T0339483

Modeling Strategic Behavior

A Graduate Introduction to Game Theory and Mechanism Design

World Scientific Lecture Notes in Economics and Policy

ISSN: 2382-6118

Series Editor: Ariel Dinar *(University of California, Riverside, USA)*

The World Scientific Lecture Notes in Economics and Policy series is aimed to produce lecture note texts for a wide range of economics disciplines, both theoretical and applied at the undergraduate and graduate levels. Contributors to the series are highly ranked and experienced professors of economics who see in publication of their lectures a mission to disseminate the teaching of economics in an affordable manner to students and other readers interested in enriching their knowledge of economic topics. The series was formerly titled World Scientific Lecture Notes in Economics.

Published:

World Scientific Lecture Notes in Economics and Policy – Vol. 6

Modeling Strategic Behavior

A Graduate Introduction to Game Theory and Mechanism Design

George J. Mailath

University of Pennsylvania, USA & Australian National University, Australia

World Scientific

NEW JERSEY · LONDON · SINGAPORE · BEIJING · SHANGHAI · HONG KONG · TAIPEI · CHENNAI · TOKYO

Published by

World Scientific Publishing Co. Pte. Ltd.

5 Toh Tuck Link, Singapore 596224

USA office: 27 Warren Street, Suite 401-402, Hackensack, NJ 07601

UK office: 57 Shelton Street, Covent Garden, London WC2H 9HE

Library of Congress Cataloging-in-Publication Data

Names: Mailath, George Joseph, author.
Title: Modeling strategic behavior: a graduate introduction to game theory and mechanism design /
    by George J Mailath (University of Pennsylvania, USA &
    Australian National University, Australia).
Description: New Jersey : World Scientific Publishing Co Pte Ltd, [2018] |
    Series: World scientific lecture notes in economics and policy ; Volume 6
Identifiers: LCCN 2018050302 | ISBN 9789813239937 (hardcover)
Subjects: LCSH: Equilibrium (Economics) | Game theory. | Microeconomics.
Classification: LCC HB145 .L43 2018 | DDC 330.01/5193--dc23
LC record available at https://lccn.loc.gov/2018050302

British Library Cataloguing-in-Publication Data

A catalogue record for this book is available from the British Library.

For any available supplementary material, please visit
https://www.worldscientific.com/worldscibooks/10.1142/10981#t=suppl

Desk Editor: Sylvia Koh

Printed in Singapore

To Loretta

Preface

These notes are based on my lecture notes for Economics 703, a first-year graduate course that I have been teaching at the Economics Department, University of Pennsylvania, for many years. It is impossible to understand modern economics without knowledge of the basic tools of game theory and mechanism design. My goal in the course (and this book) is to teach those basic tools so that students can understand and appreciate the corpus of modern economic thought, and so contribute to it.

A key theme in the course is the interplay between the formal development of the tools and their use in applications. At the same time, extensions of the results that are beyond the course, but important for context are (briefly) discussed.

While I provide more background verbally on many of the examples, I assume that students have seen some undergraduate game theory (such as covered in Osborne, 2004, Tadelis, 2013, and Watson, 2013). In addition, some exposure to intermediate microeconomics and decision making under uncertainty is helpful.

Since these are lecture notes for an introductory course, I have not tried to attribute every result or model described. The result is a somewhat random pattern of citations and references.

There is much more here than can be covered in a one semester course. I often do not cover Section 4.2 (Foundations of Nash Equilibrium), selectively cover material in Section 3.3 and Chapter 7 (Repeated Games), and never both Sections 8.2 (Coase conjecture) and 8.3 (Reputations).

I have been very lucky in my coauthors and colleagues (particularly Larry Samuelson and Andrew Postlewaite), from whom I learned a tremendous amount. Thanks to the many generations of Penn graduate students who were subjected to early versions of these notes, and made many helpful comments. Thanks also to Larry Samuelson and Tilman Börgers for helpful comments. Finally, thanks to Ashwin Kambhampati and Changhwa Lee, who did an outstanding job proofreading these notes.

Game Theory in Economics

Game theory studies the strategic interactions of agents (often called "players" or decision-makers). An example of a *strategic* interaction is the pricing behavior of two petrol (gas) stations on the same intersection. Each station, in choosing its price, will both respond to the current price of the other station and to how it believes the other station will respond to its price.

To study strategic interactions, it is useful to use parlor games such as chess and poker as examples. The first game theorists were mathematicians, and viewed the study of strategic interactions (beginning with parlor games) as applied mathematics.[1] The goal was to calculate a *solution*, which was a prediction: How would rational players behave? A solution can also be a recommendation: How should a rational player behave (assuming the other player is rational)?

The perspective of these notes is that of an economist. Economists are social scientists and, as such, want to understand social behavior. Any model simplifies the situation being modeled, and models of strategic interactions in an economic or social context are no different. The resulting modeling choices make the use of game theory within economics very different from game theory as applied mathematics (which takes the rules of the game as given). In particular, the modeling choices reflect the modeler's judgment as to what the players treat as strategically relevant. This judgment determines both the choices of strategy spaces (the actions and information players have), and the choice of solution concept. There is often a subtle interplay between the question under investigation, the modeling choices (including that of the solution concept), and the resulting analysis.

---

[1]Leonard (2010) gives a fascinating history of the birth of game theory.

Contents

Chapter 1

Normal and Extensive Form Games

1.1 Normal Form Games

Most introductions to game theory start with the *prisoner's dilemma*.[1]
Two suspects (I and II) are separately interrogated. The prosecutors have
sufficient evidence to convict each of a minor offence, but wish to convict
them of a major offence. The potential results of the interrogation are
illustrated in Figure 1.1.1. Clearly, no matter what the other suspect does,
it is always better to confess than not confess.

This game is often interpreted as a partnership game, in which two
partners simultaneously choose between exerting effort and shirking. Effort
E produces an output of 6 at a cost of 4, while shirking S yields no output
at no cost. Total output is shared equally. The result is given in Figure
1.1.2. With this formulation, no matter what the other partner does (E or
S), the partner maximizes his/her payoff by shirking.

The scenarios illustrated in Figures 1.1.1 and 1.1.2 are examples of normal form games.

Definition 1.1.1. *An n-player* normal (or strategic) form game G *is an
n-tuple* $\{(S_1, U_1), \ldots, (S_n, U_n)\}$, *where for each player i,*

- S_i *is a nonempty set, i's strategy space, with typical element s_i, and*

- $U_i : \prod_{k=1}^{n} S_k \to \mathbb{R}$, *$i$'s payoff function.*

---

[1]Poundstone (1993) discusses the prisoner's dilemma and its role in the early history
of game theory in the context of the cold war.

$$II$$

	Confess	Don't confess
I Confess	$-6, -6$	$0, -9$
Don't confess	$-9, 0$	$-1, -1$

Figure 1.1.1: The prisoner's dilemma, with the numbers describing length of sentence (the minus signs indicate that longer sentences are less desirable). In each cell, the first number is player *I*'s sentence, while the second is player *II*'s.

	E	*S*
E	$2, 2$	$-1, 3$
S	$3, -1$	$0, 0$

Figure 1.1.2: The prisoner's dilemma, as a partnership game. In each cell, the first number is the row player's payoff, while the second number is the column player's payoff.

The normal form game G is finite *if $n < \infty$ and $|S_i| < \infty$ for all i.*

Notation: The set of strategy profiles is $S := \prod_{k=1}^{n} S_k$, with a *strategy profile* denoted by $s := (s_1, \ldots, s_n) \in S$. The strategy profile omitting player i's strategy is $s_{-i} := (s_1, \ldots, s_{i-1}, s_{i+1}, \ldots, s_n) \in S_{-i} := \prod_{k \neq i} S_k$. Finally, $(s_i', s_{-i}) := (s_1, \ldots, s_{i-1}, s_i', s_{i+1}, \ldots, s_n) \in S$, so that $s = (s_i, s_{-i})$.

We sometimes write payoffs as a vector-valued function $U : S \to \mathbb{R}^n$, or when S is finite, as the vector $U \in \mathbb{R}^{n|S|}$ (recall that a vector $x \in \mathbb{R}^k$ can be viewed as the function $x : \{1, \ldots, k\} \to \mathbb{R}$, and conversely, a function from the finite set $\{1, \ldots, k\}$ to a set Y can be viewed as a vector in Y^k).

Example 1.1.1 (Sealed-bid second-price auction). Two bidders simultaneously submit bids (in sealed envelopes) for an object. A bid is a nonnegative number, with i's bid denoted by $b_i \in \mathbb{R}_+$. Bidder i's value for the object (reservation price, willingess to pay) is denoted by v_i. The object is awarded to the highest bidder, who pays the second highest bid. Ties are resolved by a fair coin toss. Then, $n = 2, S_i = \mathbb{R}_+$, and (taking expectations)

$$U_i(b_1, b_2) = \begin{cases} v_i - b_j, & \text{if } b_i > b_j, \\ \frac{1}{2}(v_i - b_j), & \text{if } b_i = b_j, \\ 0, & \text{if } b_i < b_j. \end{cases} \qquad \bigstar$$

Example 1.1.2 (Sealed bid first price auction). The only change from Example 1.1.1 is that the winning bidder pays their own bid. So, $n = 2, S_i = \mathbb{R}_+$, and

$$U_i(b_1, b_2) = \begin{cases} v_i - b_i, & \text{if } b_i > b_j, \\ \frac{1}{2}(v_i - b_i), & \text{if } b_i = b_j, \\ 0, & \text{if } b_i < b_j. \end{cases} \qquad \bigstar$$

Example 1.1.3 (Cournot duopoly). Two firms simultaneously choose quantities of goods that are perfect substitutes. The market clearing price is given by $P(Q) = \max\{a - Q, 0\}$, where $Q = q_1 + q_2$. Firm i's cost function is $C(q_i) = cq_i$, $0 < c < a$. Finally, $n = 2$.

Quantity competition: $S_i = \mathbb{R}_+$, and $U_i(q_1, q_2) = (P(q_1 + q_2) - c)q_i$. $\bigstar$

Example 1.1.4 (Bertrand duopoly). The economic environment is as for example 1.1.3, but now firms engage in price competition (i.e., firms simultaneously choose prices). Since goods are perfect substitutes, the lowest pricing firm gets the whole market (with the market split in the event of a tie). We again have $S_i = \mathbb{R}_+$, but now

$$U_1(p_1, p_2) = \begin{cases} (p_1 - c) \max\{a - p_1, \, 0\}, & \text{if } p_1 < p_2, \\ \frac{1}{2}(p_1 - c) \max\{a - p_1, \, 0\}, & \text{if } p_1 = p_2, \\ 0, & \text{if } p_1 > p_2. \end{cases} \qquad \bigstar$$

Example 1.1.5 (Voting by veto). Three outcomes: $x, y,$ and z. Player 1 first vetoes an outcome, and then player 2 vetoes one of the remaining outcomes. The non-vetoed outcome results. Suppose 1 ranks outcomes as $x \succ y \succ z$ (i.e., $u_1(x) = 2, u_1(y) = 1, u_1(z) = 0$), and 2 ranks outcomes as $y \succ x \succ z$ (i.e., $u_2(x) = 1, u_2(y) = 2, u_2(z) = 0$).

Player 1's strategy is an uncontingent veto, so $S_1 = \{x, y, z\}$.

Player 2's strategy is a contingent veto, so $S_2 = \{(abc) : a \in \{y, z\}, b \in \{x, z\}, c \in \{x, y\}\}$. The interpretation of abc is that player 2 vetoes a if 1 has vetoed x, b if y, and c if z.

The normal form is given in Figure 1.1.3. $\bigstar$

2

		(yxx)	(yxy)	(yzx)	(yzy)	(zxx)	(zxy)	(zzx)	(zzy)
	x	0,0	0,0	0,0	0,0	1,2	1,2	1,2	1,2
1	y	0,0	0,0	2,1	2,1	0,0	0,0	2,1	2,1
	z	1,2	2,1	1,2	2,1	1,2	2,1	1,2	2,1

Figure 1.1.3: The normal form of the voting by veto game in Example 1.1.5.

Definition 1.1.2. *A strategy s_i' strictly dominates another strategy s_i'' if for all $s_{-i} \in S_{-i}$,*

$$U_i(s_i', s_{-i}) > U_i(s_i'', s_{-i}).$$

A strategy s_i is a strictly dominant strategy if it strictly dominates every strategy $s_i'' \neq s_i, s_i'' \in S_i$.

If player i has a strictly dominant strategy, then

$$\arg\max_{s_i} U_i(s_i, s_{-i}) = \left\{ s_i : U_i(s_i, s_{-i}) = \max_{s_i'} U_i(s_i', s_{-i}) \right\}$$

is a singleton and is independent of s_{-i}.

Remark 1.1.1. The definitions of a normal form game and strict dominance make no assumption about the knowledge that players have about the game. We will, however, typically assume (at least) that players know the strategy spaces, and their own payoffs as a function of strategy profiles. However, as the large literature on learning and evolutionary game theory in biology suggests (briefly sketched in Section 4.2), this is not necessary.

The assertion that a player will not play a strictly dominated strategy is compelling when the player knows the strategy spaces and his/her own payoffs. Or rather, a player's payoffs capture the game *as perceived by that player*. This provides an important perspective on experimental studies of behavior is games. It is sometimes argued that in practice, people do not behave as "predicted" by game theory, since a significant fraction of experimental subjects will play the strictly dominated action in the prisoner's dilemma. This is, however, a misinterpretation of the evidence. In a typical experiment, the subjects are asked to choose an action in a game whose *monetary* payoffs are consistent with a prisoner's dilemma structure. But the players' true payoffs may well (typically do) depend upon other factors than money, such as fairness. The observation that a player deliberately chose an action that was strictly dominated by another action in monetary terms is clear evidence that the player cared about other considerations, and so the game is *not* in fact a a prisoner's dilemma. We assume that payoffs in the game capture all of these considerations. ◆

Definition 1.1.3. *A strategy s_i' (weakly) dominates another strategy s_i'' if for all $s_{-i} \in S_{-i}$,*

$$U_i(s_i', s_{-i}) \geq U_i(s_i'', s_{-i}),$$

and there exists $s_{-i}' \in S_{-i}$ such that

$$U_i(s_i', s_{-i}') > U_i(s_i'', s_{-i}').$$

Definition 1.1.4. *A strategy is said to be* strictly *(or, respectively, weakly)* undominated *if it is not strictly (or, resp., weakly) dominated (by any other strategy). A strategy is* weakly dominant *if it weakly dominates every other strategy.*

If the adjective is omitted from dominated (or undominated), weak is typically meant (but not always, unfortunately).

Lemma 1.1.1. *If a weakly dominant strategy exists, it is unique.*

Proof. Suppose s_i is a weakly dominant strategy. Then for all $s_i' \in S_i$, there exists $s_{-i} \in S_{-i}$, such that

$$U_i(s_i, s_{-i}) > U_i(s_i', s_{-i}).$$

But this implies that s_i' cannot weakly dominate s_i, and so s_i is the only weakly dominant strategy. ∎

Remark 1.1.2. Warning! There is also a notion of *dominant* strategy:

Definition 1.1.5. *A strategy s_i' is a* dominant *strategy if for all $s_i'' \in S_i$ and all $s_{-i} \in S_{-i}$,*

$$U_i(s_i', s_{-i}) \geq U_i(s_i'', s_{-i}).$$

If s_i' is a dominant strategy for i, then $s_i' \in \arg\max_{s_i} U_i(s_i,, s_{-i})$, for all s_{-i}; but $\arg\max_{s_i} U_i(s_i,, s_{-i})$ need not be a singleton and it need not be independent of s_{-i} (construct an example to make sure you understand). If i has only one dominant strategy, then that strategy weakly dominates every other strategy, and so is weakly dominant.

Dominant strategies have played an important role in mechanism design and implementation (see Remark 1.1.3 and Section 11.2), but not otherwise (since a dominant strategy–when it exists–will typically weakly dominate every other strategy, as in Example 1.1.6). ◆

Remark 1.1.3 (Strategic behavior is ubiquitous). Consider a society consisting of a finite number n of members and a finite set of outcomes Z. Suppose each member of society has a strict preference ordering of Z, and let $\mathcal{P}$ be the set of all possible strict orderings on Z. A profile $(P_1, \ldots, P_n) \in \mathcal{P}^n$ describes a particular society (a preference ordering for each member).

A *social choice rule* or *function* is a mapping $\xi : \mathcal{P}^n \to Z$. A social choice rule ξ is *dictatorial* if there is some i such that for all $(P_1, \ldots, P_n) \in \mathcal{P}^n$, $[xP_iy, \ \forall y \neq x] \implies x = \xi(P_1, \ldots, P_n)$.

The *direct mechanism* is the normal form game in which all members of society simultaneously announce a preference ordering and the outcome is determined by the social choice rule as a function of the announced preferences.

Theorem 1.1.1 (Gibbard-Satterthwaite). *Suppose $|\xi(\mathcal{P}^n)| \geq 3$. Announcing truthfully in the direct mechanism is a dominant strategy for all preference profiles if, and only if, the social choice rule is dictatorial.*

A social choice rule is said to be *strategy proof* if announcing truthfully in the direct mechanism is a dominant strategy for all preference profiles. It is trivial that for any dictatorial social choice rule, it is a dominant strategy to always truthfully report in the direct mechanism. The surprising result, proved in Section 11.2,[2] is the converse.   ◆

Example 1.1.6 (Continuation of example 1.1.1). In the second price auction (also called a *Vickrey auction*, after Vickrey, 1961), each player has a weakly dominant strategy, given by $b_1 = v_1$.

It is sufficient to show this for player 1. First, we argue that bidding v_1 is a best response for 1, no matter what bid 2 makes (i.e., it is a dominant strategy). Recall that payoffs are given by

$$U_1(b_1, b_2) = \begin{cases} v_1 - b_2, & \text{if } b_1 > b_2, \\ \frac{1}{2}(v_1 - b_2), & \text{if } b_2 = b_1, \\ 0, & \text{if } b_1 < b_2. \end{cases}$$

Two cases:

1. $b_2 < v_1$: Then, $U_1(v_1, b_2) = v_1 - b_2 \geq U_1(b_1, b_2)$ for any $b_1 \in \mathbb{R}_+$. [Bidding more than v_1 does not change anything, and bidding less only changes the outcome if player 1 bids less than b_2 and loses. But player 1 wants to win the auction in this situation.]

---

[2]It is stated there in an equivalent form as Theorem 11.2.2.

2. $b_2 \geq v_1$: Then, $U_1(v_1, b_2) = 0 \geq U_1(b_1, b_2)$ for any $b_1 \in \mathbb{R}_+$. [Player 1 does not want to win if this involves paying than v_1.]

Thus, bidding v_1 is optimal.

Bidding v_1 also weakly dominates every other bid (and so v_1 is weakly dominant). Suppose $b_1 < v_1$ and $b_1 < b_2 < v_1$. Then $U_1(b_1, b_2) = 0 < v_1 - b_2 = U_1(v_1, b_2)$. If $b_1 > v_1$ and $b_1 > b_2 > v_1$, then $U_1(b_1, b_2) = v_1 - b_2 < 0 = U_1(v_1, b_2)$. ★

Example 1.1.7 (Provision of a public good). Society consists of n agents, with agent i valuing the public good at r_i. The total cost of the public good is C.

Suppose costs are shared uniformly and utility is linear, so agent i's net utility is $v_i := r_i - \frac{1}{n}C$.

Efficient provision requires that the public good be provided if, and only if, $0 \leq \sum v_i$, i.e., $C \leq \sum r_i$.

Suppose the social planner does not know agents' preferences and so must elicit them. One possibility is to ask each agent to announce (report) his/her valuation (denote the reported valuation by $\hat{v}_i$) and provide the public good if $\sum \hat{v}_i \geq 0$. The problem with such a scheme is that it gives agents an incentive to overstate if $v_i > 0$ and understate if $v_i < 0$.

The *Groves mechanism* addresses this incentive as follows: If the public good is provided, pay agent i an amount $\sum_{j \neq i} \hat{v}_j$ (tax if negative). This implies

$$
\text{payoff to agent } i = \begin{cases} v_i + \sum_{j \neq i} \hat{v}_j, & \text{if } \hat{v}_i + \sum_{j \neq i} \hat{v}_j \geq 0, \\ 0, & \text{if } \hat{v}_i + \sum_{j \neq i} \hat{v}_j < 0, \end{cases}
$$

where v_i is agent i's valuation and $\hat{v}_i$ is i's *reported* valuation. Note that there is no budget balance requirement. It is a weakly dominant strategy for i to announce $\hat{v}_i = v_i$: If $v_i + \sum_{j \neq i} \hat{v}_j > 0$, announcing $\hat{v}_i = v_i$ ensures good is provided, while if $v_i + \sum_{j \neq i} \hat{v}_j < 0$, announcing $\hat{v}_i = v_i$ ensures good is not provided. Moreover, conditional on provision, announcement does not affect payoff—note similarity to second price auction.

A difficulty with the mechanism is that while there are no payments if no provision, payments can be large if provision: Payments to the agents when the public good is provided total $\sum_i \left(\sum_{j \neq i} \hat{v}_j \right) = (n-1) \sum_i \hat{v}_i$. This total can be reduced without altering incentives by taxing agent i by an amount independent of i's behavior. In particular, taxing agent i the amount $\max\{\sum_{j\neq i}\hat{v}_j, 0\}$ yields

$$
\text{payoff to agent } i = \begin{cases}
v_i, & \text{if } \sum_j \hat{v}_j \geq 0 \text{ and } \sum_{j\neq i} \hat{v}_j \geq 0, \\
v_i + \sum_{j\neq i}\hat{v}_j, & \text{if } \sum_j \hat{v}_j \geq 0 \text{ and } \sum_{j\neq i}\hat{v}_j < 0, \\
-\sum_{j\neq i}\hat{v}_j, & \text{if } \sum_j \hat{v}_j < 0 \text{ and } \sum_{j\neq i}\hat{v}_j \geq 0, \\
0, & \text{if } \sum_j \hat{v}_j < 0 \text{ and } \sum_{j\neq i}\hat{v}_j < 0.
\end{cases}
$$

This is the *pivot mechanism*. Note that i only pays a tax if i changes social decision. Moreover, total taxes are no larger than $\sum_i \max\{\hat{v}_i, 0\}$ if the good is provided and no larger than $\sum_i \max\{-\hat{v}_i, 0\}$ if the good is not provided. For more on this, see Section 11.3.   ★

Example 1.1.8 (Continuation of Example 1.1.3, Cournot). There are no weakly dominating quantities in the Cournot duopoly: Suppose $q_2 < a - c$. Then $\arg\max_{q_1} U_1(q_1, q_2) = \arg\max(a - c - q_1 - q_2)q_1$. The first order condition implies $a - c - 2q_1 - q_2 = 0$ or

$$
q_1(q_2) = \frac{a - c - q_2}{2}.
$$

Since $\arg\max U_1$ is unique and a nontrivial function of q_2, there is no weakly dominating quantity.   ★

1.2   Iterated Deletion of Dominated Strategies

Example 1.2.1. In the game displayed in Figure 1.2.1, neither player has an obvious way to play, in the sense that neither has a strictly dominant strategy.

But, R is strictly dominated, and so will not be played. But if the row player knows this then, in the game obtained by deleting R, column's strategy of B is now strictly dominated, and so will not be played. If the column player knows that the row player will not play B, then the column player will not play L. Note that this chain of logic involves the column player knowing that the row player knows the column player's payoffs (as well as knowing the row player's payoffs).

The result of iteratively deleting strictly dominated strategies leads to (T, M).   ★

	L	C	R
T	1, 0	1, 2	0, 1
B	0, 3	0, 1	2, 0

Figure 1.2.1: A game to illustrate the iterative deletion of strictly dominated strategies, Example 1.2.1.

2

(z, z, x)

1	y	2,1
	z	1,2

Figure 1.2.2: The result of iterative deleting weakly dominated strategies in the voting-by-veto game.

Example 1.2.2 (Continuation of Example 1.1.5). Apply iterative deletion of weakly dominated strategies to the veto game. After one round of deletions, only (z, z, x) remains for player 2 (both y and z remain for player 1), and so 1 vetoes y, and not z! See Figure 1.2.2.   ★

Remark 1.2.1. For finite games, the order of removal of strictly dominated strategies is irrelevant (see Problem 1.4.2(a)). This is not true for weakly dominated strategies, as illustrated by the game in Figure 1.2.3.

Both TL and BL can be obtained as the *singleton* profile that remains from the iterative deletion of weakly dominated strategies. In addition, $\{TL, TM\}$ results from a different sequence, and $\{BL, BR\}$ from yet another sequence. (The order does *not* matter for games like the veto game of example 1.1.5, see Theorem 2.2.2. See Marx and Swinkels, 1997, for more general conditions that imply the order does not matter.)

Similarly, the order of elimination may matter for infinite games (see Problem 1.4.2(b)).

Because of this, the procedure of the *iterative deletion of weakly (or strictly) dominated strategies* is often understood to *require* that at each stage, all weakly (or strictly) dominated strategies be deleted. These notes follow that understanding (unless explicitly stated otherwise). With that

	L	M	R
T	1,1	1,1	0,0
B	1,1	0,0	1,1

Figure 1.2.3: A game illustrating that the order matters when iteratively deleting weakly dominated strategies (Remark 1.2.1).

understanding, the iterated deletion of weakly dominated strategies in this example leads to $\{TL, BL\}$.   ♦

Remark 1.2.2. The plausibility of the iterated deletion of dominated strategies requires something like players knowing the structure of the game (including that other players know the game), not just their own payoffs. For example, in Example 1.2.1, in order for the column player to delete L at the third round, the column player needs to know that the row player will not play B, which requires the column player to know that the row player knows that the column player will not play R.[3]

As illustrated in Section 4.2, this kind of iterated knowledge is not *necessary* for the plausibility of the procedure (though in many contexts it does provide the most plausible foundation).   ♦

1.3   Extensive Form Games

Game trees look like decision trees. The role of the definition is to make clear who does what, when, knowing what. It provides a complete description of the environment.

Definition 1.3.1. *A finite extensive form* consists of the following:

1. *A set of players* $\{1, \ldots, n\}$ *and nature, denoted player 0.*

2. *A game tree* $(T, \prec)$, *where* $(T, \prec)$ *is an* arborescence: T *is a finite set of nodes and* $\prec$ *is a binary relation on* T *denoting "precedence" satisfying*

   (a) $\prec$ *is asymmetric* $(t \prec t' \implies t' \not\prec t)$,[4]

---

[3]For more on the epistemic foundations of these procedures, see Remark 2.4.3.
[4]Note that this implies that $\prec$ is irreflexive: $t \not\prec t$ for all $t \in T$.

(b) *transitive (for all $t, t', t'' \in T, t \prec t', t' \prec t'' \implies t \prec t''$),*[5]

(c) *if $t \prec t''$ and $t' \prec t''$ then either $t \prec t'$ or $t' \prec t$, and finally,*

(d) *there is a unique* initial *node, $t_0 \in T$, i.e., $\{t_0\} = \{t \in T : \nexists t' \in T, t' \prec t\}$.*

Let $p(t) \in T$ *denote the* immediate predecessor *of t, i.e., $p(t) \prec t$ and there is no t' for which $p(t) \prec t' \prec t$. The* path *to a node t is the sequence $t_0 = p^k(t), p^{k-1}(t), \ldots, p(t), t$, where $p^\ell(t) = p(p^{\ell-1}(t))$ for $\ell \geq 2$.[6] Requirement (d) implies that for every noninitial node t, there is a* unique *path from the initial node to t.*

The set $s(t) := \{t' \in T : t \prec t'$ and $\nexists t'', t \prec t'' \prec t'\} = \{t' \in T : p(t') = t\}$ *is the set of* immediate successors *of t.*

The set $Z := \{t \in T : \nexists t' \in T, t \prec t'\}$ *is the set of* terminal *nodes, with element z.*

The set $X := T \setminus Z$ *is the set of* decision *nodes, with element x.*

3. *Assignment of players to decision nodes, $\iota : X \to \{0, 1, \ldots n\}$. Define $X_j := \iota^{-1}(j) = \{t \in X : \iota(t) = j\}, \forall j \in \{0, 1, \ldots, n\}$.*

4. *Actions: Actions lead to (label) immediate successors, i.e., there is a set A and a mapping*

$$\alpha : T \setminus \{t_0\} \to A,$$

such that $\alpha(t') \neq \alpha(t'')$ for all $t', t'' \in s(t)$. Define $A(t) := \alpha(s(t))$, the set of actions available at $t \in X$.

5. *Information sets: H_i is a partition of X_i for all $i \neq 0$.[7] For all $t, t' \in h$,*

(a) *$t \not\prec t', t' \not\prec t$,[8]*

(b) *$A(t) = A(t') := A(h)$, and*

(c) *perfect recall (every player knows whatever he knew previously, including own previous actions).[9] Two extensive forms that violate perfect recall are in Figure 1.3.1.*

---

[5]A binary relation satisfying 2(a) and 2(b) is called a strict partial order.

[6]Note that the equality $t_0 = p^k(t)$ *determines* the value of k, that is, the path has k steps from t to the initial node.

[7]A *partition* H_i is a collection of subsets of X_i such that (i) for all $t \in X_i$, there exists $h \in H_i$ such that $t \in h$, and (ii) for all $h, h' \in H_i$, if $h \neq h'$ then $h \cap h' = \varnothing$.

[8]This is often viewed as one of the restrictions implied by perfect recall.

[9]Denote the information set containing $t \in T_i$ by $h(t) \in H_i$. The formal requirement is:

If $t, t', t'' \in T_i$, $t \prec t'$, and $t'' \in h(t')$, then there exists $t^\dagger \in h(t)$ with $t^\dagger \prec t''$. Moreover, if ℓ and m are defined by $t = p^\ell(t')$ and $t^\dagger = p^m(t'')$, then $\alpha(p^{\ell-1}(t')) = \alpha(p^{m-1}(t''))$.

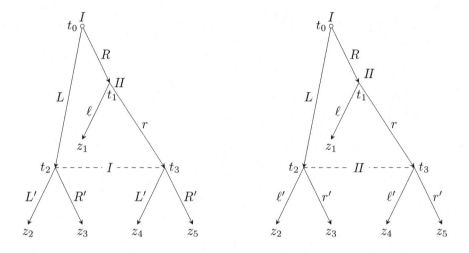

Figure 1.3.1: Two extensive forms without perfect recall. An arrow con-
nects a node with its *immediate* predecessor, and is labelled
by the action associated with that node.

Definition 1.3.2. *A finite extensive form game* Γ *consists of is a finite
extensive form with*

1. *payoffs,* $u_i : Z \to \mathbb{R}$, *and*

2. *a probability distribution for nature,* $\rho : X_0 \to \bigcup_{t \in X_0} \Delta(A(t))$ *such
    that* $\rho(t) \in \Delta(A(t))$.

Definition 1.3.3. *An* extensive form strategy *for player i is a function*

$$s_i : H_i \to \bigcup_h A(h) \text{ such that } s_i(h) \in A(h), \ \forall h \in H_i.$$

The set of player i's strategies is i's strategy space, *denoted* S_i, *and as
before, a* strategy profile *is* $(s_1, \ldots, s_n)$.

Definition 1.3.4. *Suppose there are no moves of nature.*

The outcome path *is the sequence of nodes reached by strategy profile,
or equivalently, the sequence of specified actions.*

The outcome *is the unique terminal node reached by the strategy profile
s, denoted* $z(s)$. *In this case, the normal form representation is given by*
$\{(S_1, U_1), \ldots, (S_n, U_n)\}$, *where* S_i *is the set of i's extensive form strategies,
and*

$$U_i(s) = u_i(z(s)).$$

If there is one or more moves of nature, actions by players results in a probability distribution over terminal nodes. In this case, we assume players are *expected utility* maximizers. The payoff function u_i is called a *Bernoulli utility function*, and the payoff from any distribution over terminal nodes is the expected value of the Bernoulli utility function (the resulting payoff function is called the *von-Neumann-Morgenstern utility function*).

Definition 1.3.5. *When there are moves of nature, the* outcome *is the implied probability distribution over terminal nodes, denoted* $\mathbf{P}^s \in \Delta(Z)$. *In this case, in the normal form representation given by* $\{(S_1, U_1), \ldots, (S_n, U_n)\}$, *where* S_i *is the set of* i's *extensive form strategies, we have*

$$U_i(s) = \sum_z \mathbf{P}^s(z) u_i(z).$$

For an example, see Example 3.1.1. Section 5.1 explores the structure of $\mathbf{P}^s$ in more detail.

The assumption that players are expected utility maximizers allows players to be risk averse. See Kreps (1988), Kreps (1990, Chapter 3), or Mas-Colell, Whinston, and Green (1995, Chapter 6) for much more on the theory of choice and how the von-Neumann-Morgenstern theory captures players' attitudes towards risk.

Example 1.3.1. The extensive form for Example 1.1.5 is described as follows: The game starts at the node t_0, owned by player 1, ($\iota(t_0) = 1$), and t_0 has three immediate successors, t_1, t_2, t_3 reached by vetoing x, y, and z (so that $\alpha(t_1) = x$, and so on). These immediate successors are all owned by player 2 (i.e., $\iota(t_j) = 2$ for $j = 1, 2, 3$). At each of these nodes, player 2 vetoes one of the remaining outcomes, with each veto leading to a distinct node. See Figure 1.3.2 for a graphical representation.

The result of the iterative deletion of weakly dominated strategies is (y, zzx), implying the outcome (terminal node) t_7.

Note that this outcome also results from the profiles (y, yzx) and (y, zzy), and (y, yzy).   ★

Definition 1.3.6. *A game has* perfect information *if all information sets are singletons.*

Example 1.3.2 (Simultaneous moves). The extensive form for the prisoner's dilemma is illustrated in Figure 1.3.3.   ★

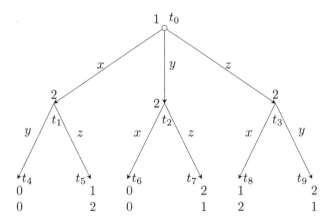

Figure 1.3.2: The extensive form for Example 1.1.5.

1.3.1 The Reduced Normal Form

Example 1.3.3. Consider the extensive form in Figure 1.3.4. Note that payoffs have not been specified, just the terminal nodes. The normal form is displayed in Figure 1.3.5. Observe that I's strategies Stop,Stop$_1$ and Stop,Go$_1$ are equivalent (in that the rows have identical entries), as are II's strategies Stop,Stop$_1$ and Stop,Go$_1$. ★

Definition 1.3.7. *Two strategies $s_i, s_i' \in S_i$ are* strategically equivalent *if $U_j(s_i, s_{-i}) = U_j(s_i', s_{-i})$ for all $s_{-i} \in S_{-i}$ and all j.*

In the (pure strategy) reduced normal form *of a game, every set of strategically equivalent strategies is treated as a single strategy.*

Example 1.3.3 (continued). The reduced normal form of Figure 1.3.5 is displayed in Figure 1.3.6. The strategy Stop for player I, for example, in the reduced normal form should be interpreted as the *equivalence class* of extensive form strategies {Stop,Stop$_1$, Stop,Go$_1$}, where the equivalence relation is given by strategic equivalence.

Notice that this reduction occurs for *any* specification of payoffs at the terminal nodes. ★

The importance of the reduced normal form arises from the reduction illustrated in Example 1.3.3: Whenever a player's strategy specifies an action that precludes another information set *owned by the same player*, that strategy is strategically equivalent to other strategies that only differ

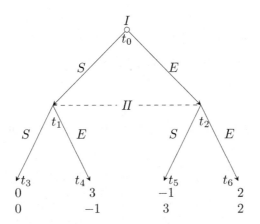

Figure 1.3.3: The extensive form for the prisoner's dilemma. Player II's information set is indicated by the dashed line.

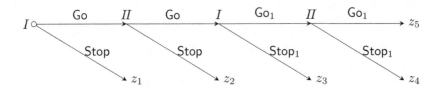

Figure 1.3.4: A simple extensive form game.

	Stop, $Stop_1$	Stop, Go_1	Go, $Stop_1$	Go, Go_1
Stop,$Stop_1$	$u(z_1)$	$u(z_1)$	$u(z_1)$	$u(z_1)$
Stop,Go_1	$u(z_1)$	$u(z_1)$	$u(z_1)$	$u(z_1)$
Go,$Stop_1$	$u(z_2)$	$u(z_2)$	$u(z_3)$	$u(z_3)$
Go,Go_1	$u(z_2)$	$u(z_2)$	$u(z_4)$	$u(z_5)$

Figure 1.3.5: The normal form of the extensive form in Figure 1.3.4, where $u(z) = (u_1(z), u_2(z))$.

	Stop	Go, Stop_1	Go, Go_1
Stop	$u(z_1)$	$u(z_1)$	$u(z_1)$
Go,Stop_1	$u(z_2)$	$u(z_3)$	$u(z_3)$
Go,Go_1	$u(z_2)$	$u(z_4)$	$u(z_5)$

Figure 1.3.6: The reduced normal form of the normal form in Figure 1.3.5.

at such information sets. As in Example 1.3.3, this is true for *arbitrary* specifications of payoffs at the terminal nodes.

When developing intuition about the relationship between the normal and extensive form, it is often helpful to assume that extensive form payoffs have no ties, i.e., for all players i and all pairs of terminal nodes, $z, z' \in Z$, $u_i(z) \neq u_i(z')$. Under that restriction, changing payoffs at terminal nodes does not change the set of reduced normal form strategies for any player.

However, if for example, $u_i(z) = u_i(z')$ for all players i and all pairs of terminal nodes, $z, z' \in Z$, then the reduced normal form consists of *one* strategy for each player. Changing payoffs now does change the set of reduced normal form strategies.

When describing the normal form representation of an extensive form, it is common to (and we will typically) use the reduced normal form.

An extensive form strategy *always* has the form given by Definition 1.3.3, while a normal form strategy may represent an equivalence class of extensive form strategies (and a reduced normal form strategy always does).

1.4   Problems

1.4.1. Consider the Cournot duopoly example (Example 1.1.3).

   (a) Characterize the set of strategies that survive iterated deletion of strictly dominated strategies.

   (b) Formulate the game when there are $n \geq 3$ firms and identify the set of strategies surviving iterated deletion of strictly dominated strategies.

1.4.2. (Note that this question does not follow the convention of Remark 1.2.1 that all possible deletions are done at each round.)

   (a) Prove that the order of deletion does not matter for the process of iterated deletion of strictly dominated strategies in a finite game (Remark 1.2.1 shows that *strictly* cannot be replaced by *weakly*).

(b) Show that the order of deletion matters for the process of iterated deletion of strictly dominated strategies for the following *infinite* game: $S_1 = S_2 = [0, 1]$ and payoffs

$$u_i(s_1, s_2) = \begin{cases} s_i, & \text{if } s_i < 1, \\ 0, & \text{if } s_i = 1, s_j < 1, \\ 1, & \text{if } s_i = s_j = 1. \end{cases}$$

1.4.3. Suppose $T = \{a, b, c, d, e, f, g, h\}$ and $\prec$ is given by (i) $a \prec b, c, d, e, f, g, h$, (ii) $b \prec c, e$, and (iii) $d \prec f, g$.

(a) Draw the implied tree. Be sure to label all nodes.

(b) Suppose this is a two player game, with player 2 owning b and d. Specify an action labeling for the game, and a pair of strategies for each of players 1 and 2 with the property that the four resulting strategy profiles have precisely c, f, g as outcomes.

(c) Suppose now that player 2 cannot distinguish between the two nodes b and d. Describe player 2's information set(s). Is it possible to specify an action labeling for the game, and a pair of strategies for each of players 1 and 2 with the property that the four resulting strategy profiles have precisely c, f, g as outcomes? Why or why not?

(d) What is the implied tree if (ii) is given by $b \prec c, d, e$?

1.4.4. Define the "follows" relation $\prec^*$ on information sets by: $h' \prec^* h$ if there exists $x \in h$ and $x' \in h'$ such that $x' \prec x$.

(a) Prove that each player i's information sets H_i are strictly partially ordered by the relation $\prec^*$ (recall footnote 5 on page 11).

(b) Perfect recall implies that the set of i's information sets satisfy Property 2(c) of Definition 1.3.1, i.e., for all $h, h', h'' \in H_i$, if $h \prec^* h''$ and $h' \prec^* h''$ then either $h \prec^* h'$ or $h' \prec^* h$. Give an intuitive argument.

(c) Give an example showing that the set of all information sets is not similarly strictly partially ordered by $\prec^*$.

(d) Prove that if $h' \prec^* h$ for $h, h' \in H_i$, then for all $x \in h$, there exists $x' \in h'$ such that $x' \prec x$. (In other words, an individual player's information is refined through play in the game. Prove should really be in quotes, since this is trivial.)

1.4.5. Consider a sealed bid second price auction with two bidders. Suppose bidder i's payoffs from owning the object being sold after paying an amount p is given by $U_i(p)$, where U_i is continuous and strictly decreasing on $\mathbb{R}$, $U_i(0) > 0$, and there exists P such that $U_i(P) < 0$. What is the unique weakly dominant strategy for bidder i?

1.4.6. A sealed bid third price auction with n bidders is a sealed bid auction in which the highest bidder wins the auction, and pays the *third* highest bid. As in Example 1.1.2, let v_i denote bidder i's reservation price for the object.

   (a) Carefully describe the normal form of the three bidder version of this game.

   (b) Do bidders in this game have a weakly dominant strategy? If so, prove it. If not, provide a counterexample.

Chapter 2

A First Look at Equilibrium

2.1 Nash Equilibrium

Most games do not have an "obvious" way to play.

Example 2.1.1 (Battle of the sexes). The game displayed in Figure 2.1.1 has no strategy iteratively dominating all the other strategies.   ★

The standard solution concept for games is Nash equilibrium (Section 2.1.1 addresses why this is the standard solution concept).

Definition 2.1.1. *The strategy profile $s^* \in S$ is a* Nash equilibrium *of $G = \{(S_1, u_1), \ldots, (S_n, u_n)\}$ if for all i and for all $s_i \in S_i$,*

$$u_i(s_i^*, s_{-i}^*) \geq u_i(s_i, s_{-i}^*).$$

The battle of the sexes in Figure 2.1.1 has two Nash equilibria, (Opera, Opera) and (Ballet, Ballet).

Example 2.1.2. Consider the simple extensive form in Figure 2.1.2. The strategy spaces are $S_1 = \{L, R\}$, $S_2 = \{\ell\ell', \ell r', r\ell', rr'\}$.

Payoffs are $U_j(s_1, s_2) = u_j(z)$, where z is terminal node reached by (s_1, s_2).

The normal form is given in Figure 2.1.3.

Two Nash equilibria: $(L, \ell\ell')$ and $(R, \ell r')$. Though $\ell r'$ is a best reply to L, $(L, \ell r')$ is not a Nash equilibrium.

Note that the equilibria are strategy profiles, *not* outcomes. The outcome path for $(L, \ell\ell')$ is $L\ell$, while the outcome path for $(R, \ell r')$ is Rr'. In examples where the terminal nodes are not separately labeled, it is common to also refer to the outcome path as simply the outcome—recall that

Sheila

		Opera	Ballet
Bruce	Opera	3, 1	0, 0
	Ballet	0, 0	1, 3

Figure 2.1.1: A Battle of the Sexes.

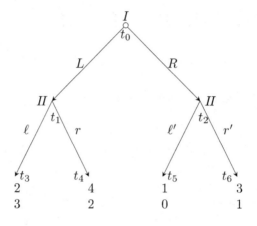

Figure 2.1.2: A simple extensive form.

every outcome path reaches a unique terminal node, and conversely, every terminal node is reached by a unique sequence of actions (and moves of nature).

NOTE: (R, rr') is *not* a Nash equilibrium, even though the outcome path associated with it, Rr', is a Nash outcome path.

★

Denote the collection of all subsets of Y by $\wp(Y)$.

A function $\phi : X \to \wp(Y) \backslash \{\varnothing\}$ is a *correspondence* from X to Y, sometimes written $\phi : X \rightrightarrows Y$.

Note that $\phi(x)$ is simply a nonempty subset of Y. If $f : X \to Y$ is a function, then $\phi(x) = \{f(x)\}$ is a correspondence and a singleton-valued correspondence can be naturally viewed as a function.

$$II$$

		$\ell\ell'$	$\ell r'$	$r\ell'$	rr'
I	L	2,3	2,3	4,2	4,2
	R	1,0	3,1	1,0	3,1

Figure 2.1.3: The normal form of the extensive form in Figure 2.1.2.

Definition 2.1.2. *The* best reply correspondence *for player i is*

$$\phi_i(s_{-i}) := \arg\max_{s_i \in S_i} u_i(s_i, s_{-i})$$
$$= \{s_i \in S_i : u_i(s_i, s_{-i}) \geq u_i(s_i', s_{-i}), \forall s_i' \in S_i\}.$$

Without assumptions on S_i and u_i, ϕ_i need not be well defined. When it is well defined everywhere, $\phi_i : S_{-i} \rightrightarrows S_i$.

If $\phi_i(s_{-i})$ is a singleton for all s_{-i}, then ϕ_i is i's *reaction function*.

Remark 2.1.1. Defining $\phi : S \rightrightarrows S$ by

$$\phi(s) := \prod_i \phi_i(s_{-i}) = \phi_1(s_{-1}) \times \cdots \times \phi_n(s_{-n}),$$

the profile s^* is a Nash equilibrium if, and only if,

$$s^* \in \phi(s^*).$$ ◆

Example 2.1.3 (Continuation of Example 1.1.8, Cournot)**.** Recall that, if $q_2 < a - c$, $\arg\max u_i(q_1, q_2)$ is unique and given by

$$q_1(q_2) = \frac{a - c - q_2}{2}.$$

More generally, i's *reaction* function is

$$\phi_i(q_j) = \max\{\frac{1}{2}(a - c - q_j), 0\}.$$

A Nash equilibrium profile (q_1^*, q_2^*) solves

$$q_1^* = \phi_1(q_2^*)$$

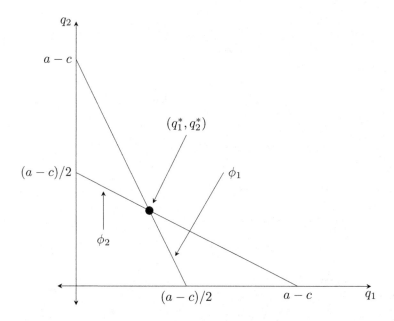

Figure 2.1.4: The reaction (or best reply) functions for the Cournot game.

and

$$q_2^* = \phi_2(q_1^*).$$

This is illustrated in Figure 2.1.4.

So (temporarily ignoring the boundary conditions),

$$
\begin{aligned}
q_1^* &= \frac{1}{2}(a - c - q_2^*) \\
&= \frac{1}{2}(a - c - \frac{1}{2}(a - c - q_1^*)) \\
&= \frac{1}{2}(a - c) - \frac{1}{4}(a - c) + \frac{q_1^*}{4} \\
&= \frac{1}{4}(a - c) + \frac{q_1^*}{4}
\end{aligned}
$$

and so

$$q_1^* = \frac{1}{3}(a - c).$$

Thus,

$$q_2^* = \frac{1}{2}(a - c) - \frac{1}{6}(a - c) = \frac{1}{3}(a - c),$$

and the boundary conditions are not binding.

The market-clearing price is given by

$$P = a - q_1^* - q_2^* = a - \frac{2}{3}(a - c) = \frac{1}{3}a + \frac{2}{3}c > 0.$$

Note also that there is no equilibrium with price equal to zero.   ★

Example 2.1.4 (Example 1.1.6 cont., sealed-bid second-price auction).
Suppose $v_1 < v_2$, and the valuations are commonly known. There are many
Nash equilibria: Bidding v_i for each i is a Nash equilibrium (of course?).
But so is any bidding profile (b_1, b_2) satisfying $b_1 < b_2$, $b_1 \leq v_2$ and $v_1 \leq b_2$.
And so is any bidding profile (b_1, b_2) satisfying $b_2 \leq v_1 < v_2 \leq b_1$! (Why?
Make sure you understand why some inequalities are weak and some are
strict).   ★

Example 2.1.5 (Example 1.1.2 cont., sealed-bid first-price auction).
Suppose $v_1 = v_2 = v$, and the valuations are commonly known. The unique
Nash equilibrium is for both bidders to bid $b_i = v$. But this equilibrium is
in *weakly dominated strategies*. But what if bids are in pennies?   ★

2.1.1   Why Study Nash Equilibrium?

Nash equilibrium is based on two principles:

1. each player is optimizing given beliefs/predictions about the behavior
   of the other players; and

2. these beliefs/predictions are correct.

Optimization is not in principle troubling, since a player chooses an
action he/she believes is optimal. For if he/she believed another action was
optimal, (surely) it would have been chosen. This reflects our assumption
that a player's payoffs capture the game as perceived by that player (recall
Remark 1.1.1).[1]

In contrast, consistency of beliefs with the actual behavior is a strong
assumption. Several justifications have been suggested:

---

[1]This is a little cavalier (but perhaps appropriately so at this point). For a more
nuanced discussion, see Section 4.2.

1. Self-fulfilling prophecy.

   Suppose an advisor (oracle) recommends (publicly predicts) a pattern
   of behavior for all the players. If the announcement is not a Nash
   equilibrium and the players believe others will play as announced,
   then there is at least one player who has a strict incentive *not* do so,
   undermining the recommendation.

2. Preplay communication.

   Before playing, the players discuss and reach an agreement on what
   to play. Such an agreement needs to be a Nash equilibrium, since
   the agreement needs to be a self-fulfilling prophecy, otherwise at least
   one player will not honor the agreement. But, the game with the
   preplay communication should be viewed as a (bigger) game, since
   players may disagree as to the preferred play (Example 2.1.1) and so
   bargaining and renegotiation becomes relevant (see Section 2.5.2),

3. Obvious or natural way to play.

   These are often called focal points, see Example 2.5.3.

4. Learning.

   Games are played within a (social) context with a history. Players
   may learn how others have played from either their own previous
   experience (individual learning) or from observing how others have
   played (social learning). A brief introduction to learning foundations
   is given in Section 4.2.

5. Provides an important discipline on modeling.

   In the absence of consistency, many seemingly inconsistent patterns
   of behavior are consistent with optimal play. For example, in the
   battle of the sexes (Example 2.1.1), (Opera, Ballet) is consistent with
   each player optimally best responding to beliefs about the play of the
   other player (indeed, this profile is rationalizable). In the absence
   of consistency, it is too easy to "explain" seemingly perverse group
   behavior by the assertion that each individual's behavior is optimal
   with respect to some beliefs. The difficulty is that in the absence of
   consistency, beliefs are effectively untethered from the actual play in
   the game being studied.

Nash equilibrium is typically viewed (particularly in applications) as a
necessary condition for a strategy profile to be worthy of investigation. It is
not sufficient. In many settings, there are multiple Nash equilibria and not
all Nash equilibria are equally plausible. The next two sections introduce

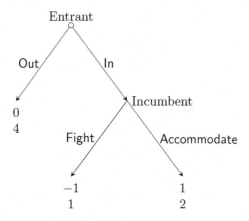

Figure 2.2.1: An entry game.

some considerations that can eliminate some Nash equilibria (Section 2.5 expands on this discussion).

2.2 Credible Threats and Backward Induction

Example 2.2.1 (Entry deterrence). The entry game illustrated in Figure 2.2.1 has two Nash equilibria: (In, Accommodate) and (Out, Fight). The latter profile is implausible. Suppose the incumbent firm threatens to fight entry (while the possibility of announcing such a threat is outside the model, the thought exercise is useful nonetheless). The entrant knows that when the incumbent is faced with the fait accompli of entry, the incumbent will not carry out the threat (it is not credible), and so the entrant ignores the threat.   ★

In the (Out, Fight) equilibrium of Example 2.2.1, the incumbent is indifferent between Accommodate and Fight; because the entrant chooses Out, the incumbent's information set is off-the-path-of-play, and so the incumbent's choice is irrelevant. The credibility logic essentially breaks (resolves) the indifference by having the incumbent evaluate the payoff implications of his choice conditional on the information set being reached. In Example 2.2.1, this evaluation is trivial, since that information set is the last information set *and* is a singleton information set (i.e., the information set contains a single node).

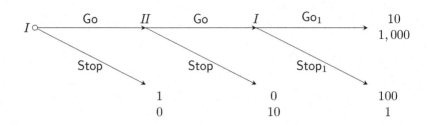

Figure 2.2.2: A short centipede game.

More generally, if there are many singleton information sets, it is natural to make such evaluations at the last information sets. Such evaluations determine behavior at those last information sets, resulting in a shorter game and we can then proceed recursively.

Definition 2.2.1. *Fix a finite extensive form game of perfect information. A* backward induction solution *is a strategy profile $\hat{s}$ obtained as follows:*

1. *Since the game is finite, there is at least one "last" decision node, t^*. Since the game has perfect information, the information set containing t^* is the singleton $\{t^*\}$. With only a slight abuse of notation, denote the information set by t^*. Player $\iota(t^*)$'s action choice under $\hat{s}$ at t^*, $\hat{s}_{\iota(t^*)}(t^*)$, is an action reaching a terminal node that maximizes $\iota(t^*)$'s payoff (over all terminal nodes reachable from t^*).[2]*

2. *With player $\iota(t^*)$'s action at t^* fixed, we now have a finite extensive form game with one less decision node. Now apply step 1 again. Continue till an action has been specified at* every *decision node.*

A backward induction outcome *is the terminal node reached by a backward induction solution.*

Example 2.2.2 (Rosenthal's (1981) centipede game). The perils of backward induction are illustrated in Figure 2.2.2, with reduced normal form given in Figure 2.2.3.

A longer (and more dramatic) version of the centipede is given in Figure 2.2.4.

The backward induction solution in both cases is that both players choose to stop the game at *every* decision point.   ★

Example 2.2.3 (Another cautionary example). The backward solution for the game in Figure 2.2.5 is (D_1R_2, r).

---

[2]This action is often unique, but need not to be. If it is not unique, each choice leads to a distinct backward induction solution.

	Stop	Go
Stop	$1, 0$	$1, 0$
Go,Stop$_1$	$0, 10$	$100, 1$
Go,Go$_1$	$0, 10$	$10, 1000$

Figure 2.2.3: The reduced normal form for the short centipede in Figure 2.2.2.

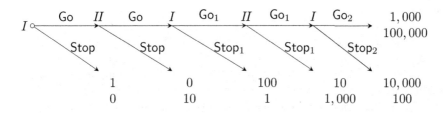

Figure 2.2.4: A long centipede game.

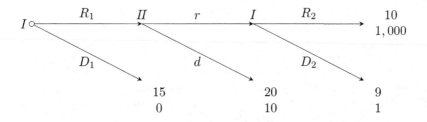

Figure 2.2.5: Another cautionary example.

But if II is asked to make a move, how confident should she be that I will play R_2 after r? After all, she already has evidence that I is not following the backward induction solution (i.e., she already has evidence that I may not be "rational," or perhaps that her beliefs about I's payoffs may be incorrect). And 9 is close to 10. But then playing d may be sensible. Of course, that action seems to justify I's original "irrational" action. ★

Since every backward induction solution is pure strategy Nash equilibrium, we immediately have the following trivial existence theorem.

Theorem 2.2.1. *A finite game of perfect information has a pure strategy Nash equilibrium.*

Schwalbe and Walker (2001) describe the role of this result (and its misattribution) in the early history of game theory.

Remark 2.2.1. Games with ties arise naturally. For example, the ultimatum game (Problem 2.6.7) has ties because all terminal nodes reached by a rejection give both players a zero payoff. Ties can result in multiple best replies at an information set. If multiple best replies do arise, then (as should be clear from Definition 2.2.1), there is a (different) backward induction solution for each best reply.

The restriction to finite games is important. If multiple best replies do arise, then there need not be a "backward induction" solution corresponding to each best reply (the ultimatum game of Problem 2.6.7 is a simple, but important, example). Moreover, there need not be any "backward induction" solution (Problem 4.3.3(c)). ♦

2.2.1   Backward Induction and Iterated Weak Dominance

"Generically in extensive form payoffs," the terminal node reached by the backward induction solution agrees with the terminal node reached by any strategy profile left after the iterated deletion of weakly dominated strategies. Given the distinction between extensive form and reduced normal form strategies, there is no hope for anything stronger.

Theorem 2.2.2 (Rochet, 1980[3]). *Suppose* Γ *is a finite extensive form game of perfect information with no ties, i.e., for all players* i *and all pairs of*

---

[3]See also Marx and Swinkels (1997).

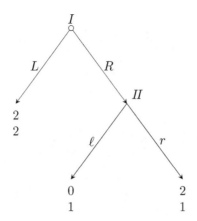

Figure 2.2.6: An example with ties. The backward induction solutions are $L\ell$, Lr, and Rr. But R is weakly dominated for I.

terminal nodes z and z', $u_i(z) \neq u_i(z')$.[4] *The game has a unique backward induction outcome. Let s be a strategy profile that remains after some maximal sequence of iterated deletions of weakly dominated strategies (i.e., the order of deletions is arbitrary and no further deletions are possible). The outcome reached by s is the backward induction outcome.*

Example 2.2.4 (An example with ties). If the game has ties, then iterated deletion of weakly dominated strategies can be stronger than backward induction (see Figure 2.2.6).   ★

2.3   Subgame Perfection

Example 2.3.1. In the game illustrated in Figure 2.3.1, the profile (L, T, r) is Nash. Is it plausible?   ★

Define $S(t) := \{t' \in T : t \prec t'\}$.

---

[4]This property is *generic in extensive form payoffs* in the following sense: Fix an extensive game tree (i.e., everything except the assignment of payoffs to terminal nodes $z \in Z$). The space of games (with that tree) is the space of all payoff assignments to the terminal nodes, $\mathbb{R}^{n|Z|}$. The space of games is thus a finite dimensional Euclidean space. The set of payoff assignments that violate the property of no ties is a subset of a closed Lebesgue measure zero subset of $\mathbb{R}^{n|Z|}$.

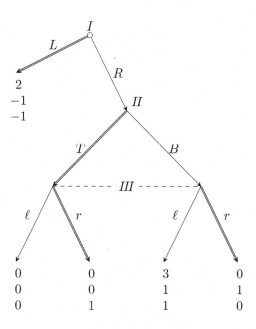

Figure 2.3.1: The profile (L, T, r) is Nash. Is it plausible?

Definition 2.3.1. *The subset $T^t := \{t\} \cup S(t)$, of T, together with payoffs, etc., restricted to T^t, is a* subgame *if for all information sets h,*

$$h \cap T^t \neq \varnothing \Rightarrow h \subset T^t.$$

Observe that the information set containing the initial node of a subgame is necessarily a singleton.

Definition 2.3.2. *A strategy profile is a* subgame perfect equilibrium *if it specifies a Nash equilibrium in every subgame.*

Example 2.3.2 (augmented PD). A (once-)repeated stage game with perfect monitoring. The *stage game* in Figure 2.3.2 is played twice, action choices in the first round are observed by both players (i.e., perfectly monitored), and payoffs are added.

Nash strategy profile: Play E in first period, and S in second period as long as opponent also cooperated (played E) in the first period, and play P if opponent didn't. Every first period action profile describes an information set for each player. Player i's strategy is

$$s_i^1 = E,$$

	E	S	P
E	$2,2$	$-1,3$	$-1,-1$
S	$3,-1$	$0,0$	$-1,-1$
P	$-1,-1$	$-1,-1$	$-2,-2$

Figure 2.3.2: The augmented prisoner's dilemma for Example 2.3.2.

	L	C	R
T	$4,6$	$0,0$	$9,0$
M	$0,0$	$6,4$	$0,0$
B	$0,0$	$0,0$	$8,8$

Figure 2.3.3: The stage game for Example 2.3.3.

$$s_i^2(a_i, a_j) = \begin{cases} S, & \text{if } a_j = E, \\ P, & \text{if } a_j \neq E. \end{cases}$$

This profile is not subgame perfect: Every first period action profile induces a subgame, on which SS must be played. But the profile specifies PS after ES, for example.

The only subgame perfect equilibrium is to always play S.   ★

Remark 2.3.1. Why was the prisoner's dilemma augmented in Example 2.3.2? The prisoner's dilemma has the property that the unique Nash equilibrium gives each player the lowest payoff that the player can guarantee him/herself, i.e., 0 is each player's security level (see Definition 2.4.2). In order to sustain E in the first period, a player who chooses S in the first period must receive a lower payoff in the second than if he/she had chosen E. But the player already receives the lowest possible payoff after E, and so the only Nash equilibrium outcome is SS in both periods.   ◆

Example 2.3.3 (A different repeated game). The stage game in Figure 2.3.3 played played twice with first round actions observed by both players.

The action profiles (T, L) and (M, C) are both Nash equilibria of the stage game. The following profile (s_1, s_2) of the once-repeated game with payoffs added is subgame perfect:

$$s_1^1 = B,$$

$$s_1^2(x, y) = \begin{cases} M, & \text{if } x = B, \\ T, & \text{if } x \neq B, \end{cases}$$

$$s_2^1 = R,$$

and

$$s_2^2(x, y) = \begin{cases} C, & \text{if } x = B, \\ L, & \text{if } x \neq B. \end{cases}$$

The outcome path induced by (s_1, s_2) is (BR, MC). The first period actions coordinate the play in the second period.

These are strategies of the extensive form of the repeated game, not of the reduced normal form. The reduced form strategies corresponding to (s_1, s_2) are

$$\hat{s}_1^1 = B,$$

$$\hat{s}_1^2(y) = M, \text{ for all } y,$$

$$\hat{s}_2^1 = R,$$

and

$$s_2^2(x) = \begin{cases} C, & \text{if } x = B, \\ L, & \text{if } x \neq B. \end{cases} \qquad \bigstar$$

Example 2.3.4. Consider the extensive form in Figure 2.3.4.

The game has three Nash equilibria: (RB, r), (LT, ℓ), and (LB, ℓ). Note that (LT, ℓ), and (LB, ℓ) are distinct extensive form strategy profiles.

The only subgame perfect equilibrium is (RB, r).

But (L, ℓ) is also subgame perfect in the extensive form in Figure 2.3.5. Both games have the same reduced normal form, given in Figure 2.3.6. $\bigstar$

Remark 2.3.2 (Equivalent representations?). A given strategic setting has both a normal form and an extensive form representation. Moreover, the extensive form apparently contains more information (since it contains information about dynamics and information). For example, in Example

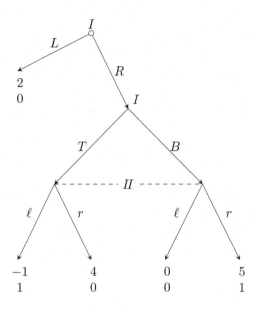

Figure 2.3.4: A game with "nice" subgames.

2.2.1, the application of weak domination to rule out the (Out, Fight) equilibrium can be argued to be less compelling than the backward induction (ex post) argument in the extensive form: faced with the fait accompli of Enter, the incumbent "must" Accommodate. As Kreps and Wilson (1982, p. 886) write: "analysis based on normal form representation inherently ignore the role of anticipated actions off the equilibrium path ... and in the extreme yields Nash equilibria that are patently implausible."

However, backward induction and the iterated deletion of weakly dominated strategies lead to the same outcomes in finite games of perfect information (Theorem 2.2.2). Motivated by this and other considerations, a "classical" argument holds that all extensive forms with the same reduced normal form representation are strategically equivalent (Kohlberg and Mertens, 1986, is a well known statement of this position; see also Elmes and Reny, 1994). Such a view implies that "good" extensive form solutions should not depend on the extensive form in the way illustrated in Example 2.3.4. For more on this issue, see van Damme (1984) and Mailath, Samuelson, and Swinkels (1993, 1997).   ♦

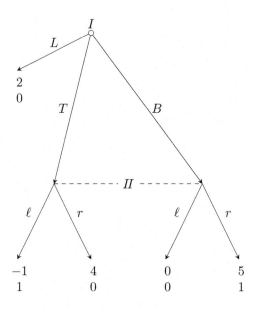

Figure 2.3.5: A game "equivalent" to the game in Figure 2.3.4 with no "nice" subgames.

	ℓ	r
L	2, 0	2, 0
T	$-1, 1$	4, 0
B	0, 0	5, 1

Figure 2.3.6: The reduced form for the games in Figures 2.3.4 and 2.3.5.

2.4   Mixing

2.4.1   Mixed Strategies and Security Levels

Example 2.4.1 (Matching Pennies). A game with no Nash equilibrium is illustrated in Figure 2.4.1.

The greatest payoff that the row player, for example, can *guarantee* himself may appear to be -1 (the unfortunate result of the column player correctly anticipating his choice).

But suppose that the row player uses a fair coin toss to determine his

	H	T
H	$1, -1$	$-1, 1$
T	$-1, 1$	$1, -1$

Figure 2.4.1: Matching Pennies.

action. Then the column player cannot anticipate the row player's choice, and so presumably the row player would be able to do better.   ★

How can we capture the possibility that a player may randomize (such as in Matching Pennies)?

Definition 2.4.1. *Suppose* $\{(S_1, U_1), \ldots, (S_n, U_n)\}$ *is an n-player normal form game. A* mixed strategy *for player i is a probability distribution over* S_i, *denoted* σ_i. *Strategies in* S_i *are called* pure *strategies. A strategy* σ_i *is* completely mixed *if* $\sigma_i(s_i) > 0$ *for all* $s_i \in S_i$.

In order for the set of mixed strategies to have a nice mathematical structure (such as being metrizable or compact), we need the set of pure strategies to also have a nice structure (often complete separable metric, i.e., Polish). For our purposes here, it will suffice to consider countable sets, or nice subsets of $\mathbb{R}^k$. More generally, a mixed strategy is a probability *measure* over the set of pure strategies. The set of probability measures over a set A is denoted $\Delta(A)$.

If S_i is countable, think of a mixed strategy σ_i as a mapping $\sigma_i : S_i \to [0, 1]$ such that $\sum_{s_i \in S_i} \sigma_i(s_i) = 1$.

Since players are expected utility maximizers, extend U_i to $\prod_{j=1}^{n} \Delta(S_j)$ by taking expected values, so that U_i is i's expected payoff under randomization.

If S_i is countable,

$$U_i(\sigma_1, \ldots, \sigma_n) = \sum_{s_1 \in S_1} \cdots \sum_{s_n \in S_n} U_i(s_1, \ldots, s_n) \sigma_1(s_1) \cdots \sigma_n(s_n).$$

Writing

$$U_i(s_i, \sigma_{-i}) = \sum_{s_{-i} \in S_{-i}} U_i(s_i, s_{-i}) \prod_{j \neq i} \sigma_j(s_j),$$

we then have

$$U_i(\sigma_i, \sigma_{-i}) = \sum_{s_i \in S_i} U_i(s_i, \sigma_{-i}) \sigma_i(s_i).$$

	L	R
T	3	0
M	2	2
B	0	3

Figure 2.4.2: In this game (payoffs are for the row player), M is not dominated by any strategy (pure or mixed) and it is the unique best reply to $\frac{1}{2} \circ L + \frac{1}{2} \circ R$.

	L	R
T	5	0
M	2	2
B	0	5

Figure 2.4.3: In this game (payoffs are for the row player), M is not dominated by T or B, it is never a best reply, and it is strictly dominated by $\frac{1}{2} \circ T + \frac{1}{2} \circ B$.

Definition 2.4.2. *Player i's security level (or payoff) is the greatest payoff that i can guarantee himself:*

$$\underline{v}_i = \sup_{\sigma_i \in \Delta(S_i)} \ \inf_{\sigma_{-i} \in \prod_{j \neq i} \Delta(S_j)} \ U_i(\sigma_i, \sigma_{-i}).$$

If σ_i^* achieves the sup, *then σ_i^* is a security strategy for i.*

In matching pennies, each player's security level is 0, guaranteed by the security strategy $\frac{1}{2} \circ H + \frac{1}{2} \circ T$.

2.4.2   Domination and Optimality

Example 2.4.2. In the game in Figure 2.4.2, M is not strictly dominated by any strategy (pure or mixed) and it is the unique best reply to $\frac{1}{2} \circ L + \frac{1}{2} \circ R$. In the game in Figure 2.4.3, M is not strictly dominated by T or B, it is never a best reply, and it is strictly dominated by $\frac{1}{2} \circ T + \frac{1}{2} \circ B$.   ★

Definition 2.4.3. *The strategy $s'_i \in S_i$ is* strictly dominated *by the mixed strategy $\sigma_i \in \Delta(S_i)$ if*

$$U_i(\sigma_i, s_{-i}) > U_i(s'_i, s_{-i}) \qquad \forall s_{-i} \in S_{-i}.$$

The strategy $s'_i \in S_i$ is weakly dominated *by the mixed strategy $\sigma_i \in \Delta(S_i)$ if*

$$U_i(\sigma_i, s_{-i}) \geq U_i(s'_i, s_{-i}) \qquad \forall s_{-i} \in S_{-i}$$

and there exists $s^*_{-i} \in S_{-i}$ such that

$$U_i(\sigma_i, s^*_{-i}) > U_i(s'_i, s^*_{-i}).$$

A strategy is admissible *if it not weakly dominated. A strategy is* inadmissible *if it is weakly dominated (by some mixed strategy).*

Henceforth, a strategy is (strictly) undominated if there is no *mixed* strategy that strictly dominates it.[5]

It is immediate that Definition 2.4.3 is equivalent to the same requirement with the displayed inequalities in Definition 2.4.3 holding with "σ_{-i}" and "for all $\sigma_{-i} \in \prod_{j \neq i} \Delta(S_j)$" replacing "$s_{-i}$" and "for all $s_{-i} \in S_{-i}$" (respectively).

Lemma 2.4.1. *Suppose $n = 2$. The strategy $s'_1 \in S_1$ is not strictly dominated by a pure or mixed strategy if, and only if, $s'_1 \in \arg\max u_1(s_1, \sigma_2)$ for some $\sigma_2 \in \Delta(S_2)$.*

For more than two players, it is possible that a strategy can fail to be a best reply to any mixed profile for the other players, and yet that strategy is *not* strictly dominated (see Problem 2.6.17 for an example).

Proof. We present the proof for finite S_i. The proof for infinite S_i is similar.

($\Longleftarrow$) If there exists $\sigma_2 \in \Delta(S_2)$ such that $s'_1 \in \arg\max U_1(s_1, \sigma_2)$, then it is straightforward to show that s'_1 is not strictly dominated by any other pure or mixed strategy (left as an exercise).

($\Longrightarrow$) We prove this direction by proving the contrapositive.[6] So, suppose s'_1 is a player 1 strategy satisfying

$$s'_1 \notin \arg\max U_1(s_1, \sigma_2) \qquad \forall \sigma_2 \in \Delta(S_2). \tag{2.4.1}$$

---

[5] Recall that a pure strategy is a special case of a mixed strategy, so an undominated strategy is not dominated by any pure strategy.

[6] The contrapositive of the conditional statement $(A \Rightarrow B)$ is $(\neg B \Rightarrow \neg A)$, and has the same truth value as the original conditional statement.

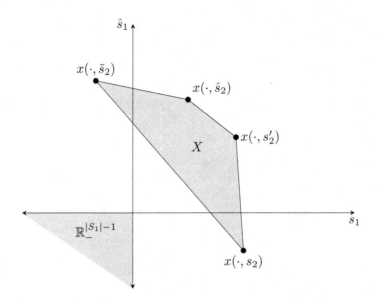

Figure 2.4.4: The sets X and $\mathbb{R}^{|S_1|-1}_-$.

We need to find a mixed strategy σ_1^* strictly dominating s_1'.

For $s_1 \neq s_1'$, define $x(s_1, s_2) = U_1(s_1, s_2) - U_1(s_1', s_2)$. Observe that for any $s_2 \in S_2$, we can represent the vector of payoff differences $\{x(s_1, s_2) : s_1 \neq s_1'\}$ as a point $x = (x_{s_1})_{s_1 \neq s_1'} \in \mathbb{R}^{|S_1|-1}$. Define

$$X := \operatorname{conv}\{x \in \mathbb{R}^{|S_1|-1} : \exists s_2 \in S_2, x_{s_1} = x(s_1, s_2) \ \forall s_1 \neq s_1'\}.$$

Denote the closed negative orthant by $\mathbb{R}^{|S_1|-1}_- := \{x \in \mathbb{R}^{|S_1|-1} : x_{s_1} \leq 0, \ \forall s_1 \neq s_1'\}$. Equation (2.4.1) implies that for all $\sigma_2 \in \Delta(S_2)$, there exists s_1 such that $\sum x(s_1, s_2)\sigma_2(s_2) > 0$, and so $\mathbb{R}^{|S_1|-1}_- \cap X = \varnothing$. Moreover, X is closed, since it is the convex hull of a finite number of vectors. See Figure 2.4.4.

By an appropriate strict separating hyperplane theorem (see, for example, Vohra, 2005, Theorem 3.7), there exists $\lambda \in \mathbb{R}^{|S_1|-1} \setminus \{0\}$ such that $\lambda \cdot x > \lambda \cdot x'$ for all $x \in X$ and all $x' \in \mathbb{R}^{|S_1|-1}_-$. Since $\mathbb{R}^{|S_1|-1}_-$ is unbounded below, $\lambda(s_1) \geq 0$ for all s_1 (otherwise making $|x'(s_1)|$ large enough for s_1 satisfying $\lambda(s_1) < 0$ ensures $\lambda \cdot x' > \lambda \cdot x$). Define

$$\sigma_1^*(s_1'') = \begin{cases} \lambda(s_1'') \Big/ \sum_{s_1 \neq s_1'} \lambda(s_1) , & \text{if } s_1'' \neq s_1', \\ 0, & \text{if } s_1'' = s_1'. \end{cases}$$

We now argue that σ_1^* is indeed a mixed strategy for 1 strictly dominating s_1': Since $0 \in \mathbb{R}_{-}^{|S_1|-1}$, we have $\lambda \cdot x > 0$ for all $x \in X$, and so

$$\sum_{s_1 \neq s_1'} \sigma_1^*(s_1) \sum_{s_2} x(s_1, s_2) \sigma_2(s_2) > 0, \qquad \forall \sigma_2,$$

i.e., for all σ_2,

$$U_1(\sigma_1^*, \sigma_2) = \sum_{s_1 \neq s_1', s_2} U_1(s_1, s_2) \sigma_1^*(s_1) \sigma_2(s_2)$$

$$> \sum_{s_1 \neq s_1', s_2} U_1(s_1', s_2) \sigma_1^*(s_1) \sigma_2(s_2) = U_1(s_1', \sigma_2).$$

∎

Remark 2.4.1. This proof requires us to strictly separate two disjoint closed convex sets (one bounded), rather than a point from a closed convex set (the standard separating hyperplane theorem). To apply the standard theorem, define $Y := \{y \in \mathbb{R}^{|S_1|-1} : \exists x \in X, y_\ell \geq x_\ell \forall \ell\}$. Clearly Y is closed, convex and $0 \notin Y$. We can proceed as in the proof, since the normal vector for the separating hyperplane must again have only nonnegative coordinates (use now the unboundedness of Y). ◆

Remark 2.4.2. There is a similar result for admissibility:

Lemma 2.4.2. *Suppose $n = 2$. The strategy $s_1' \in S_1$ is admissible if, and only if, $s_1' \in \arg\max U_1(s_1, \sigma_2)$ for some full support $\sigma_2 \in \Delta(S_2)$.*

A hint for the proof is given in Problem 2.6.15. ◆

At least for two players, iterated strict dominance is thus iterated non-best replies—the *rationalizability* notion of Bernheim (1984) and Pearce (1984). For more than two players, as illustrated in Problem 2.6.17, there is an issue of correlation: A strategy may be a best reply to a correlated strategy profile of two other players, but not a best reply to any profile of mixed strategies (which requires independent randomization across players). The rationalizability notion of Bernheim (1984) and Pearce (1984) does not allow such correlation, since it focuses on deleted strategies that are not best replies to any *mixed* profile of the other players.

Lemma 2.4.1 holds for mixed strategies (see Problem 2.6.14).

Remark 2.4.3. Fix a two player game played by Bruce and Sheila, and interpret *rationality* as not playing an action that is not a best reply to some belief over the opponent's play. Then if Sheila knows that Bruce is rational, she knows he will not play an action that is not a best reply to some belief over her play, that is, Bruce will not play a strictly dominated action. Hence, the result of the iterative deletion of strictly dominated actions can be interpreted as the result of the assumption that rationality is common knowledge between Bruce and Sheila (i.e., each is rational, each knows each is rational, each knows that each knows that each is rational, and so on). Hence the use of the term *rationalizability* by Bernheim (1984) and Pearce (1984).

What about the iterative deletion of weakly dominated actions? There is a similar interpretation, but complicated by the kinds of issues raised in Remark 1.2.1 (first raised by Samuelson, 1992).[7] See Brandenburger, Friedenberg, and Keisler (2008) for (much) more on this; their Section 2 provides a nice heuristic introduction to the issues.   ♦

2.4.3   Equilibrium in Mixed Strategies

Definition 2.4.4. *Suppose* $\{(S_1, U_1), \ldots, (S_n, U_n)\}$ *is an n-player normal form game. A Nash equilibrium in mixed strategies is a profile* $(\sigma_1^*, \ldots, \sigma_n^*)$ *such that, for all i and for all* $\sigma_i \in \Delta(S_i)$,

$$U_i(\sigma_i^*, \sigma_{-i}^*) \geq U_i(\sigma_i, \sigma_{-i}^*). \tag{2.4.2}$$

Since

$$U_i(\sigma_i, \sigma_{-i}^*) = \sum_{s_i \in S_i} U_i(s_i, \sigma_{-i}^*)\sigma_i(s_i) \leq \max_{s_i'} U_i(s_i', \sigma_{-i}^*),$$

the requirement (2.4.2) can be equivalently written as for all $s_i \in S_i$,

$$U_i(\sigma_i^*, \sigma_{-i}^*) \geq U_i(s_i, \sigma_{-i}^*).$$

Lemma 2.4.3. *Suppose* S_i *is countable. A strategy* σ_i^* *is a best reply to* σ_{-i}^* *(i.e., satisfies (2.4.2)) if, and only if,*

$$\sigma_i^*(s_i') > 0 \implies s_i' \in \arg\max_{s_i} U_i(s_i, \sigma_{-i}^*).$$

---

[7]In Figure 1.2.3, M is deleted because the column player believes that the row player may play B, and indeed if only M is deleted, then T is deleted for the row player. Of course, a similar logic (with a different conclusion!) applies if R is deleted first. But if the row player is truly convinced that the column player will play L then there is no reason to delete any strategy for the row player. This is sometimes called the "inclusion-exclusion" problem: a strategy is deleted for reasons that are subsequently excluded.

	L	R
T	2, 1	0, 0
B	0, 0	1, 1

Figure 2.4.5: The game for Example 2.4.3.

Proof. Left as an exercise (Problem 2.6.18). ∎

Corollary 2.4.1. *A strategy σ_i^* is a best reply to σ_{-i}^* (i.e., satisfies (2.4.2))*
if, and only if, for all $s_i \in S_i$,

$$U_i(\sigma_i^*, \sigma_{-i}^*) \geq U_i(s_i, \sigma_{-i}^*).$$

Example 2.4.3. Consider the game displayed in Figure 2.4.5. For this
game, $\Delta(S_1) = \Delta(S_2) = [0, 1]$, $p = \Pr(T)$, $q = \Pr(L)$.

The best reply correspondences in mixed strategies for the two players
are

$$\phi_1(q) = \begin{cases} \{1\}, & \text{if } q > \frac{1}{3}, \\ [0, 1], & \text{if } q = \frac{1}{3}, \\ \{0\}, & \text{if } q < \frac{1}{3}, \end{cases}$$

and

$$\phi_2(p) = \begin{cases} \{1\}, & \text{if } p > \frac{1}{2}, \\ [0, 1], & \text{if } p = \frac{1}{2}, \\ \{0\}, & \text{if } p < \frac{1}{2}. \end{cases}$$

The best replies are graphed in Figure 2.4.6. ★

For games like auctions, the strategy space is a continuum, often $\mathbb{R}_+$.
For such games, a mixed strategy for a player i is a probability distribution
on $\mathbb{R}_+$ (which we can denote F_i). Player 1's expected payoff from an action
b_1 is

$$\int U_1(s_1, s_2) dF_2(s_2).$$

As an aside, note that this notation covers all relevant possibilities: If the
mixed strategy of player 2 has a countable support $\{s^k\}$ with action s^k

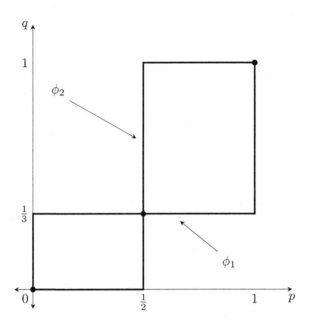

Figure 2.4.6: The best reply mappings for Example 2.4.3.

having probability $\sigma_2(s^k) > 0$ (the distribution is said to be *discrete* in this case), we have

$$\int U_1(s_1, s_2)dF_2(s_2) = \sum_{s^k} U_1(s_1, s^k)\sigma_2(s^k).$$

Of course, $\sum_{s^k} \sigma_2(s^k) = 1$. Any single action receiving strictly positive probability is called an *atom*. If the distribution function describing player 2's behavior has a density f_2 (in which case, there are no atoms), then

$$\int U_1(s_1, s_2)dF_2(s_2) = \int U_1(s_1, s_2)f_2(s_2)ds_2.$$

Finally, combinations of distributions with densities on part of the support and atoms elsewhere, as well as more esoteric possibilities (that are almost never relevant) are also covered.

Suppose F_1 is a best reply for player 1 to player 2's strategy F_2. Then

$$\iint U_1(s_1, s_2)\, dF_2(s_2)\, dF_1(s_1) = \max_{s_1} \int U_1(s_1, s_2)\, dF_2(s_2).$$

Observe first that if F_1 is discrete with support $\{s^k\}$ and action s^k having probability $\sigma_1(s^k) > 0$, then

$$\iint U_1(s_1, s_2) \, dF_2(s_2) \, dF_1(s_1) = \sum_{s^k} \int U_1(s^k, s_2) \, dF_2(s_2) \sigma_1(s^k),$$

and so we immediately have

$$\sum_{s^k} \int U_1(s^k, s_2) \, dF_2(s_2) \sigma_1(s^k) = \max_{s_1} \int U_1(s_1, s_2) \, dF_2(s_2)$$

and so, for all s^k,

$$\int U_1(s^k, s_2) \, dF_2(s_2) = \max_{s_1} \int U_1(s_1, s_2) \, dF_2(s_2).$$

This is just Lemma 2.4.3 and Corollary 2.4.1.

What is the appropriate version of this statement for general F_1? The key observation is that zero probability sets don't matter. Thus, the statement is: Let $\hat{S}_1$ be the set of actions that are suboptimal against F_2, i.e.,

$$\hat{S}_1 = \left\{ \hat{s}_1 : \int U_1(\hat{s}_1, s_2) \, dF_2(s_2) < \max_{s_1} \int U_1(s_1, s_2) \, dF_2(s_2) \right\}.$$

Then the set $\hat{S}_1$ is assigned zero probability by any best response F_1. [If such a set received positive probability under some F_1, then F_1 could not be a best reply, since expected payoffs are clearly increased by moving probability off the set $\hat{S}_1$.]

In most applications, the set $\hat{S}_1$ is disjoint from the support of F_1, in which case player 1 is indeed indifferent over all actions in his support, and each such action maximizes his payoff against F_2. However, Example 2.4.4 is an example where $\hat{S}_1$ includes one point of the support. More generally, the set $S_1 \setminus \hat{S}_1$ must be dense in the support of F_1.[8]

Example 2.4.4. Consider a first-price sealed-bid auction. There are two *risk neutral* bidders with values $v_1 < v_2$. These values are public.

We first prove that this game does not have an equilibrium in pure strategies by contradiction. So, suppose (b_1, b_2) is a pure strategy equilibrium. If $b_1 = b_2 = b$ and $b < v_2$, then bidder 2 has as a profitable deviation outbidding the other player and so avoiding the tie. If $b \geq v_2$, then bidder

---

[8]The support of a probability distribution is the smallest *closed* set receiving probability one. If $S_1 \setminus \hat{S}_1$ is not dense in the support of F_1, then F_1 must assign a probability strictly less than 1 to the closure of $S_1 \setminus \hat{S}_1$; but this is impossible if F_1 is a best response, since then F_1 would be assigning strictly positive probability to $\hat{S}_1$.

1 lowering b_1 to avoid a loss is a profitable deviation. Hence, $b_1 \neq b_2$. But then there is a single winner i with payoff $v_i - b_i$. If $v_i - b_i < 0$, then i has any losing bid as a profitable deviation. If $v_i - b_i \geq 0$, then i has any lower bid (strictly higher than b_j) as a profitable deviation. Therefore, no pure strategy pair is an equilibrium. (If bids are restricted to come from a discrete set, so that lowering a bid and still winning for sure may not be feasible, then pure strategy equilibria exist, see Problem 2.6.19.)

We now claim that the following is a mixed strategy equilibrium in which bidder 1 randomizes and bidder 2 does not. Denote by $\underline{b}$ the minimum of the support of bidder 1's bids, and by $\bar{b}$ the maximum of the support of bidder 1's bids. Any mixed strategy profile in which $v_1 \leq \bar{b} < v_2$, bidder 1 does not bid $\bar{b}$ with positive probability (i.e., $\Pr_1(b_1 = \bar{b}) = 0$),[9] and for all $b_2 \leq \bar{b}$,

$$[\Pr_1(b_1 < b_2) + \tfrac{1}{2}\Pr_1(b_1 = b_2)](v_2 - b_2) \leq v_2 - \bar{b}; \qquad (2.4.3)$$

and bidder 2 bids $\bar{b}$ with probability 1 is a Nash equilibrium. Under this profile, bidder 2 wins the auction for sure, so bidder 1 is indifferent over all bids strictly less than $\bar{b}$. Inequality (2.4.3) is the requirement that bidder 2 not find it profitable to bid less than $\bar{b}$; it is clearly suboptimal to bid more than $\bar{b}$.

There are many such equilibria, differing in the behavior of bidder 1 (which may, but need not, involve atoms at $b_1 < \bar{b}$). Note also that the only requirement on $\underline{b}$ is $\underline{b} < \bar{b}$. In particular, there is a sequence of mixed strategy equilibria in which $\underline{b} \to \bar{b}$. The limit is *not* a Nash equilibrium, but the limit outcome can be described as both bidders bid $\bar{b}$, and the tie is broken in favor of the high value bidder. Unless $\bar{b} = v_1$, the mixed strategy equilibrium involves weakly dominated strategies for bidder 1. It is worth noting that there are analogous pure strategy equilibria in the game with a discrete set of bids.

Note that if $\bar{b} > v_1$, $b_1 = \bar{b}$ is not a best reply for bidder 1 to the bid of $b_2 = \bar{b}$ by bidder 2. As discussed before the example, this does not contradict Lemma 2.4.3 since each bidder's pure strategy space is not countable.   ★

2.4.4 Behavior Strategies

How to model mixing in extensive form games? Recall Definition 1.3.3:

---

[9]Since bidder 1 is randomizing, bidder 1's bid b_1 is a random variable; its probability distribution is denoted by $\Pr_1$.

Definition 2.4.5. *A pure strategy for player i is a function*

$$s_i : H_i \to \bigcup_h A(h) \text{ such that } s_i(h) \in A(h) \; \forall h \in H_i.$$

We can think of a pure strategy for i as a book:[10] The book has $|H_i|$ pages (one page for each information set), and on each page is written the action the pure strategy specifies for the corresponding information set. The collection of i's pure strategies is denoted by S_i; of course, $|S_i| < \infty$ for finite extensive form games. The collection S_i is a library, consisting of all of i's "books." A mixed strategy for i is a probability distribution over the library:

Definition 2.4.6. A mixed strategy *for player i, σ_i, is a probability distribution over S_i, i.e., $\sigma_i \in \Delta(S_i)$.*

We now introduce a different kind of book. The book again has $|H_i|$ pages, but now there is a probability distribution written on each page, describing the randomization player i will use at each information set:

Definition 2.4.7. *A behavior strategy for player i is a function*

$$b_i : H_i \to \bigcup_h \Delta(A(h)) \text{ such that } b_i(h) \in \Delta(A(h)), \; \forall h \in H_i.$$

Write $b_i(h)(a)$ for the probability assigned to the action $a \in A(h)$ by the probability distribution $b_i(h)$. At times (particularly in the proof of Theorem 2.4.1 and Chapter 5), we assume actions have been labeled so that $A(h) \cap A(h') = \varnothing$ for all $h \neq h'$; an action uniquely identifies the information set at which it is taken. In that case, we can write $b_i(a)$, rather than the more cumbersome $b_i(h)(a)$. With this notational convention, it is straightforward to define, for any behavior strategy profile b, the induced probability distribution on Z, $\mathbf{P}^b$ (see Section 5.1).

Note that if $|H_i| = 3$ and $|A(h)| = 2 \; \forall h \in H_i$, then $|S_i| = 8$ and so $\Delta(S_i)$ is a 7-dimensional simplex. A behavior strategy in this case requires only 3 numbers (the probability on the first action in each information set).

The behavior strategy corresponding to a pure strategy s_i is given by

$$b_i(h) = \delta_{s_i(h)}, \; \forall h \in H_i,$$

where $\delta_x \in \Delta(X)$, $x \in X$, is the degenerate distribution,

$$\delta_x(y) = \begin{cases} 1, & \text{if } y = x, \\ 0, & \text{otherwise.} \end{cases}$$

---

[10]I can't remember where I first encountered this interpretation, but it is too good not to use.

Definition 2.4.8. *Two strategies for a player i are* realization equivalent *if, fixing the strategies of the other players, the two strategies induce the same distribution over outcomes (terminal nodes).*

Thus, two strategies that are realization equivalent are strategically equivalent (Definition 1.3.7).

Moreover, if two extensive form strategies only differ in the specification of behavior at an information set that one of those strategies had precluded, then the two strategies are realization equivalent. For example the strategies Stop,Stop$_1$ and Stop,Go$_1$ in Example 2.2.2 are realization equivalent.

Given a behavior strategy b_i, the realization-equivalent mixed strategy $\sigma_i \in \Delta(S_i)$ is

$$\sigma_i(s_i) = \prod_{h \in H_i} b_i(h)(s_i(h)).$$

Theorem 2.4.1 (Kuhn, 1953). *Every mixed strategy has a realization equivalent behavior strategy.*

The behavior strategy realization equivalent to the mixture σ_i can be calculated as follows: Fix an information set h for player i (i.e., $h \in H_i$). Suppose h is reached with strictly positive probability under σ_i, for *some* specification $\hat{s}_{-i}$. Then, $b_i(h)$ is the distribution over $A(h)$ implied by σ_i *conditional* on h being reached. While this calculation appears to depend on the particular choice of $\hat{s}_{-i}$, it turns out it does not. (If for all specifications s_{-i}, h is reached with zero probability under σ_i, then $b_i(h)$ can be determined arbitrarily.) The proof of Theorem 2.4.1 is in Appendix 14.1.

Using behavior strategies, mixing can be easily accommodated in subgame perfect equilibria (for an example, see Problem 2.6.21).

2.5   Dealing with Multiplicity

In many situations, the multiplicity of Nash equilibria appropriately reflects an inherent indeterminacy in the theory. Any game is an abstraction, studied out of context. It should not be surprising that, once the detail and context of a strategic interaction have been eliminated, theory often does not yield a unique equilibrium. As a trivial example, consider the interaction between two pedestrians. Passing on the right and passing on the left are both Nash equilibria. The particular equilibrium played is determined by context (people in the U.S. pass on the right, in Australia on the left).

Nonetheless, not all Nash equilibria are equally deserving of our attention. Broadly speaking, there are two approaches to identifying those equilibria: refinement and selection.

2.5.1   Refinements

As we have seen in Example 2.1.2, sometimes Nash equilibria require the "right" specification of behavior for a player who has multiple best responses. But more careful, or refined, consideration of the game may lead the analyst to conclude that the "right" specification is implausible, suggesting that the Nash equilibrium is implausible (that is, the Nash equilibrium is, in fact, *not* self-enforcing). In such a case, the Nash equilibrium can be said to have failed this more refined consideration. A Nash equilibrium *refinement* effectively imposes constraints that the choice from multiple best responses must satisfy.

A Nash equilibrium of a game with a nontrivial extensive form typically involves a selection from multiple best replies, since a player will be indifferent between strategies that only differ at unreached information sets. Requiring that the behavior at unreached information sets be optimal (credible, sequential rational) underlies many refinements. We have already seen two well-known instances of such equilibrium refinements: backward induction and subgame perfection. We explore these ideas in more detail in Chapter 5. Here, I briefly introduce two important refinements.

Forward Induction

Example 2.5.1. Consider the game in Figure 2.5.1. This game has two pure strategy subgame perfect equilibria: (RT, ℓ) and (LB, r). However, the second equilibrium fails to be "self-enforcing" in the following sense: Suppose II expects I to choose L, and then sees that I has *not* chosen L. What should II think? The only possible reason I could have for deviating is that she is expecting a larger payoff from the deviation, which leads II to infer that I intends to choose T after the deviation, which "rationalizes" the deviation.

This logic is mirrored in the iterated deletion of weakly dominated strategies in the reduced normal form of this game, which eliminates (LB, r).

The logic underlying the failure of the equilibrium in which II chooses L to be self-enforcing is an example of *forward induction* (Kohlberg and Mertens, 1986), which essentially requires that responses to deviations from equilibrium be, if possible, consistent with some rationalization of the deviation. In the words of Kohlberg and Mertens (1986, p. 1013):

> Essentially, what is involved here is an argument of "forward induction": a subgame should not be treated as a separate game, because it is preceded by a very specific form of preplay

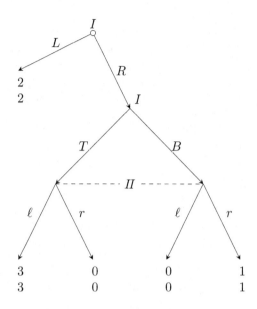

Figure 2.5.1: The game for Example 2.5.1.

communication–the play leading to the subgame. In the above example, it is common knowledge that, when player II has to play in the subgame, preplay communication (for the subgame) has effectively ended with the following message from player I to player II: "Look, I had the opportunity to get 2 for sure, and nevertheless I decided to play in this subgame, and my move is already made. And we both know that you can no longer talk to me, because we are in the game, and my move is made. So think now well, and make your decision."

While seemingly compelling, forward induction has strong and some-times surprising implications. For example, Kohlberg and Mertens's (1986) implementation of forward induction implies separation in canonical signal-ing games (Section 6.2.3). For more on this, see Kohlberg (1990), Osborne and Rubinstein (1994, Section 6.6.2), van Damme (2002), and Hillas and Kohlberg (2002).

Trembling Hand Perfection in the Normal Form

Example 2.5.2. The game in Figure 2.5.2 has two Nash equilibria (UL and DR), but only DR is plausible. The other equilibrium is in weakly

	L	R
U	2, 1	0, 0
D	2, 0	1, 1

Figure 2.5.2: The game for Example 2.5.2.

dominated strategies. ★

It is natural to require that equilibria be "robust" to small mistakes. The following notion captures the minimal requirement that behavior should be robust at least to some mistakes. The stronger requirement that behavior be robust to all mistakes is too strong, in the sense that typically, no behavior satisfies that requirement, and so is rarely imposed (see Problem 2.6.24 for an example).

Definition 2.5.1. *An equilibrium σ of a finite normal from game G is (normal form) trembling hand perfect if there exists a sequence $\{\sigma^k\}_k$ of completely mixed strategy profiles converging to σ such that σ_i is a best reply to every σ^k_{-i} in the sequence.*

This is *not* the standard definition in the literature, but is equivalent to it (see Subsection 14.2.1).

Every finite normal form game has a trembling hand perfect equilibrium (see Subsection 14.2.1).

Weakly dominated strategies cannot be played in a trembling hand perfect equilibrium:

Theorem 2.5.1. *If a strategy profile in a finite normal form game is trembling hand perfect then it is a Nash equilibrium in weakly undominated strategies. If there are only two players, every Nash equilibrium in weakly undominated strategies is trembling hand perfect.*

Proof. The proof of the first statement is straightforward and left as an exercise (Problem 2.6.22). A proof of the second statement can be found in van Damme (1991, Theorem 3.2.2). ∎

Problem 2.6.23 illustrates the two statements of Theorem 2.5.1.

Note that normal form trembling hand perfection does not imply subgame perfection (Problem 2.6.25). We explore the role of trembles in extensive form games in Section 5.3.

Section 14.2 contains additional material on trembling hand perfect equilibria.

	A	a	b
A	2,2	0,0	0,0
a	0,0	0,0	2,2
b	0,0	2,2	0,0

Figure 2.5.3: A game with three pure strategy equilibria, but only one, AA, is focal.

	ℓ	r
T	2,2	0,0
B	0,0	1,1

Figure 2.5.4: The game for Example 2.5.4.

2.5.2   Selection

Even when a game has multiple strict Nash equilibria, we may be able to select between them on various grounds. Unlike equilibrium refinements, an equilibrium selection argument often invokes a broader context. Learning is one context (i.e., it may be that some Nash equilibria can be learned, while others cannot), which we briefly discuss in Section 4.2.

Example 2.5.3 (Focal Points). The game in Figure 2.5.3 has three pure strategy equilibria. If players engage in preplay communication, we can be confident the players would agree on one of these equilibria, but in the absence of other considerations, not know which one. But the distinguished nature of AA is one such consideration, and so AA is *focal*. Even in the absence of preplay communication, the distinguished nature of AA makes it the natural prediction for this game.   ★

Example 2.5.4 (Payoff Dominance). One source of focalness is payoffs. The game in Figure 2.5.4 has two pure strategy Nash equilibria. The profile $T\ell$ is a Nash equilibrium that *payoff dominates* the other equilibrium, and indeed dominates every other cell. If players can engage in preplay communication, then this increases the persuasiveness of payoff dominance.   ★

Example 2.5.5 (Renegotiation). The stage game is displayed in Figure 2.5.5.

	L	C	R
T	$4,4$	$0,0$	$9,0$
M	$0,0$	$6,6$	$0,0$
B	$0,0$	$0,0$	$8,8$

Figure 2.5.5: The stage game for Example 2.5.5.

	A	B
A	$9,9$	$0,5$
B	$5,0$	$7,7$

Figure 2.5.6: The stag hunt.

The profiles (T, L) and (M, C) are both Nash equilibria of the stage game. The following profile (s_1, s_2) of the once-repeated game with payoffs added is subgame perfect:

$$s_1^1 = B,$$

$$s_1^2(x, y) = \begin{cases} M, & \text{if } x = B, \\ T, & \text{if } x \neq B, \end{cases}$$

$$s_2^1 = R,$$

$$s_2^2(x, y) = \begin{cases} C, & \text{if } x = B, \text{ and} \\ L, & \text{if } x \neq B. \end{cases}$$

The outcome path induced by (s_1, s_2) is (BR, MC). But TL is Pareto dominated by MC, and so players may renegotiate (which presumably involves some preplay communication) from TL to MC after 1's deviation. But then 1 has no incentive not to play T in the first period.

It is worth contrasting this example with Example 2.3.3.   ★

Example 2.5.6 (Stag hunt game; illustration of risk dominance). Figure 2.5.6 is essentially Rousseau's (1984) stag hunt, where A is hunting the stag and B is catching the hare.

The profile AA is the efficient profile (or is *payoff dominant*). But B is "less risky" than A: Technically, it is *risk dominant* since B is the unique best reply to the uniform lottery over $\{A, B\}$, i.e., to the mixture

$$\frac{1}{2} \circ A + \frac{1}{2} \circ B.$$

It turns out that the risk dominant equilibrium is the only stochastically stable equilibrium, and so selected by stochastic evolutionary arguments (Kandori, Mailath, and Rob, 1993).   ★

2.6 Problems

2.6.1. Suppose $\{(S_i, U_i)_{i=1}^n\}$ is a normal form game, and $\hat{s}_1 \in S_1$ is a weakly dominated strategy for player 1. Let $S_1' = S_1 \setminus \{\hat{s}_1\}$, and $S_i' = S_i$ for $i \neq 1$. Suppose s is a Nash equilibrium of $\{(S_i', U_i)_{i=1}^n\}$. Prove that s is a Nash equilibrium of $\{(S_i, U_i)_{i=1}^n\}$.

2.6.2. Suppose $\{(S_i, U_i)_{i=1}^n\}$ is a normal form game, and s is a Nash equilibrium of $\{(S_i, U_i)_{i=1}^n\}$. Let $\{(S_i', U_i)_{i=1}^n\}$ be the normal form game obtained by the iterated deletion of some or all strictly dominated strategies. Prove that s is a Nash equilibrium of $\{(S_i', U_i)_{i=1}^n\}$. (Of course, you must first show that $s_i \in S_i'$ for all i.) Give an example showing that this is false if *strictly* is replaced by *weakly*.

2.6.3. Consider (again) the Cournot example (Example 1.1.3). What is the Nash Equilibrium of the n-firm Cournot oligopoly? [**Hint:** To calculate the equilibrium, first solve for the total output in the Nash equilibrium.] What happens to both individual firm output and total output as n approaches infinity?

2.6.4. Consider now the Cournot duopoly where inverse demand is $P(Q) = a - Q$ but firms have asymmetric marginal costs: c_i for firm i, $i = 1, 2$.

   (a) What is the Nash equilibrium when $0 < c_i < a/2$ for $i = 1, 2$? What happens to firm 2's equilibrium output when firm 1's costs, c_1, increase? Can you give an intuitive explanation?

   (b) What is the Nash equilibrium when $c_1 < c_2 < a$ but $2c_2 > a+c_1$?

2.6.5. Consider the following Cournot duopoly game: The two firms are identical. The cost function facing each firm is denoted by $C(q)$, is continuously differentiable with $C(0) = 0, C'(0) = 0, C'(q) > 0\ \forall q > 0$. Firm i chooses $q_i, i = 1, 2$. Inverse demand is given by $p = P(Q)$, where $Q = q_1 + q_2$ is total supply. Suppose P is continuous and there exists $\overline{Q} > 0$ such that $P(Q) > 0$ for $Q \in [0, \overline{Q})$ and $P(Q) = 0$ for $Q \geq \overline{Q}$. Assume firm i's profits are strictly concave in q_i for all $q_j, j \neq i$.

(a) Prove that for each value of q_j, firm i ($i \neq j$) has a unique profit maximizing choice. Denote this choice $R_i(q_j)$. Prove that $R_i(q) = R_j(q)$, i.e., the two firms have the same reaction function. Thus, we can drop the subscript of the firm on R.

(b) Prove that $R(0) > 0$ and that $R(\overline{Q}) = 0 < \overline{Q}$.

(c) We know (from the maximum theorem) that R is a continuous function. Use the Intermediate Value Theorem to argue that this Cournot game has at least one symmetric Nash equilibrium, i.e., a quantity q^*, such that (q^*, q^*) is a Nash equilibrium. [**Hint:** Apply the Intermediate Value Theorem to the function $f(q) = R(q) - q$. What does $f(q) = 0$ imply?]

(d) Give some conditions on C and P that are sufficient to imply that firm i's profits are strictly concave in q_i for all $q_j, j \neq i$.

2.6.6. In the canonical Stackelberg model, there are two firms, I and II, producing the same good. Their inverse demand function is $P = 6 - Q$, where Q is market supply. Each firm has a constant marginal cost of \$4 per unit and a capacity constraint of 3 units (the latter restriction will not affect optimal behavior, but assuming it eliminates the possibility of negative prices). Firm I chooses its quantity first. Firm II, knowing firm I's quantity choice, then chooses its quantity. Thus, firm I's strategy space is $S_1 = [0, 3]$ and firm II's strategy space is $S_2 = \{\tau_2 \mid \tau_2 : S_1 \to [0, 3]\}$. A strategy profile is $(q_1, \tau_2) \in S_1 \times S_2$, i.e., an action (quantity choice) for I and a specification for *every* quantity choice of I of an action (quantity choice) for II.

(a) What are the outcome and payoffs of the two firms implied by the strategy profile (q_1, τ_2)?

(b) Show that the following strategy profile does not constitute a Nash equilibrium: $(\frac{1}{2}, \tau_2)$, where $\tau_2(q_1) = (2 - q_1)/2$. Which firm(s) is (are) not playing a best response?

(c) Prove that the following strategy profile constitutes a Nash equilibrium: $(\frac{1}{2}, \hat{\tau}_2)$, where $\hat{\tau}_2(q_1) = \frac{3}{4}$ if $q_1 = \frac{1}{2}$ and $\hat{\tau}_2(q_1) = 3$ if $q_1 \neq \frac{1}{2}$, i.e., II threatens to flood the market unless I produces exactly $\frac{1}{2}$. Is there any other Nash equilibrium which gives the outcome path $(\frac{1}{2}, \frac{3}{4})$? What are the firms' payoffs in this equilibrium?

(d) Prove that the following strategy profile constitutes a Nash equilibrium: $(0, \tilde{\tau}_2)$, where $\tilde{\tau}_2(q_1) = 1$ if $q_1 = 0$ and $\tilde{\tau}_2(q_1) = 3$ if $q_1 \neq 0$, i.e., II threatens to flood the market unless I produces exactly 0. What are the firms' payoffs in this equilibrium?

(e) Given $q_1 \in [0, 2]$, specify a Nash equilibrium strategy profile in which I chooses q_1. Why is it not possible to do this for $q_1 \in (2, 3]$?

(f) What is the unique backward induction solution of this game?

2.6.7. Player 1 and 2 must agree on the division of a pie of size 1. They are playing an ultimatum game (sometimes called a take-it-or-leave-it-offer game): Player 1 makes an offer x from a set $S_1 \subset [0, 1]$, which player 2 accepts or rejects. If player 2 accepts, the payoffs are $1 - x$ to player 1 and x to player 2; if player 2 rejects, both players receive a zero payoff.

(a) Describe the strategy spaces for both players.

(b) Suppose $S_1 = \left\{ 0, \frac{1}{n}, \ldots, \frac{n-1}{n}, 1 \right\}$, for some positive integer $n \geq 2$. Describe all the backward induction solutions.

(c) Suppose $S_1 = [0, 1]$. Describe the unique backward induction solution. (While the game is not a finite game, it is a game of perfect information and has a finite horizon, and so the notion of backward induction applies in an "obvious" way.) Why is it unique?

2.6.8. Consider the extensive form in Figure 2.6.1.

(a) What is the normal form of this game?

(b) Describe the pure strategy Nash equilibrium strategies *and* outcomes of the game.

(c) Describe the pure strategy subgame perfect equilibria (there may only be one).

2.6.9. Consider the following game: Player 1 first chooses between A or B, with A giving a payoff of 1 to each player, and B giving a payoff of 0 to player 1 and 3 to player 2. After player 1 has publicly chosen between A and B, the two players play the game G (with 1 being the row player) in Figure 2.6.2.

Payoffs in the overall game are given by the sum of payoffs from 1's initial choice and the bimatrix game.

(a) What is the extensive form of G?

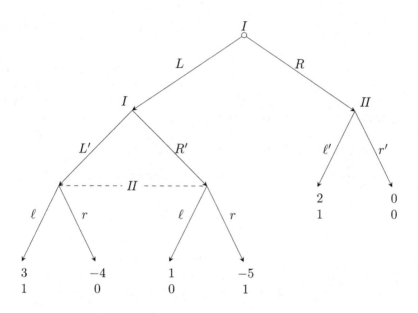

Figure 2.6.1: The game for Problem 2.6.8.

	L	R
U	1, 1	0, 0
D	0, 0	3, 3

Figure 2.6.2: The game G for Problem 2.6.9.

(b) Describe a subgame perfect equilibrium strategy profile in pure strategies in which 1 chooses B.

(c) What is the reduced normal form of G?

(d) What is the result of the iterated deletion of weakly dominated strategies?

2.6.10. Suppose s is a pure strategy Nash equilibrium of a finite extensive form game, Γ. Suppose Γ' is a subgame of Γ that is on the path of play of s. Prove that s prescribes a Nash equilibrium on Γ'. (It is probably easier to first consider the case where there are no moves of nature. The result is also true for mixed strategy Nash equilibria, though the proof is more notationally intimidating.)

2.6.11. Prove that player i 's security level (Definition 2.4.2) is also given by

$$\underline{v}_i = \sup_{\sigma_i \in \Delta(S_i)} \quad \inf_{s_{-i} \in \prod_{j \neq i} S_j} \quad u_i(\sigma_i, s_{-i}).$$

Prove that

$$\underline{v}_i \geq \sup_{s_i \in S_i} \quad \inf_{s_{-i} \in \prod_{j \neq i} S_j} \quad u_i(s_i, s_{-i}),$$

and give an example illustrating that the inequality can be strict.

2.6.12. Suppose the 2×2 normal form game G has a unique Nash equilibrium, and each player's Nash equilibrium strategy and security strategy are both completely mixed.

(a) Describe the implied restrictions on the payoffs in G.

(b) Prove that each player's security level is given by his/her Nash equilibrium payoff.

(c) Give an example showing that (in spite of part (b)), the Nash equilibrium profile need not agree with the strategy profile in which each player is playing his or her security strategy. (This is not possible for zero-sum games, see Problem 4.3.2.)

(d) For games like you found in part (c), which is the better prediction of play, security strategy or Nash equilibrium?

2.6.13. Suppose $\{(S_1, U_1), \ldots, (S_n, U_n)\}$ is a finite normal form game. Prove that if $s'_1 \in S_1$ is strictly dominated in the sense of Definition 2.4.3, then it is not a best reply to any belief over S_{-i}. [While you can prove this by contradiction, try to obtain the direct proof, which is more informative.] (This is the contrapositive of the "straightforward" direction of Lemma 2.4.1.)

$$\begin{array}{c|c|c|}
 & L & R \\
\hline
T & 5,0 & 0,1 \\
\hline
C & 2,6 & 4,0 \\
\hline
B & 0,0 & 5,1 \\
\hline
\end{array}$$

Figure 2.6.3: The game for Problem 2.6.14(b).

2.6.14. (a) Prove that Lemma 2.4.1 also holds for mixed strategies, i.e., prove that $\sigma_1 \in \Delta(S_1)$ is strictly dominated by some other strategy σ_1' (i.e., $U_1(\sigma_1', s_2) > U_1(\sigma_1, s_2)$ for all $s_2 \in S_2$) if and only if σ_1 is not a best reply to any mixture $\sigma_2 \in \Delta(S_2)$.

(b) For the game illustrated in Figure 2.6.3, prove that $\frac{1}{2} \circ T + \frac{1}{2} \circ B$ is not a best reply to any mixture over L and R. Describe a strategy that strictly dominates it.

2.6.15. Prove Lemma 2.4.2. [As for Lemma 2.4.1, one direction is trivial. For the other, prove directly that if a strategy is admissible, then it is a best response to some full support strategy of player 2. This strategy is delivered by an application of a separating hyperplane theorem.]

2.6.16. Suppose $\{(S_1, U_1), (S_2, U_2)\}$ is a two player finite normal form game and let $\widehat{S}_2$ be a strict subset of S_2. Suppose $s_1' \in S_1$ is not a best reply to any strategy σ_2 with support $\widehat{S}_2$. Prove that there exists $\varepsilon > 0$ such that s_1' is not a best reply to any strategy $\sigma_2 \in \Delta(S_2)$ satisfying $\sigma_2(\widehat{S}_2) > 1 - \varepsilon$. Is the restriction to two players important?

2.6.17. Consider the three player game in Figure 2.6.4 (only player 3's payoffs are presented).

(a) Prove that player 3's strategy of M is not strictly dominated.

(b) Prove that player 3's strategy of M is not a best reply to any mixed strategy profile $(\sigma_1, \sigma_2) \in \Delta(S_1) \times \Delta(S_2)$ for players 1 and 2. (The algebra is a little messy. Given the symmetry, it suffices to show that L yields a higher expected payoff than M for all mixed strategy profiles satisfying $\sigma_1(t) \geq \frac{1}{2}$.)

2.6.18. Prove Lemma 2.4.3.

2.6.19. (a) Prove that the game in Example 2.4.4 has no other Nash equilibria than those described.

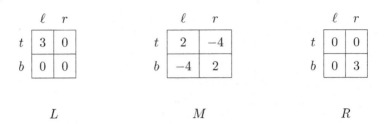

$$L \qquad\qquad\qquad M \qquad\qquad\qquad R$$

Figure 2.6.4: The game for Problem 2.6.17. Player 1 chooses rows (i.e.,
$s_1 \in \{t, b\}$), player 2 chooses columns (i.e., $s_2 \in \{\ell, r\}$), and
player 3 chooses matrices (i.e., $s_3 \in \{L, M, R\}$). Only player
3's payoffs are given.

(b) Suppose that in the game in Example 2.4.4, bidders are restricted
to a discrete set of bids, $\{0, \frac{1}{n}, \frac{2}{n}, \dots\}$, with $n > 2/(v_2 - v_1)$.
The game with the discrete set of possible bids does have pure
strategy Nash equilibria. What are they?

2.6.20. Consider the following variant of a sealed bid auction: There are two
bidders who each value the object at v, and simultaneously submit
bids. As usual, the highest bid wins and in the event of a tie, the
object is awarded on the basis of a fair coin toss. But now *all* bidders
pay their bid. (This is an *all-pay auction*.)

(a) Formulate this auction as a normal form game.

(b) Show that there is no equilibrium in pure strategies.

(c) This game has an equilibrium in mixed strategies. What is it?
(You should verify that the strategies do indeed constitute an
equilibrium).

2.6.21. Suppose player 1 must publicly choose between the game on the left
and the game on the right in Figure 2.6.5 (where player 1 is choosing
rows and player 2 is choosing columns). Prove that this game has
no Nash equilibrium in pure strategies. What is the unique subgame
perfect equilibrium (the equilibrium is in behavior strategies)?

2.6.22. Is it necessary to assume that σ is a Nash equilibrium in the definition of normal form trembling hand perfection (Definition 2.5.1)?
Prove that every trembling hand perfect equilibrium of a finite normal
form game is a Nash equilibrium in weakly undominated strategies.

2.6.23. Two examples to illustrate Theorem 2.5.1.

	t	h
t	$1, -1$	$-1, 1$
h	$-1, 1$	$1, -1$

	T	H
T	$3, -1$	$-1, 1$
H	$-1, 1$	$1, -1$

Figure 2.6.5: The game for Problem 2.6.21.

	L	C	R
T	$2, 2$	$1, 1$	$0, 0$
B	$2, 0$	$0, 0$	$4, 2$

Figure 2.6.6: The game for Problem 2.6.23(a).

(a) For the game in Figure 2.6.6, verify that TL is trembling hand perfect by explicitly describing the sequence of trembles.

(b) The game in Figure 2.6.7 has an undominated Nash equilibrium that is not trembling hand perfect. What is it?

2.6.24. Say a profile σ is *robust to all trembles* if for all sequences $\{\sigma^k\}_k$ of completely mixed strategy profiles converging to σ, σ_i is eventually a best reply to every σ_{-i}^k in the sequence.[11]

(a) Prove that no profile in the game in Figure 2.6.8 is robust to all trembles.

(b) There is an extensive form with a nontrivial subgame that has as its normal the game in Figure 2.6.8. This extensive form game has two subgame perfect equilibria. What are they? Compare with your analysis of part (a).

2.6.25. It is not true that every trembling hand perfect equilibrium of a normal form game induces subgame perfect behavior in an extensive game with that normal form. Illustrate using the game in Figure 2.6.9.

---

[11] By eventually, I mean there exists K such that σ_i is a best reply to σ_{-i}^k for all $k > K$. Note that K depends on the sequence.

	ℓ	r
T	$1,1,1$	$1,0,1$
B	$1,1,1$	$0,0,1$

L

	ℓ	r
T	$1,1,0$	$0,0,0$
B	$0,1,0$	$1,0,0$

R

Figure 2.6.7: The game for Problem 2.6.23(b). Player 1 chooses a row (T or B), player 2 chooses a column (ℓ or r), and player 3 chooses a matrix (L or R). In each cell, the first payoff is player 1's, the second is player 2's, and the third is player 3's.

	L	M	R
T	$1,5$	$2,3$	$0,0$
B	$1,5$	$0,0$	$3,2$

Figure 2.6.8: The game for Problem 2.6.24.

	L	M	R
T	$1,5$	$2,6$	$0,0$
B	$1,5$	$0,3$	$3,2$

Figure 2.6.9: The game for Problem 2.6.25.

Chapter 3

Games with Nature

3.1 An Introductory Example

Example 3.1.1 (A variant of Example 1.1.3). Firm 1's costs are private information, while firm 2's are public. Nature determines the costs of firm 1 at the beginning of the game, with $\Pr(c_1 = c_L) = \theta \in (0,1)$. The terminal nodes of the extensive form are given by (c_1, q_1, q_2). As in Example 1.1.3, firm i's *ex post* profit is

$$\pi_i(c_1, q_1, q_2) = [(a - q_1 - q_2) - c_i]q_i,$$

where c_i is firm i's cost.

Assume $c_L, c_H, c_2 < a/2$. A strategy for player 2 is a quantity q_2. A strategy for player 1 is a function $q_1 : \{c_L, c_H\} \to \mathbb{R}_+$. For simplicity, write q_L for $q_1(c_L)$ and q_H for $q_1(c_H)$.

Note that for any strategy profile $((q_L, q_H), q_2)$, the associated outcome distribution is

$$\theta \circ (c_L, q_L, q_2) + (1 - \theta) \circ (c_H, q_H, q_2),$$

that is, with probability θ, the terminal node (c_L, q_L, q_2) is realized and with probability $1 - \theta$, the terminal node (c_H, q_H, q_2) is realized.

To find a Nash equilibrium, we must solve for three numbers q_L^*, q_H^*, and q_2^*.

Assume interior solution. We must have

$$(q_L^*, q_H^*) = \underset{q_L, q_H}{\arg\max} \ \theta[(a - q_L - q_2^*) - c_L]q_L$$

$$+ (1 - \theta)[(a - q_H - q_2^*) - c_H]q_H.$$

This implies pointwise maximization, i.e.,

$$q_L^* = \arg\max_{q_1} \; [(a - q_1 - q_2^*) - c_L]q_1$$

$$= \frac{1}{2}(a - q_2^* - c_L).$$

and, similarly,

$$q_H^* = \frac{1}{2}(a - q_2^* - c_H).$$

We must also have

$$q_2^* = \arg\max_{q_2} \; \theta[(a - q_L^* - q_2) - c_2]q_2$$

$$+ (1 - \theta)[(a - q_H^* - q_2) - c_2]q_2$$

$$= \arg\max_{q_2} \; [(a - \theta q_L^* - (1 - \theta)q_H^* - q_2) - c_2]q_2$$

$$= \frac{1}{2}(a - c_2 - \theta q_L^* - (1 - \theta)q_H^*).$$

Solving,

$$q_H^* = \frac{a - 2c_H + c_2}{3} + \frac{\theta}{6}(c_H - c_L),$$

$$q_L^* = \frac{a - 2c_L + c_2}{3} - \frac{(1 - \theta)}{6}(c_H - c_L),$$

and

$$q_2^* = \frac{a - 2c_2 + \theta c_L + (1 - \theta)c_H}{3}. \qquad \bigstar$$

3.2   Purification

The assumption that players can randomize is sometimes criticized on three grounds:

1. players don't randomize;

2. there is no reason for a player randomize with just the right probability, when the player is indifferent over all possible randomization probabilities (including 0 and 1); and

$$
\begin{array}{c|c|c|}
 & A & B \\
\hline
A & 9,9 & 0,5 \\
\hline
B & 5,0 & 7,7 \\
\hline
\end{array}
$$

Figure 3.2.1: The benchmark game for Example 3.2.1.

   3. a randomizing player is subject to *ex post regret*.

   Purification (to be described soon) provides a persuasive response to the first two criticisms. Turning to the third, a player is said to be subject to *ex post regret* if after all uncertainty is resolved, a player would like to change his/her decision (i.e., has regret). In a game with no moves of nature, no player has ex post regret in a pure strategy equilibrium. This is clearly false in an equilibrium with nontrivial mixing, since players are best responding to the randomizing behavior of other players.

   But any pure strategy equilibrium of a game with moves of nature will typically also have ex post regret. Ex post regret should not be viewed as a criticism of mixing, but rather a caution to modelers. If a player has ex post regret, then that player has an incentive to change his/her choice. Whether a player is able to do so depends upon the scenario being modeled. If the player cannot do so, then there is no issue. If, however, the player can do so, then that option should be included in the game description.

   Player i's mixed strategy σ_i of a game G is said to be *purified* if in an "approximating" version of G with private information (with player i's private information given by T_i), that player's behavior can be written as a pure strategy $s_i : T_i \to A_i$ such that

$$
\sigma_i(a_i) \approx \Pr\{s_i(t_i) = a_i\},
$$

where Pr is given by the prior distribution over T_i (and so describes player $j \neq i$ beliefs over T_i).

Example 3.2.1. The game in Figure 3.2.1 has two strict pure strategy Nash equilibria and one symmetric mixed strategy Nash equilibrium. Let $p = \Pr\{A\}$, then (using Lemma 2.4.3)

$$
\begin{aligned}
9p &= 5p + 7(1-p) \\
\Longleftrightarrow \quad 9p &= 7 - 2p \\
\Longleftrightarrow \quad 11p &= 7 \iff p = 7/11.
\end{aligned}
$$

   Trivial purification: Give player i *payoff-irrelevant* information t_i, where $t_i \sim \mathcal{U}([0,1])$, and t_1 and t_2 are independent. This is a game with private

	A	B
A	$9 + \varepsilon t_1, 9 + \varepsilon t_2$	$0, 5$
B	$5, 0$	$7, 7$

Figure 3.2.2: The game $G(\varepsilon)$ for Example 3.2.1. Player i has type t_i, with $t_i \sim \mathcal{U}([0,1])$, and t_1 and t_2 are independent.

information, where player i learns t_i before choosing his or her action. The mixed strategy equilibrium is purified by many pure strategy equilibria in the game with private information, such as

$$s_i(t_i) = \begin{cases} B, & \text{if } t_i \leq 4/11, \\ A, & \text{if } t_i \geq 4/11. \end{cases}$$

Harsanyi (1973) purification: Consider the game $G(\varepsilon)$ displayed in Figure 3.2.2. This is again a game with private information, where player i learns t_i before choosing his or her action, $t_i \sim \mathcal{U}([0,1])$ and t_1 and t_2 are independent. But now the information *is* payoff relevant.

A pure strategy for player i is $s_i : [0,1] \to \{A, B\}$. Suppose 2 is following a cutoff strategy,

$$s_2(t_2) = \begin{cases} A, & t_2 \geq \bar{t}_2, \\ B, & t_2 < \bar{t}_2, \end{cases}$$

with $\bar{t}_2 \in (0,1)$.

Type t_1 expected payoff from A is

$$\begin{aligned} U_1(A, t_1, s_2) &= (9 + \varepsilon t_1) \Pr\{s_2(t_2) = A\} \\ &= (9 + \varepsilon t_1) \Pr\{t_2 \geq \bar{t}_2\} \\ &= (9 + \varepsilon t_1)(1 - \bar{t}_2), \end{aligned}$$

while from B is

$$\begin{aligned} U_1(B, t_1, s_2) &= 5 \Pr\{t_2 \geq \bar{t}_2\} + 7 \Pr\{t_2 < \bar{t}_2\} \\ &= 5(1 - \bar{t}_2) + 7\bar{t}_2 \\ &= 5 + 2\bar{t}_2. \end{aligned}$$

Thus, A is optimal if and only if

$$(9 + \varepsilon t_1)(1 - \bar{t}_2) \geq 5 + 2\bar{t}_2,$$

i.e.,

$$t_1 \geq \frac{11\bar{t}_2 - 4}{\varepsilon(1 - \bar{t}_2)}.$$

Thus the best reply to the cutoff strategy s_2 is a cutoff strategy with $\bar{t}_1 = (11\bar{t}_2 - 4)/\varepsilon(1 - \bar{t}_2)$.[1] Since the game is symmetric, we conjecture there is a symmetric equilibrium: $\bar{t}_1 = \bar{t}_2 = \bar{t}$. In such an equilibrium, the common cutoff value $\bar{t}$ solves

$$\bar{t} = \frac{11\bar{t} - 4}{\varepsilon(1 - \bar{t})},$$

or

$$\varepsilon\bar{t}^2 + (11 - \varepsilon)\bar{t} - 4 = 0. \tag{3.2.1}$$

Let $t(\varepsilon)$ denote the value of $\bar{t}$ satisfying (3.2.1). Note first that $t(0) = 4/11$. Writing (3.2.1) as $g(\bar{t}, \varepsilon) = 0$, we can apply the implicit function theorem (since $\partial g/\partial \bar{t} \neq 0$ at $\varepsilon = 0$) to conclude that for $\varepsilon > 0$ but close to 0, the cutoff value of $\bar{t}$, $t(\varepsilon)$, is close to $4/11$, the probability of the mixed strategy equilibrium in the unperturbed game.[2] In other words, for ε small, there is a symmetric equilibrium in cutoff strategies, with interior cutoff $\bar{t} \in (0,1)$. This equilibrium is not only pure, but almost everywhere strict.

The interior cutoff equilibrium of $G(\varepsilon)$ approximates the mixed strategy equilibrium of $G(0)$ in the following sense: Let $p(\varepsilon)$ be the probability assigned to A by the symmetric cutoff equilibrium strategy of $G(\varepsilon)$. Then, $p(0) = 7/11$ and $p(\varepsilon) = 1 - t(\varepsilon)$. Since we argued in the previous paragraph that $t(\varepsilon) \to 4/11$ as $\varepsilon \to 0$, we have that for all $\eta > 0$, there exists $\varepsilon > 0$ such that

$$|p(\varepsilon) - p(0)| < \eta. \qquad\qquad \bigstar$$

Harsanyi's (1973) purification theorem provides a compelling justification for mixed equilibria in finite normal form games. Intuitively, it states that for almost all finite normal form games, and *any* sequence of games in which each player's payoffs are subject to independent non-atomic private shocks (i.e, known only to the player whose payoffs are perturbed), converging to the complete-information normal form game, the following is true: every equilibrium (pure or mixed) of the original game is the limit of equilibria of these close-by games with payoff shocks. Moreover, in these games with private information, players have essentially strict best replies,

---

[1] Indeed, even if player 2 were not following a cutoff strategy, payer 1's best reply is a cutoff strategy.

[2] Since (3.2.1) is a quadratic, we could have obtained the same conclusion by solving the equation for $\bar{t}$. The advantage of the approach in the text is that it makes clear that the precise form of (3.2.1) is irrelevant for the conclusion; see the discussion after this example.

and so will not randomize. Consequently, a mixed strategy equilibrium can be viewed as a pure strategy equilibrium of *any* close-by game with private payoff shocks.

Govindan, Reny, and Robson (2003) provide a modern exposition and generalization of Harsanyi (1973). A brief introduction can also be found in Morris (2008).

3.3   Auctions and Related Games

An auction (or a similar environment) is said to have *private values* if each buyer's (private) information is sufficient to determine his value (i.e., it is a sufficient statistic for the other buyers' information). The values are *independent* if each buyer's private information is stochastically independent of every other bidder's private information.

An auction (or a similar environment) is said to have *interdependent values* if the value of the object to the buyers is unknown at the start of the auction, and if a bidder's (expectation of the) value can be affected by the private information of other bidders. If all bidders have the same value, then we have the case of *common value*.

Example 3.3.1 (First-price sealed-bid auction—private values).

Bidder i's value for the object, v_i is known only to i. Nature chooses v_i, $i = 1, 2$ at the beginning of the game, with v_i being independently drawn from the interval $[\underline{v}_i, \bar{v}_i]$, with distribution F_i and density f_i. Bidders know F_i (and so f_i).

This is an example of *independent private values*.

The set of possible bids is $\mathbb{R}_+$.

Bidder i's *ex post* payoff as a function of bids b_1 and b_2, and values v_1 and v_2:

$$u_i(b_1, b_2, v_1, v_2) = \begin{cases} 0, & \text{if } b_i < b_j, \\ \frac{1}{2}(v_i - b_i), & \text{if } b_i = b_j, \\ v_i - b_i, & \text{if } b_i > b_j. \end{cases}$$

Suppose bidder 2 uses a strategy $\sigma_2 : [\underline{v}_2, \bar{v}_2] \to \mathbb{R}_+$. Then, bidder 1's expected (or *interim*) payoff from bidding b_1 at v_1 is

$$U_1(b_1, v_1; \sigma_2) = \int u_1(b_1, \sigma_2(v_2), v_1, v_2) \, dF_2(v_2)$$

$$= \frac{1}{2}(v_1 - b_1) \Pr\{\sigma_2(v_2) = b_1\}$$

$$+ \int_{\{v_2 : \sigma_2(v_2) < b_1\}} (v_1 - b_1) f_2(v_2) \, dv_2.$$

Player 1's *ex ante* payoff from the strategy σ_1 is given by

$$\int U_1(\sigma_1(v_1), v_1; \sigma_2)\, dF_1(v_1),$$

and so for an optimal strategy σ_1, the bid $b_1 = \sigma_1(v_1)$ must maximize $U_1(b_1, v_1; \sigma_2)$ for almost all v_1.

I proceed by "guess and verify," that is, I impose a sequence of increasingly demanding conditions on the strategy of player 2, and prove that there is a best reply for player 1 satisfying these conditions. I begin by supposing σ_2 is strictly increasing, so that $\Pr\{\sigma_2(v_2) = b_1\} = 0$. Without loss of generality, I can restrict attention to bids b_1 in the range of σ_2,[3] so that

$$
\begin{aligned}
U_1(b_1, v_1; \sigma_2) &= \int_{\{v_2 : \sigma_2(v_2) < b_1\}} (v_1 - b_1) f_2(v_2)\, dv_2 \\
&= E[v_1 - b_1 \mid \text{winning}]\Pr\{\text{winning}\} \\
&= (v_1 - b_1)\Pr\{\sigma_2(v_2) < b_1\} \\
&= (v_1 - b_1)\Pr\{v_2 < \sigma_2^{-1}(b_1)\} \\
&= (v_1 - b_1)F_2(\sigma_2^{-1}(b_1)).
\end{aligned}
$$

Assuming σ_2 is, moreover, differentiable, and that the bid $b_1 = \sigma_1(v_1)$ is an interior maximum, the first order condition is

$$0 = -F_2\left(\sigma_2^{-1}(b_1)\right) + (v_1 - b_1) f_2\left(\sigma_2^{-1}(b_1)\right) \frac{d\sigma_2^{-1}(b_1)}{db_1}.$$

But

$$\frac{d\sigma_2^{-1}(b_1)}{db_1} = \frac{1}{\sigma_2'(\sigma_2^{-1}(b_1))},$$

so

$$F_2\left(\sigma_2^{-1}(b_1)\right)\sigma_2'\left(\sigma_2^{-1}(b_1)\right) = (v_1 - b_1) f_2\left(\sigma_2^{-1}(b_1)\right),$$

i.e.,

$$\sigma_2'\left(\sigma_2^{-1}(b_1)\right) = \frac{(v_1 - b_1) f_2\left(\sigma_2^{-1}(b_1)\right)}{F_2\left(\sigma_2^{-1}(b_1)\right)}.$$

To proceed further, assume the environment is symmetric, so that $F_1 = F_2$, and suppose the equilibrium is symmetric, so that $\sigma_1 = \sigma_2 = \tilde\sigma$, and

---

[3]It is clearly suboptimal for player 1 to submit a bid that is strictly greater than any bid submitted by player 2. Moreover, any bid strictly less than any bid submitted by player 2 yields the same payoff (of 0) as bidding the smallest bid submitted by player 2 (since the probability of a tie is zero).

$b_1 = \sigma_1(v)$ implies $v = \sigma_2^{-1}(b_1)$. Then, dropping subscripts,

$$\tilde{\sigma}'(v) = \frac{(v - \tilde{\sigma}(v))f(v)}{F(v)},$$

or

$$\tilde{\sigma}'(v)F(v) + \tilde{\sigma}(v)f(v) = vf(v). \tag{3.3.1}$$

But

$$\frac{d}{dv}\tilde{\sigma}(v)F(v) = \sigma'(v)F(v) + \tilde{\sigma}(v)f(v),$$

so

$$\tilde{\sigma}(\hat{v})F(\hat{v}) = \int_{\underline{v}}^{\hat{v}} vf(v)dv + k,$$

where k is a constant of integration. Moreover, evaluating both sides at $\hat{v} = \underline{v}$ shows that $k = 0$, and so

$$\tilde{\sigma}(\hat{v}) = \frac{1}{F(\hat{v})}\int_{\underline{v}}^{\hat{v}} vf(v)dv = E[v \mid v \le \hat{v}]. \tag{3.3.2}$$

Each bidder bids the expectation of the other bidder's valuation, conditional on that valuation being less than his (i.e., conditional on his value being the highest). As will be clear later (in particular, from Problem 12.5.2), the form of (3.3.2) is not an accident.

Summarizing the calculations till this point, I have shown that *if* $(\tilde{\sigma}, \tilde{\sigma})$ is a Nash equilibrium in which $\tilde{\sigma}$ is a strictly increasing and differentiable function, and $\tilde{\sigma}(v)$ is interior (which here means strictly positive), *then* it is given by (3.3.2). Note that $E[v \mid v \le \hat{v}]$ is increasing in $\hat{v}$ and lies in the interval $[\underline{v}, Ev]$.

It remains to verify the hypotheses. It is immediate that the function in (3.3.2) is strictly increasing and differentiable. Moreover, for $v > \underline{v}$, $\tilde{\sigma}(v)$ is strictly positive. It remains to verify the optimality of bids (i.e., we still need to verify that bidding $b_1 = \tilde{\sigma}(v_1)$ is optimal when bidder 2 is bidding according to $\tilde{\sigma}$; given the symmetry, this is enough).

Showing optimality is a little more involved (see Problem 3.6.3), but is easily done for a special case: suppose $\bar{v} = \underline{v} + 1$ and the values are uniformly distributed on $[\underline{v}, \underline{v} + 1]$. Then

$$\tilde{\sigma}(v) = \frac{1}{2}(v + \underline{v})$$

and bidder 1's interim payoff is (since $\tilde{\sigma}(v_2) \in [\underline{v}, \underline{v} + \frac{1}{2}]$ for all $v_2 \in [\underline{v}, \underline{v}+1]$)

$$U_1(b_1, v_1; \tilde{\sigma}_2) = (v_1 - b_1)F_2(\sigma_2^{-1}(b_1))$$

$$= \begin{cases} v_1 - b_1, & \text{if } b_1 > \underline{v} + \frac{1}{2}, \\ (v_1 - b_1)(2b_1 - 2\underline{v}), & \text{if } b_1 \in [\underline{v}, \underline{v} + \frac{1}{2}], \\ 0, & \text{if } b_1 < \underline{v}. \end{cases}$$

This function is strictly concave for $b_1 \in [\underline{v}, \underline{v} + \frac{1}{2}]$, and it is clearly not optimal to bid outside the interval, and so bidding $b_1 = \tilde{\sigma}(v_1)$ is optimal.

As an illustration of the kind of arguments that are useful, I now argue that *every* Nash equilibrium must be in nondecreasing strategies.

Lemma 3.3.1. *Suppose (σ_1, σ_2) is a Nash equilibrium of the first-price sealed-bid auction with independent private values. Suppose type v_i' wins the auction with positive probability. Then, $\sigma_i(v_i'') \geq \sigma_i(v_i')$ for all $v_i'' > v_i'$.*

Proof. Without loss of generality, consider $i = 1$. Suppose the lemma is false. Then there exists $v_1'' > v_1'$ with $\sigma_1(v_1') =: b_1' > b_1'' := \sigma_1(v_1'')$.

Since σ_1 is a best response to σ_2,

$$U_1(b_1', v_1'; \sigma_2) \geq U_1(b_1'', v_1'; \sigma_2)$$

$$\text{and} \qquad U_1(b_1', v_1''; \sigma_2) \leq U_1(b_1'', v_1''; \sigma_2).$$

Note that

$$U_1(b_1, v_1'; \sigma_2) - U_1(b_1, v_1''; \sigma_2) =$$

$$\frac{1}{2}(v_1' - v_1'') \Pr\{\sigma_2(v_2) = b_1\} + (v_1' - v_1'') \Pr\{\sigma_2(v_2) < b_1\}.$$

Subtracting the second from the first inequality gives

$$U_1(b_1', v_1'; \sigma_2) - U_1(b_1', v_1''; \sigma_2) \geq U_1(b_1'', v_1'; \sigma_2) - U_1(b_1'', v_1''; \sigma_2),$$

and so substituting,

$$\frac{1}{2}(v_1' - v_1'') \Pr\{\sigma_2(v_2) = b_1'\} + (v_1' - v_1'') \Pr\{\sigma_2(v_2) < b_1'\} \geq$$

$$\frac{1}{2}(v_1' - v_1'') \Pr\{\sigma_2(v_2) = b_1''\} + (v_1' - v_1'') \Pr\{\sigma_2(v_2) < b_1''\},$$

and simplifying (and dividing by $(v_1' - v_1'') < 0$) we get

$$0 \geq \Pr\{b_1'' \leq \sigma_2(v_2) < b_1'\}$$

$$+ \frac{1}{2}\left(\Pr\{\sigma_2(v_2) = b_1'\} - \Pr\{\sigma_2(v_2) = b_1''\}\right)$$

$$= \Pr\{b_1'' < \sigma_2(v_2) < b_1'\}$$

$$+ \frac{1}{2}\left(\Pr\{\sigma_2(v_2) = b_1'\} + \Pr\{\sigma_2(v_2) = b_1''\}\right).$$

This implies

$$0 = \Pr\{\sigma_2(v_2) = b_1'\},$$
$$0 = \Pr\{\sigma_2(v_2) = b_1''\},$$
$$\text{and} \qquad 0 = \Pr\{b_1'' < \sigma_2(v_2) < b_1'\}.$$

That is, bidder 2 does not make a bid between b_1'' and b_1', and there are no ties at b_1' or b_1''. A bid of b_1' and b_1'' therefore wins with the same probability. But this implies a contradiction: Since b_1' wins with positive probability, v_1' strictly prefers to win with the same probability at the strictly lower bid of b_1''. ∎

Example 3.3.2 (An example of mixed strategies). Maintaining the assumptions of Example 3.3.1, except that the value for each bidder is drawn independently from a uniform distribution on the two point set $\{\underline{v}, \bar{v}\}$, with $\underline{v} < \bar{v}$. For similar reasons to Example 2.4.4, this game has no equilibrium in pure strategies (see Problem 3.6.6). It is also worth noting that rather than working directly with *mixed* strategies (which are randomizations over pairs of bids (b', b''), where b' is the bid of $\underline{v}$ and b'' is the bid of $\bar{v}$), it is convenient to work with *behavior* strategies.[4] In a behavior strategy, the randomizations over bids are specified *separately* for the bidder of value $\underline{v}$ and $\bar{v}$.

Suppose bidder 2 follows the strategy σ_2 of bidding $\underline{v}$ if $v_2 = \underline{v}$ (why is that a reasonable "guess"?), and of bidding according to the distribution function $F_2(b)$ if $v_2 = \bar{v}$. Then, assuming there are no atoms in F_2, player 1 has interim payoffs from $b > \underline{v}$ given by

$$\begin{aligned} U_1(b, \bar{v}; \sigma_2) &= \frac{1}{2}(\bar{v} - b) + \frac{1}{2}(\bar{v} - b)F_2(b) \\ &= \frac{1}{2}(\bar{v} - b)(1 + F_2(b)). \end{aligned}$$

Note that the minimum of the support of F_2 is given by $\underline{v}$ (why?). Denote the maximum of the support by $\bar{b}$.

Suppose 1 is also randomizing over the set of bids, $(\underline{v}, \bar{b}]$. The indifference condition requires that 1 is indifferent over all $b \in (\underline{v}, \bar{b}]$. The bid $b = \underline{v}$ is excluded because there is a positive probability of a tie at $\underline{v}$ (from the

---

[4]Moreover, behavior strategies are also more natural when viewing the game as one of incomplete information (Section 3.4).

low-value bidder) and so it cannot be optimal for $\bar{v}$ to bid $\underline{v}$. That is, for all small $\varepsilon > 0$,

$$\frac{1}{2}(\bar{v} - b)(1 + F_2(b)) = U_1(\underline{v} + \varepsilon, \bar{v}; \sigma_2),$$

and

$$U_1(\underline{v} + \varepsilon, \bar{v}, ; \sigma_2) = \frac{1}{2}(\bar{v} - \underline{v} - \varepsilon)(1 + F_2(\underline{v} + \varepsilon)).$$

Since $\lim_{\varepsilon \to 0} F_2(\underline{v} + \varepsilon) = F_2(\underline{v}) = 0$ (where the first equality follows from the continuity of probabilities and the second equality follows from the assumption of no atoms), and so

$$(\bar{v} - b)(1 + F_2(b)) = \bar{v} - \underline{v},$$

yielding

$$F_2(b) = \frac{b - \underline{v}}{\bar{v} - b}, \qquad \forall b \in (\underline{v}, \bar{b}].$$

Note that $F_2(\underline{v}) = 0$.

Moreover, $F_2(\bar{b}) = 1$ implies $\bar{b} = (\bar{v} + \underline{v})/2$. An alternative derivation of the value of $\bar{b}$ is to note that bidder 1 should receive the payoff $\frac{1}{2}(\bar{v} - \underline{v})$ from bidding $\bar{b}$ (at which 1 wins for sure, since there are no atoms), or

$$\frac{1}{2}(\bar{v} - \underline{v}) = \bar{v} - \bar{b},$$

and then solving for $\bar{b}$.

It is straightforward to verify that the symmetric profile in which each bidder bids $\underline{v}$ if $v = \underline{v}$, and bids according to the distribution function $F(b) = (b - \underline{v})/(\bar{v} - b)$ if $v = \bar{v}$ is a Nash equilibrium.

Reminiscent of Example 2.4.4, the high-value bidder is *not* indifferent over every bid in the support $[\underline{v}, \bar{b}]$:

$$U_1(b, \bar{v}; \sigma_2) = \begin{cases} \bar{v} - b, & \text{if } b > \bar{b}, \\ \frac{1}{2}(\bar{v} - \underline{v}), & \text{if } b \in (\underline{v}, \bar{b}], \\ \frac{1}{4}(\bar{v} - \underline{v}), & \text{if } b = \underline{v}, \\ 0, & \text{if } b < \underline{v}. \end{cases}$$

Figure 3.3.1 graphs this payoff function.   ★

Remark 3.3.1. Given the discontinuities in the payoff functions of a first-price auction, it should not be surprising that, when the distribution over

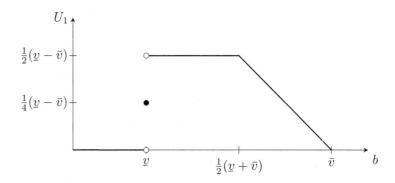

Figure 3.3.1: The payoff function U_1 at equilibrium for Example 3.3.2.

valuations is discrete, equilibrium requires randomization. Example 3.3.1 illustrates a version of purification: even in the presence of discontinuities, if the private information is continuously distributed, pure strategy equilibria exist and mixed strategies can be "purified" without perturbing the game (Aumann, Katznelson, Radner, Rosenthal, and Weiss, 1983; Milgrom and Weber, 1985)   ♦

Example 3.3.3 (independent private values, symmetric n bidders). Suppose now there n identical bidders, with valuations v_i independently distributed on $[\underline{v}, \bar{v}]$ according to F with density f.

We are interested in characterizing the symmetric Nash equilibrium (if it exists). Let σ be the symmetric strategy, and suppose it is strictly increasing. Consequently, the probability of a tie is zero, and so bidder i's interim payoff from the bid b_i is

$$U_i(b_i, v_i; \sigma) = E[v_i - b_i \mid \text{winning}] \Pr\{\text{winning}\}$$
$$= (v_i - b_i) \Pr\{v_j < \sigma^{-1}(b_i), \ \forall j \neq i\}$$
$$= (v_i - b_i) \prod_{j \neq i} \Pr\{v_j < \sigma^{-1}(b_i)\}$$
$$= (v_i - b_i) F^{n-1}(\sigma^{-1}(b_i)).$$

As before, assuming σ is differentiable, and an interior solution, the first order condition is

$$0 = -F^{n-1}(\sigma^{-1}(b_i))$$
$$+ (v_i - b_i)(n-1)F^{n-2}(\sigma^{-1}(b_i))f(\sigma^{-1}(b_i))\frac{d\sigma^{-1}(b_i)}{db_i},$$

and simplifying (similarly to (3.3.1)), we get

$$\sigma'(v)F^{n-1}(v) + \sigma(v)(n-1)F^{n-2}(v)f(v) = v(n-1)F^{n-2}(v)f(v),$$

that is,

$$\frac{d}{dv}\sigma(v)F^{n-1}(v) = v(n-1)F^{n-2}(v)f(v),$$

or (where the constant of integration is zero, since $F(\underline{v}) = 0$),

$$\sigma(v) = \frac{1}{F^{n-1}(v)} \int_{\underline{v}}^{v} x \, dF^{n-1}(x).$$

Remark 3.3.2 (Order statistics). Given n independent draws from a common distribution F, denoted $v_1, \ldots, v_n$, let $y_{(1)}^n, y_{(2)}^n, \ldots, y_{(n)}^n$ denote the rearrangement satisfying $y_{(1)}^n \leq y_{(2)}^n \leq \ldots \leq y_{(n)}^n$. The statistic $y_{(k)}^n$ is the kth-order statistic from the sample of n draws.

The distribution of $y_{(n)}^n$ is $\Pr\{y_{(n)}^n \leq y\} = F^n(y)$.   ◆

If σ is a symmetric Nash equilibrium, then

$$\sigma(v) = E[y_{(n-1)}^{n-1} \mid y_{(n-1)}^{n-1} \leq v].$$

That is, each bidder bids the expectation of the maximum of all the other bidders' valuation, conditional on that valuation being less than his (i.e., conditional on his value being the highest). Equivalently, the bidder bids the expected value of the $(n-1)^{th}$ order statistic of values, conditional on his value being the nth order statistic.   ★

Example 3.3.4 (First-price sealed-bid auction—common values). Each bidder receives a private signal about the value of the object, t_i, with $t_i \in T_i = [0,1]$, uniformly independently distributed. The common (to both players) value of the object is $v = t_1 + t_2$.

Ex post payoffs are given by

$$u_i(b_1, b_2, t_1, t_2) = \begin{cases} t_1 + t_2 - b_i, & \text{if } b_i > b_j, \\ \frac{1}{2}(t_1 + t_2 - b_i), & \text{if } b_i = b_j, \\ 0, & \text{if } b_i < b_j. \end{cases}$$

Suppose bidder 2 uses strategy $\sigma_2 : T_2 \to \mathbb{R}_+$. Suppose σ_2 is strictly increasing. Then, t_1's expected payoff from bidding b_1 is

$$U_1(b_1, t_1; \sigma_2) = \dot{} \, E[t_1 + t_2 - b_1 \mid \text{winning}] \Pr\{\text{winning}\}$$

$$
\begin{aligned}
&= E[t_1 + t_2 - b_1 \mid t_2 < \sigma_2^{-1}(b_1)] \Pr\{t_2 < \sigma_2^{-1}(b_1)\} \\
&= (t_1 - b_1)\sigma_2^{-1}(b_1) + \int_0^{\sigma_2^{-1}(b_1)} t_2 \, dt_2 \\
&= (t_1 - b_1)\sigma_2^{-1}(b_1) + (\sigma_2^{-1}(b_1))^2/2.
\end{aligned}
$$

If σ_2 is differentiable, the first order condition is

$$
0 = -\sigma_2^{-1}(b_1) + (t_1 - b_1)\frac{d\sigma_2^{-1}(b_1)}{db_1} + \sigma_2^{-1}(b_1)\frac{d\sigma_2^{-1}(b_1)}{db_1},
$$

and so

$$
\sigma_2^{-1}(b_1)\sigma_2'(\sigma_2^{-1}(b_1)) = (t_1 + \sigma_2^{-1}(b_1) - b_1).
$$

Assume now that the environment is symmetric, i.e., $F_1 = F_2$, and suppose the equilibrium is symmetric, so that $\sigma_1 = \sigma_2 = \sigma$, and $b_1 = \sigma_1(t)$ implies $t = \sigma_2^{-1}(b_1)$. Then,

$$
t\sigma'(t) = 2t - \sigma(t).
$$

Integrating,

$$
t\sigma(t) = t^2 + k,
$$

where k is a constant of integration. Evaluating both sides at $t = 0$ shows that $k = 0$, and so

$$
\sigma(t) = t.
$$

Note that this is *not* the profile that results from the analysis of the private value auction when $\underline{v} = 1/2$ (the value of the object in the common value auction, conditional on *only* t_1, is $E[t_1 + t_2 \mid t_1] = t_1 + 1/2$). In particular, letting $v' = t + \frac{1}{2}$, we have

$$
\sigma_{\text{private value}}(t) = \tilde{\sigma}(v') = \frac{v' + 1/2}{2} = \frac{t + 1}{2} > t = \sigma_{\text{common value}}(t).
$$

This illustrates the *winner's curse*: $E[v \mid t_1] > E[v \mid t_1, \text{winning}]$. In particular, in the equilibrium just calculated,

$$
\begin{aligned}
E[v \mid t_1, \text{winning}] &= E[t_1 + t_2 \mid t_1, t_2 < t_1] \\
&= t_1 + \frac{1}{\Pr\{t_2 < t_1\}} \int_0^{t_1} t_2 \, dt_2 \\
&= t_1 + \frac{1}{t_1}\left[(t_2)^2/2\right]_0^{t_1} = \frac{3t_1}{2},
\end{aligned}
$$

while $E[v \mid t_1] = t_1 + \frac{1}{2} > 3t_1/2$ (recall $t_1 \in [0,1]$).   ★

Example 3.3.5 (War of attrition).

Two players compete for a prize by engaging in a costly game of waiting. The first player to drop out loses to the other player. The action spaces are $S_i = \mathbb{R}_+$, where $s_i \in S_i$ is the time at which player i stops competing (drops out) and so loses. Both players incur the same disutility of competing: staying until s_i costs s_i.

The value of the prize to player i, denoted by $t_i \in T_i = \mathbb{R}_+$, is private information, with distribution F_i. Assume F_i has a density f_i and f_i is strictly positive on T_i.

To simplify calculations, assume that in the event of a tie (both players drop out at the same time), neither player wins. *Ex post* payoffs (effectively a second-price all-pay auction, see Problem 2.6.20) are given by

$$u_i(s_1, s_2, t_1, t_2) = \begin{cases} t_i - s_j, & \text{if } s_j < s_i, \\ -s_i, & \text{if } s_j \geq s_i. \end{cases}$$

Suppose player 2 uses strategy $\sigma_2 : T_2 \to S_2$. Then, t_1's expected (or *interim*) payoff from stopping at s_1 is

$$U_1(s_1, t_1; \sigma_2) = \int u_1(s_1, \sigma_2(t_2), t)\, dF_2(t_2)$$

$$= -s_1 \Pr\{\sigma_2(t_2) \geq s_1\}$$

$$+ \int_{\{t_2 : \sigma_2(t_2) < s_1\}} (t_1 - \sigma_2(t_2))\, dF_2(t_2).$$

Every Nash equilibrium is *sequentially rational* on the equilibrium path. In the current context, this can be restated as: Suppose σ_1 is a best reply to σ_2, $\tau < \sigma_1(t_1)$, and $\Pr\{\sigma_2(t_2) > \tau\} > 0$ (so that play reaches τ with positive probability when player 1 has value t_1). Suppose player 1 has value t_1 and τ is reached (i.e., 2 has not yet dropped out). Then, stopping at $\sigma_1(t_1)$ should still be optimal. I provide a direct proof by contradiction. Suppose that, conditional on play reaching τ, the stopping time $\hat{s}_1 > \tau$ yields a higher payoff than the original stopping time $s_1 = \sigma_1(t_1)$, i.e.,

$$E_{t_2}[u_1(s_1, \sigma_2(t_2), t) \mid \sigma_2(t_2) > \tau] < E_{t_2}[u_1(\hat{s}_1, \sigma_2(t_2), t) \mid \sigma_2(t_2) > \tau]$$

Then,

$$U_1(s_1, t_1; \sigma_2) = E_{t_2}[u_1(s_1, \sigma_2(t_2), t) \mid \sigma_2(t_2) \leq \tau] \Pr\{\sigma_2(t_2) \leq \tau\}$$
$$+ E_{t_2}[u_1(s_1, \sigma_2(t_2), t) \mid \sigma_2(t_2) > \tau] \Pr\{\sigma_2(t_2) > \tau\}$$
$$< E_{t_2}[u_1(s_1, \sigma_2(t_2), t) \mid \sigma_2(t_2) \leq \tau] \Pr\{\sigma_2(t_2) \leq \tau\}$$
$$+ E_{t_2}[u_1(\hat{s}_1, \sigma_2(t_2), t) \mid \sigma_2(t_2) > \tau] \Pr\{\sigma_2(t_2) > \tau\}$$

$$= E_{t_2}[u_1(\hat{s}_1, \sigma_2(t_2), t) \mid \sigma_2(t_2) \leq \tau] \Pr\{\sigma_2(t_2) \leq \tau\}$$
$$+ E_{t_2}[u_1(\hat{s}_1, \sigma_2(t_2), t) \mid \sigma_2(t_2) > \tau] \Pr\{\sigma_2(t_2) > \tau\}$$
$$= U_1(\hat{s}_1, t_1; \sigma_2),$$

and so s_1 cannot have been the unconditionally optimal stopping time. This is an application of the principle of Problem 2.6.10 to an infinite game.

Define $\bar{s}_i := \inf\{s_i : \Pr\{\sigma_i(t_i) \leq s_i\} = 1\} = \inf\{s_i : \Pr\{\sigma_i(t_i) > s_i\} = 0\}$, where $\inf\{\varnothing\} = \infty$.

Lemma 3.3.2. *Suppose σ_2 is a best reply to σ_1. If $\bar{s}_1, \bar{s}_2 > 0$, then $\bar{s}_1 \leq \bar{s}_2$.*

The proof is left as an exercise (Problem 3.6.10). The intuition is that if $\bar{s}_2 < \bar{s}_1$, for sufficiently large types for player 2, there are late stopping times that are profitable deviations.

Lemma 3.3.2 implies that in in any Nash equilibrium with $\bar{s}_1, \bar{s}_2 > 0$, $\bar{s}_1 = \bar{s}_2$.

Lemma 3.3.3. *Suppose (σ_1, σ_2) is a Nash equilibrium profile. Then, σ_i is nondecreasing for $i = 1, 2$.*

Proof. We use a standard revealed preference argument (similar to the proof Lemma 3.3.1). Let $s'_1 = \sigma_1(t'_1)$ and $s''_1 = \sigma_1(t''_1)$, with $s'_1, s''_1 \leq \bar{s}_1$. If σ_1 is a best reply to σ_2,

$$U_1(s'_1, t'_1; \sigma_2) \geq U_1(s''_1, t'_1; \sigma_2)$$

and

$$U_1(s''_1, t''_1; \sigma_2) \geq U_1(s'_1, t''_1; \sigma_2).$$

Thus,

$$U_1(s'_1, t'_1; \sigma_2) - U_1(s'_1, t''_1; \sigma_2) \geq U_1(s''_1, t'_1; \sigma_2) - U_1(s''_1, t''_1; \sigma_2).$$

Since,

$$U_1(s_1, t'_1; \sigma_2) - U_1(s_1, t''_1; \sigma_2) = (t'_1 - t''_1) \Pr\{t_2 : \sigma_2(t_2) < s_1\}$$

we have

$$(t'_1 - t''_1) \Pr\{t_2 : \sigma_2(t_2) < s'_1\} \geq (t'_1 - t''_1) \Pr\{t_2 : \sigma_2(t_2) < s''_1\},$$

i.e.,

$$(t'_1 - t''_1)[\Pr\{t_2 : \sigma_2(t_2) < s'_1\} - \Pr\{t_2 : \sigma_2(t_2) < s''_1\}] \geq 0.$$

Suppose $t'_1 > t''_1$. Then, $\Pr\{t_2 : \sigma_2(t_2) < s'_1\} \geq \Pr\{t_2 : \sigma_2(t_2) < s''_1\}$. If $s'_1 < s''_1$, then $\Pr\{t_2 : s'_1 \leq \sigma_2(t_2) < s''_1\} = 0$. That is, 2 does not stop between s'_1 and s''_1.

The argument to this point has only used the property that σ_1 is a best reply to σ_2. To complete the argument, we appeal to Lemma 3.3.2, i.e., that $\bar{s}_1 \leq \bar{s}_2$ (an implication of σ_2 being a best reply to σ_1), which implies $\Pr\{\sigma_2(t_2) \geq s_1''\} > 0$, and so stopping earlier (at a time $s_1 \in (s_1', s_1'')$) is a profitable deviation for t_1''. Thus, $s_1' = \sigma_1(t_1') \geq s_1'' = \sigma_1(t_1'')$. $\blacksquare$

It can also be shown that in any Nash equilibrium, σ_i is a strictly increasing and continuous function. Thus,

$$U_1(s_1, t_1; \sigma_2)$$

$$= -s_1 \Pr\left\{t_2 \geq \sigma_2^{-1}(s_1)\right\} + \int_{\{t_2 < \sigma_2^{-1}(s_1)\}} (t_1 - \sigma_2(t_2)) f_2(t_2) \, dt_2$$

$$= -s_1 \left(1 - F_2\left(\sigma_2^{-1}(s_1)\right)\right) + \int_0^{\sigma_2^{-1}(s_1)} (t_1 - \sigma_2(t_2)) f_2(t_2) \, dt_2.$$

Assuming σ_2 is, moreover, differentiable, the first-order condition is

$$0 = -\left(1 - F_2\left(\sigma_2^{-1}(s_1)\right)\right)$$
$$+ s_1 f_2\left(\sigma_2^{-1}(s_1)\right) \frac{d\sigma_2^{-1}(s_1)}{ds_1} + (t_1 - s_1) f_2\left(\sigma_2^{-1}(s_1)\right) \frac{d\sigma_2^{-1}(s_1)}{ds_1}.$$

But

$$\frac{d\sigma_2^{-1}(s_1)}{ds_1} = 1/\sigma_2'\left(\sigma_2^{-1}(s_1)\right),$$

so

$$\left[1 - F_2\left(\sigma_2^{-1}(s_1)\right)\right] \sigma_2'\left(\sigma_2^{-1}(s_1)\right) = t_1 f_2\left(\sigma_2^{-1}(s_1)\right),$$

i.e.,

$$\sigma_2'\left(\sigma_2^{-1}(s_1)\right) = \frac{t_1 f_2\left(\sigma_2^{-1}(s_1)\right)}{1 - F_2\left(\sigma_2^{-1}(s_1)\right)}.$$

Assume the environment is symmetric, i.e., $F_1 = F_2$, and suppose the equilibrium is symmetric, so that $\sigma_1 = \sigma_2 = \sigma$, and $s_1 = \sigma_1(t)$ implies $t = \sigma_2^{-1}(s_1)$. Then,

$$\sigma'(t) = \frac{t f(t)}{1 - F(t)}.$$

Since $\sigma(0) = 0$,

$$\sigma(t) = \int_0^t \frac{\tau f(\tau)}{1 - F(\tau)} \, d\tau.$$

If $f(t) = e^{-t}$, then $F(t) = 1 - e^{-t}$, and

$$\sigma(t) = \int_0^t \frac{\tau e^{-\tau}}{e^{-\tau}} \, d\tau = t^2/2.$$

I leave as an exercise that the first order condition is indeed sufficient for optimality, when $\sigma(t) = t^2/2$ (the hint to Problem 3.6.3 also works here).

Note that $\sigma(t) > t$ for $t > 2$. That is, it is possible for a player to have a negative ex post payoff. This does not contradict sequential rationality on the equilibrium path: at time t, the player with value t has a strictly positive *continuation* value (i.e., value from continuing) since the cost of staying until t is sunk and so irrelevant for incentives. But, of course, each player's interim payoff is strictly positive for all t.[5] A straightforward calculation shows that

$$U(t^2/2, t; \sigma) = t + e^{-t} - 1 > 0.$$

If we extend the strategy space to allow for never stopping, i.e., $S_i = \mathbb{R}_+ \cup \{\infty\}$ and allow payoffs to take on the value $-\infty$, then there are also two asymmetric equilibria, in which one player drops out immediately, and the other never drops out.   ★

Example 3.3.6 (Double Auction). A seller owns a good that a (potential) buyer is interested in purchasing. The seller's valuation of the good is v_s, and the buyer's valuation is v_b. The valuations are private information, with each being independently and uniformly distributed on $[0, 1]$. If a sale occurs at price p, the buyer receives a payoff of $v_b - p$ and the seller receives a payoff of $p - v_s$.

In a double auction, the seller and buyer simultaneously propose ask and bid prices $p_s \in [0, 1]$ and $p_b \in [0, 1]$, respectively. Trade occurs at a price $\frac{1}{2}(p_s + p_b)$ if $p_s \le p_b$; otherwise, there is no trade.

The buyer's strategy is a function $\widetilde{p}_b : [0, 1] \to [0, 1]$ and the seller's strategy is a function $\widetilde{p}_s : [0, 1] \to [0, 1]$. We check for interim optimality (i.e., optimality of a strategy conditional on a value). (Recall that with a continuum of strategies, ex ante optimality requires only that the strategy is optimal for *almost all* vales.)

Fix the seller's strategy, $\widetilde{p}_s : [0, 1] \to [0, 1]$, the buyer's valuation v_b and her bid price p_b. The buyer's (conditional) expected payoff is:

$$
\begin{aligned}
U_b(p_b, v_b; \widetilde{p}_s) &= \int\limits_{\{v_s : p_b \ge \widetilde{p}_s(v_s)\}} \left(v_b - \frac{1}{2}(\widetilde{p}_s(v_s) + p_b) \right) dv_s \\
&= \Pr\left(\{v_s : p_b \ge \widetilde{p}_s(v_s)\}\right) \\
&\quad \times \left(v_b - \frac{1}{2}p_b - \frac{1}{2}E\left(\widetilde{p}_s(v_s) \,|\, \{v_s : p_b \ge \widetilde{p}_s(v_s)\}\right) \right).
\end{aligned}
$$

---

[5]Why of course? If there were a player valuation with negative interim payoff, that value player would have a profitable deviation to $s = 0$.

Suppose the seller's strategy is linear in his valuation, i.e. $\widetilde{p}_s(v_s) = a_s + c_s v_s$, with $a_s \geq 0$, $c_s > 0$ and $a_s + c_s \leq 1$. Then,

$$\Pr\left(\{v_s : p_b \geq \widetilde{p}_s(v_s)\}\right) = \Pr\left(\left\{v_s : v_s \leq \frac{p_b - a_s}{c_s}\right\}\right).$$

So,

$$\Pr\left(\{v_s : p_b \geq \widetilde{p}_s(v_s)\}\right) = \begin{cases} 0, & \text{if } p_b \leq a_s, \\ \frac{p_b - a_s}{c_s}, & \text{if } a_s \leq p_b \leq a_s + c_s, \\ 1, & \text{if } p_b \geq a_s + c_s, \end{cases}$$

and so

$$E\left(\widetilde{p}_s(v_s) \mid \{v_s : p_b \geq \widetilde{p}_s(v_s)\}\right) = a_s + c_s E\left(v_s \left| \left\{v_s : v_s \leq \frac{p_b - a_s}{c_s}\right\}\right.\right)$$

(and if $a_s \leq p_b \leq a_s + c_s$)

$$= a_s + c_s \frac{1}{2} \frac{p_b - a_s}{c_s} = \frac{p_b + a_s}{2}.$$

So,

$$E\left(\widetilde{p}_s(v_s) \mid \{v_s : p_b \geq \widetilde{p}_s(v_s)\}\right) = \begin{cases} \text{not defined}, & \text{if } p_b < a_s, \\ \frac{p_b + a_s}{2}, & \text{if } a_s \leq p_b \leq a_s + c_s, \\ a_s + \frac{1}{2} c_s, & \text{if } p_b \geq a_s + c_s, \end{cases}$$

and

$$U_b(p_b, v_b; \widetilde{p}_s) = \begin{cases} 0, & \text{if } p_b \leq a_s, \\ \left(\frac{p_b - a_s}{c_s}\right)\left(v_b - \frac{3}{4}p_b - \frac{1}{4}a_s\right), & \text{if } a_s \leq p_b \leq a_s + c_s, \\ v_b - \frac{1}{2}p_b - \frac{1}{2}a_s - \frac{1}{4}c_s, & \text{if } p_b \geq a_s + c_s. \end{cases}$$

The interior expression is maximized by solving the first-order condition

$$0 = \frac{1}{c_s}\left[v_b - \frac{3}{4}p_b - \frac{1}{4}a_s - \frac{3}{4}p_b + \frac{3}{4}a_s\right],$$

and so

$$\frac{6}{4}p_b = v_b - \frac{a_s}{4} + \frac{3}{4}a_s$$

$$= v_b + \frac{1}{2}a_s$$

$$\Longrightarrow \ p_b = \frac{1}{3}a_s + \frac{2}{3}v_b.$$

Thus any bid less than a_s is optimal if $v_b \leq a_s$, $\frac{1}{3}a_s + \frac{2}{3}v_b$ is the unique optimal bid if $a_s < v_b \leq a_s + \frac{3}{2}c_s$ and $a_s + c_s$ is the unique optimal bid if $v_b \geq a_s + \frac{3}{2}c_s$. Thus, the strategy $\widetilde{p}_b(v_b) = \frac{1}{3}a_s + \frac{2}{3}v_b$ is a best response to $\widetilde{p}_s(v_s) = a_s + c_s v_s$ as long as $1 \leq a_s + \frac{3}{2}c_s$.

A symmetric argument shows that if $\widetilde{p}_b(v_s) = a_b + c_b v_b$, then the seller's optimal bid (if interior) is $\widetilde{p}_s(v_s) = \frac{2}{3}v_s + \frac{1}{3}(a_b + c_b)$. Thus a linear equilibrium must have $a_b = \frac{1}{3}a_s$, $a_s = \frac{1}{3}(a_b + c_b)$, $c_b = \frac{2}{3}$ and $c_s = \frac{2}{3}$, so $a_s = \frac{1}{4}$ and $a_b = \frac{1}{12}$. There is then a linear equilibrium with

$$\widetilde{p}_s(v_s) = \frac{1}{4} + \frac{2}{3}v_s$$

and

$$\widetilde{p}_b(v_b) = \frac{1}{12} + \frac{2}{3}v_b.$$

Ex post efficient trade requires trade if $v_s < v_b$, and no trade if $v_s > v_b$.

Under the linear equilibrium, trade occurs if, and only if, $\widetilde{p}_s(v_s) \leq \widetilde{p}_b(v_b)$, which requires

$$v_s + \frac{1}{4} \leq v_b.$$

Thus, for valuations in the set $\left\{(v_s, v_b) \mid v_s < v_b < v_s + \frac{1}{4}\right\}$, trade is efficient but does not occur in equilibrium.

Not only is there is no equilibrium of this game with ex post efficient trade, there is *no game* respecting the private information of the players with an ex post efficient equilibrium (Section 12.3.1). This linear equilibrium does maximize the ex ante gains from trade, across *all games* respecting the private information of the players (Section 12.3.2).

Note: There are other equilibria (see Problem 3.6.12).   ★

3.4   Games of Incomplete Information

Example 3.4.1. Suppose payoffs of a two player two action game are given by one of the two bimatrices in Figure 3.4.1.

Suppose first that $x = 0$. Either player II has a strictly dominant strategy to play L or a strictly dominant strategy to play R. Suppose that II knows his own payoffs but player I thinks there is a probability α that payoffs are given by the left matrix, and a probability $1 - \alpha$ that they are

	L	R
T	$1,1$	$0,0$
B	$0,1$	$1,0$

	L	R
T	$1,x$	$0,1$
B	$0,0$	$1,1$

Figure 3.4.1: The payoff matrices for Example 3.4.1.

given by the right matrix. Clearly, "equilibrium" must have: II plays L if payoffs are given by the left matrix, R if by the right matrix; I plays T if $\alpha > \frac{1}{2}$, B if $\alpha < \frac{1}{2}$.

But suppose now $x = 2$. Should player II still feel comfortable playing R if the payoffs are the right matrix? The optimality of II's action choice of R depends on his believing that I will play B, which only occurs if I assigns probability at most $\frac{1}{2}$ to the right matrix. But suppose II does not know I's beliefs α. Then, II has beliefs over I's beliefs and so II finds R optimal if he assigns probability at least $\frac{1}{2}$ to I assigning probability at least $\frac{1}{2}$ to the right matrix.

But, we are not done: how confident is I that II will play R in the right matrix? Now we need to deal with I's beliefs over II's beliefs. Player I will only play B if he assigns probability at least $\frac{1}{2}$ to II assigning probability at least $\frac{1}{2}$ to I assigning probability at least $\frac{1}{2}$ to the right matrix.

This leads us into an infinite regress, since now I's beliefs about II's beliefs about I's beliefs become relevant for II!   ★

So, how can we analyze such problems in general?

Definition 3.4.1 (Harsanyi (1967, 1968a,b)). *A game of incomplete information or Bayesian game is the collection $\{(A_i, T_i, p_i, u_i)_{i=1}^{n}\}$, where*

- *A_i is i's action space,*

- *T_i is i's type space,*

- *$p_i : T_i \to \Delta\left(\prod_{j \neq i} T_j\right)$ is i's subjective beliefs about the other players' types, given i's type and*

- *$u_i : \prod_j A_j \times \prod_j T_j \to \mathbb{R}$ is i's payoff function.*

We think of a player's type t_i as describing *everything* that player i knows that is not common knowledge (including player i's beliefs). The function p_i is common knowledge, so that player j (say) knows how player i's beliefs depend upon player i's type t_i. But player j may not know t_i, and so may not know player i's beliefs (as illustrated in the next section). A strategy for i is

$$s_i : T_i \to A_i.$$

Let $s(t) := (s_1(t_1), \ldots, s_n(t_n))$, etc.

Definition 3.4.2. *The profile* $(\hat{s}_1, \ldots, \hat{s}_n)$ *is a* Bayes-Nash *(or* Bayesian Nash*) equilibrium if, for all i and all $t_i \in T_i$,*

$$E_{t_{-i}}[u_i(\hat{s}(t), t)] \geq E_{t_{-i}}[u_i(a_i, \hat{s}_{-i}(t_{-i}), t)], \quad \forall a_i \in A_i, \qquad (3.4.1)$$

where the expectation over t_{-i} is taken with respect to the probability distribution $p_i(t_i)$.

If the type spaces are finite, then the probability i assigns to the vector $t_{-i} \in \prod_{j \neq i} T_j =: T_{-i}$ when his type is t_i can be denoted $p_i(t_{-i}; t_i)$, and (3.4.1) can be written as

$$\sum_{t_{-i}} u_i(\hat{s}(t), t) p_i(t_{-i}; t_i) \geq \sum_{t_{-i}} u_i(a_i, \hat{s}_{-i}(t_{-i}), t) p_i(t_{-i}; t_i), \quad \forall a_i \in A_i.$$

Example 3.4.2 (Revisiting Example 3.1.1)**.** The Cournot game is represented as a game of incomplete information, as follows: The action spaces are $A_i = \mathbb{R}_+$. Firm 1's type space is $T_1 = \{t_1', t_1''\}$ while firm 2's type space is a singleton $T_2 = \{t_2\}$. The belief mapping p_1 for firm 1 is trivial: both types assign probability one to the type t_2 (since T_2 is a singleton, there is no alternative), while the belief mapping for firm 2 is

$$p_2(t_2) = \theta \circ t_1' + (1 - \theta) \circ t_1'' \in \Delta(T_1).$$

Finally, payoffs are

$$u_1(q_1, q_2, t_1, t_2) = \begin{cases} [(a - q_1 - q_2) - c_H]q_1, & \text{if } t_1 = t_1', \\ [(a - q_1 - q_2) - c_L]q_1, & \text{if } t_1 = t_1'', \end{cases}$$

and

$$u_2(q_1, q_2, t_1, t_2) = [(a - q_1 - q_2) - c_2]q_2.$$

In this example, it is of course more natural to denote the type t_1' by c_H and t_1'' by c_L. See Problem 3.6.11 for a version of this game capturing "higher order" beliefs of the kind hinted at in the end of Example 3.4.1.   ★

Remark 3.4.1. The idea that games of incomplete information (as defined in Definition 3.4.1) formally capture hierarchies of beliefs can be made precise, and leads to the notion of the *universal type space* (which describes all possible hierarchies). This is clearly well beyond the scope of this course. For a (relatively) gentle introduction to this, see Myerson (1991, Section 2.9).   ♦

The perspective of a game of incomplete information is *interim*: the beliefs of player i are specified *type by type*. Nonetheless, it is possible to specify a prior for player i such that the interim subjective beliefs can be interpreted as the conditional beliefs derived from that prior.

To simplify the discussion, suppose the type spaces are finite or countably infinite. Let $\hat{q}_i$ be an arbitrary full support distribution on T_i, i.e., $\hat{q}_i \in \Delta(T_i)$. Then, defining

$$q_i(t) := \hat{q}_i(t_i) p_i(t_{-i} \mid t_i) \qquad \forall t$$

generates a prior $q_i \in \Delta(T)$ for player i with the property that $p_i(\cdot \mid t_i) \in \Delta(T_{-i})$ is the belief on t_{-i} conditional on t_i. Indeed, there are many priors consistent with the subjective beliefs (since $\hat{q}_i$ is arbitrary). In general, these priors have no inherent meaning (as reflected in their multiplicity).

There is one special case when these priors do have inherent meaning.

Definition 3.4.3. *The subjective beliefs are* consistent *or satisfy the* Common Prior Assumption *(*CPA*) if there exists a single probability distribution* $p \in \Delta\left(\prod_i T_i\right)$ *such that, for each* i, $p_i(t_i)$ *is the probability distribution on* T_{-i} *conditional on* t_i *implied by* p.

If the type spaces are finite, this is equivalent to

$$p_i(t_{-i}; t_i) = p(t_{-i}|t_i) = \frac{p(t)}{\sum_{t'_{-i}} p\left(t'_{-i}, t_i\right)}.$$

If beliefs are consistent, a Bayesian game can be interpreted as having an initial move by nature, which selects $t \in T$ according to p. The Common Prior Assumption is controversial, sometimes viewed as a mild assumption (Aumann, 1987) and sometimes not (Gul, 1998). Nonetheless, in applications it is standard to assume it. One common justification for the CPA is in its absence, too much "crazy" behavior is permitted.[6]

---

[6]For example, if Bruce and Sheila have inconsistent beliefs about the probability of heads on a coin toss, say Bruce assigns probability 0.4 and Sheila assigns probability 0.6 to heads, and these beliefs are *commonly known*, then the two agents are happy to take the opposite sides of a bet, and even happy to pay a small fee to do so. This trade looks Pareto improving and yet, this trade would seem to have no real economic value. See Gilboa, Samuelson, and Schmeidler (2014) for a discussion and a notion of Pareto dominance in the absence of the CPA.

	t_2'	t_2''
$p_1(t_1')$	$\frac{5}{9}$	$\frac{4}{9}$
$p_1(t_1'')$	$\frac{2}{3}$	$\frac{1}{3}$

	$p_2(t_2')$	$p_2(t_2'')$
t_1'	$\frac{5}{7}$	$\frac{4}{5}$
t_1''	$\frac{2}{7}$	$\frac{1}{5}$

	t_2'	t_2''
t_1'	$\frac{5}{12}$	$\frac{4}{12}$
t_1''	$\frac{2}{12}$	$\frac{1}{12}$

Figure 3.4.2: The subject beliefs for Example 3.4.3, and the common prior consistent with these subjective beliefs.

Example 3.4.3. Suppose $n = 2$ and $T_i = \{t_i', t_i''\}$. The subjective beliefs are given in the first two matrices of Figure 3.4.2, with the CPA consistent with them on the right.

Consider now a slight perturbation: Change $p_2(t_2')$ to the subjective belief assigning probability $\frac{6}{7}$ to t_1' (and $\frac{1}{7}$ to t_1''). The CPA is now *not* satisfied. The easiest way to see this is to observe that, in this example, a common prior is an element of the 3-dimensional simplex (i.e., there are three degrees of freedom). This implies that the collection of conditional beliefs can also only have three degrees of freedom. In particular, in this example, the values of $p_1(t_1')$, $p_1(t_1'')$, and $p_2(t_2'')$ determine the common prior and so $p_2(t_2')$. We are not free to perturb the latter while keeping a common prior. ★

While there is a unique common prior in Example 3.4.3, if the subjective beliefs do not have full support, there may be multiple common priors. For example, modify the subjective beliefs in Figure 3.4.2 so that $p_i(t_i') = 1 \circ t_j'$ and $p_i(t_i'') = 1 \circ t_j''$. Then, any common prior $q \circ (t_1't_2') + (1-q) \circ (t_1''t_2'')$, $q \in (0,1)$, is consistent with these beliefs.

We continue with our supposition that type spaces are finite. Viewed as a game of complete information, a profile $\hat{s}$ is a Nash equilibrium if, for all i,

$$\sum_t u_i(\hat{s}(t), t)p(t) \geq \sum_t u_i(s_i(t_i), \hat{s}_{-i}(t_{-i}), t)p(t), \quad \forall s_i : T_i \to A_i.$$

This inequality can be rewritten as (where $p_i^*(t_i) := \sum_{t_{-i}} p(t_{-i}, t_i)$)

$$\sum_{t_i} \left\{ \sum_{t_{-i}} u_i(\hat{s}(t), t)\, p_i(t_{-i}; t_i) \right\} p_i^*(t_i) \geq$$

$$\sum_{t_i} \left\{ \sum_{t_{-i}} u_i(s_i(t_i), \hat{s}_{-i}(t_{-i}), t)\, p_i(t_{-i}; t_i) \right\} p_i^*(t_i),$$

	A	B
A	θ,θ	$\theta-9,5$
B	$5,\theta-9$	$7,7$

Figure 3.5.1: The game for Example 3.5.1. For $\theta = 9$, this is the game studied in examples 3.2.1 and 2.5.6.

$$\forall s_i : T_i \rightarrow A_i.$$

If $p_i^*(t_i) \neq 0$, this is then equivalent to the definition of a Bayes-Nash equilibrium.

Remark 3.4.2. To reiterate, a game of incomplete information is a game as described in Definition 3.4.1. If the CPA is satisfied (i.e., there is a common prior), then the game of incomplete information has a complete information *representation* (where the game starts explicitly with a move of nature). But the game as described in Definition 3.4.1 is still a game of incomplete information. The games in the earlier sections of this chapter are all more naturally described as games of incomplete information satisfying CPA.

A Bayes-Nash equilibrium is *always* the equilibrium (in the sense of Definition 3.4.2) of a Bayesian game, i.e., of a game of incomplete information.

While the literature appears to have settled on the term *game of incomplete information* for games as described in Definition 3.4.1, the alternative *Bayesian game* seems more appropriate, since the notion involves the explicit description of beliefs. These notes will follow the literature and typically use game of incomplete information (though this does leave us with the quandary of what to call games with truly incomplete information and unspecified beliefs).   ◆

3.5   Higher Order Beliefs and Global Games

Example 3.5.1. Consider the game in Figure 3.5.1. For $\theta = 9$, this is the game studied in Examples 2.5.6 and 3.2.1.

Suppose, as in example 3.2.1, that there is incomplete information about payoffs. However, now the information will be correlated. In particular, suppose $\theta \in \{4, 9\}$, with prior probability $\Pr\{\theta = 9\} > 7/9$.

If players have no information about θ, then there are two pure strategy Nash equilibria, with (A, A) Pareto dominating (B, B).

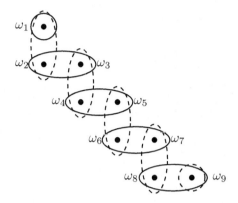

Figure 3.5.2: The solid lines describe player 1's information sets, while the dashed lines describe player 2's. Player 1 knows the row and player 2 knows the column containing the realized state.

In state ω_3, both players know $\theta = 9$, player 2 knows that player 1 knows that $\theta = 9$, but player 1 assigns probability $1/2$ to ω_2, and so to the event that player 2 does not know that $\theta = 9$.

In state ω_4, both players know $\theta = 9$, both players know that both players know that $\theta = 9$, player 1 knows that player 2 knows that player 1 knows that $\theta = 9$, but player 2 does not know that player 1 knows that player 2 knows that $\theta = 9$.

Now suppose players have some private information represented as follows: There is an underlying state space $\Omega := \{\omega_1, \omega_2, \ldots, \omega_M\}$, where M is odd. In state ω_1, we have $\theta = 4$, while in all other states ω_ℓ, $\ell \geq 2$, we have $\theta = 9$. Player 1's information is described by the partition on Ω given by $\{\{\omega_1\}, \{\omega_2, \omega_3\}, \ldots, \{\omega_{M-1}, \omega_M\}\}$; we denote $\{\omega_1\}$ by t_1^0, and $\{\omega_{2\ell}, \omega_{2\ell+1}\}$ by t_1^ℓ for $1 \leq \ell \leq (M-1)/2$. Player 2's information is described by the partition $\{\{\omega_1, \omega_2\}, \{\omega_3, \omega_4\}, \ldots, \{\omega_{M-2}, \omega_{M-1}\}, \{\omega_M\}\}$, and we denote $\{\omega_{2\ell+1}, \omega_{2\ell+2}\}$ by t_2^ℓ for $0 \leq \ell < (M-1)/2$, and $\{\omega_M\}$ by $t_2^{(M-1)/2}$. Each player knows only his own type (the element of his information partition that contains the realized ω). Finally, the probability distribution on Ω is uniform. Figure 3.5.2 illustrates Ω for $M = 9$.

Say that a player *knows* θ if the player assigns probability 1 to a particular value of θ (this clearly depends on ω). So, player 1 knows the value of θ at every ω, while player 2 only knows θ for $\omega \notin \{\omega_1, \omega_2\} = t_2^0$. Write $K_i E$ if player i *knows* the event E. Then, at ω_1, we only have $K_1\{\theta = 4\}$, and at ω_2, we only have $K_1\{\theta = 9\}$. At ω_3, we have $K_1\{\theta = 9\}$ and $K_2\{\theta = 9\}$

(both players know $\theta = 9$) and $K_2 K_1 \{\theta = 9\}$ (player 2 knows that player 1 knows $\theta = 9$), but no more. Note that moving from ω_2 to ω_3 does not change what player 1 knows, since both states are in the same element of player 1's partition. As a final illustration, consider ω_4. Since ω_3 and ω_4 are in the same element of 2's partition, player 2's knowledge does not change, and so for 2 we only have $K_2 \{\theta = 9\}$ and $K_2 K_1 \{\theta = 9\}$. But player 1 knows what player 2 knows at ω_3, so in addition to $K_1 \{\theta = 9\}$, we also have $K_1 K_2 \{\theta = 9\}$ and $K_1 K_2 K_1 \{\theta = 9\}$.[7]

A pure strategy for player 1 is

$$s_1 : \{t_1^0, t_1^1, \ldots, t_1^{(M-1)/2}\} \to \{A, B\},$$

while a pure strategy for player 2 is

$$s_2 : \{t_2^1, t_2^2, \ldots, t_2^{(M-1)/2}\} \to \{A, B\}.$$

Before we begin examining equilibrium behavior, note that $\Pr\{\theta = 9\} = (M-1)/M$, and this converges to 1 as $M \to \infty$.

This game of asymmetric information has a unique equilibrium.

Claim 3.5.1. *This game has a unique Nash equilibrium $(\hat{s}_1, \hat{s}_2)$, and in this equilibrium, both players necessarily choose B, i.e., $\hat{s}_1(t_1) = \hat{s}_2(t_2) = B$ for all t_1 and t_2.*

Proof. (by induction) Let s^* be a Nash equilibrium. Note first that $s_2^*(t_2^0) = B$ (since B is the unique best response at t_2^0).

Suppose $s_2^* \left(t_2^{\ell-1}\right) = B$ for $1 \le \ell < (M-1)/2$. Then, since

$$\Pr\{t_2^{\ell-1}|t_1^\ell\} = \Pr\{\{\omega_{2\ell-1}, \omega_{2\ell}\} \mid \{\omega_{2\ell}, \omega_{2\ell+1}\}\}$$
$$= \frac{\Pr\{\omega_{2\ell}\}}{\Pr\{\omega_{2\ell}, \omega_{2\ell+1}\}} = \frac{1/M}{2/M} = \frac{1}{2},$$

$s_1^*\left(t_1^\ell\right) = B$ (the probability that 2 plays B is at least $1/2$). Moreover, if $s_1^*\left(t_1^\ell\right) = B$, then since

$$\Pr\{t_1^\ell|t_2^\ell\} = \Pr\{\{\omega_{2\ell}, \omega_{2\ell+1}\} \mid \{\omega_{2\ell+1}, \omega_{2\ell+2}\}\}$$
$$= \frac{\Pr\{\omega_{2\ell+1}\}}{\Pr\{\omega_{2\ell+1}, \omega_{2\ell+2}\}} = \frac{1}{2},$$

we also have $s_2^*\left(t_2^\ell\right) = B$.   ∎

---

[7] This is more formally described as follows: An *event* is simply a subset of Ω. A player *knows* an event (such as the value of θ) if the player assigns probability 1 to that event. The event consisting of all states at which player i assigns probability 1 to the event E is denoted by $K_i E$. The event $K_i E$ is a union of elements of player i's partition. The event E could itself be the event $K_j F$ for some other event F. In particular, $K_1 \{\theta = 4\} = \{\omega_1\}$, $K_1 \{\theta = 9\} = \{\omega_2 \ldots, \omega_M\}$, $K_2 \{\theta = 4\} = \varnothing$, $K_2 \{\theta = 9\} = K_2 K_1 \{\theta = 9\} = \{\omega_3, \ldots, \omega_M\}$, and $K_1 K_2 \{\theta = 9\} = K_1 K_2 K_1 \{\theta = 9\} = \{\omega_4, \ldots, \omega_M\}$.

The proof actually proves something a little stronger, that the only profile that survives the iterated deletion of strictly dominated strategies involves both players always choosing B. Since B is the unique best response at t_2^0, any strategy s_2' satisfying $s_2'(t_2^0) = A$ is strictly dominated by the strategy $\hat{s}_2'$ given by $\hat{s}_2'(t_2^0) = B$ and $\hat{s}_2'(t_2) = s'(t_2)$ for all other t_2. We now proceed by iteratively deleting strictly dominated strategies. ★

Remark 3.5.1. Rubinstein's (1989) email game is essentially the infinite version ($M = \infty$) of Example 3.5.1 (with slightly different payoffs): if $\theta = 9$, player 1 sends a message to player 2 that $\theta = 9$ (no message is sent if $\theta = 4$). This message is exogenous. Moreover, the message does not arrive with some probability $\varepsilon > 0$. Players then exchange confirmations, with each confirming message having the same exogenous probability of not arriving. When a message does not arrive (which will occur with probability one), communication stops and players choose. Each player's type (information) is determined by the number of messages that player sent, so player i's type if he sent ℓ messages is t_i^ℓ. A similar argument to the proof of Claim 3.5.1 shows that, for ε small, this game also has a unique Nash equilibrium, and in this game both players choose B at all types. This implies that equilibria are not lower hemicontinuous in levels of knowledge: AA is a Nash equilibrium when $\theta = 9$ is common knowledge, and yet no player will play A in the email game, no matter how many messages were exchanged. One interpretation of this example is that, from a strategic point of view, "almost common knowledge" is not captured by "mutual knowledge of an arbitrary high, but finite, level." Monderer and Samet (1989) describe a way of approximating common knowledge (by common p-belief) that maintains the lower hemicontinuity of equilibria. ◆

Remark 3.5.2. As another illustration of the explicit representation of games with the incomplete information, we represent the game of Example 3.5.1 as follows. The action spaces for both players are $\{A, B\}$. The types spaces are $T_1 = \{t_1^0, t_1^1, \ldots, t_1^{(M-1)/2}\}$ and $T_2 = \{t_2^0, t_2^1, \ldots, t_2^{(M-1)/2}\}$. From the proof of Claim 3.5.1, the belief mappings are

$$p_1(t_1^\ell) = \begin{cases} \frac{1}{2} \circ t_2^{\ell-1} + \frac{1}{2} \circ t_2^\ell, & \text{if } \ell \geq 1, \\ 1 \circ t_2^0, & \text{if } \ell = 0, \end{cases}$$

and

$$p_2(t_2^\ell) = \begin{cases} \frac{1}{2} \circ t_1^\ell + \frac{1}{2} \circ t_1^{\ell+1}, & \text{if } \ell < (M-1)/2, \\ 1 \circ t_1^{(M-1)/2}, & \text{if } \ell = (M-1)/2. \end{cases}$$

Finally, payoffs are given by, for $i = 1, 2$,

$$u_i(a_1, a_2, t_1, t_2) = \begin{cases} 9, & \text{if } a = AA, \text{ and } t_1 \neq t_1^0, \\ 4, & \text{if } a = AA, \text{ and } t_1 = t_1^0, \\ 0, & \text{if } a_i = A, \, a_j = B, \text{ and } t_1 \neq t_1^0, \\ -5, & \text{if } a_i = A, \, a_j = B, \text{ and } t_1 = t_1^0, \\ 5, & \text{if } a_i = B, \, a_j = A, \\ 7, & \text{if } a = BB. \end{cases}$$ ◆

Remark 3.5.3 (Payoff and belief types). In many situations (such as Example 3.4.1 and 3.5.1), there is common knowledge of the environment, but beliefs are not common knowledge. It is becoming common (particularly in the study of robust mechanism design, see Bergemann and Morris, 2005) to distinguish between a player's *payoff type* and a player's *belief type*. Each player has a payoff and a belief type, with each player's payoffs depending only on (possibly all) players' payoff types. In Example 3.1.1, firm 1 has two payoff types, and firm 2 has one payoff type (since firm 2's payoffs are common knowledge). Each firm has only one belief type, since firm 2's beliefs over firm 1's payoff type are common knowledge.

In Example 3.5.1, the set of payoff types is $\Theta := \Theta_1 \times \Theta_2 := \{4, 9\} \times \{\theta_2\}$, and both players payoffs depend on the actions chosen and the vector of payoff types (note that player 2's payoff type space is degenerate[8]). We add to the type space representation from Remark 3.5.2 the additional mappings $\hat{\theta}_i : T_i \to \Theta_i$ for $i = 1, 2$ given by

$$\hat{\theta}_1(t_1^\ell) = \begin{cases} 9, & \ell > 0, \\ 4, & \ell = 0, \end{cases}$$

and

$$\hat{\theta}_2(t_1^\ell) = \theta_1, \quad \forall \ell.$$

Player i's payoff type is $\hat{\theta}_i(t_i)$. Player i's belief type is $p_i(t_i) \in \Delta(T_{-i})$. ◆

The complete information version of the game with $\theta = 9$ has two strict equilibria. Nonetheless, by making a small perturbation to the game by introducing a particular form of incomplete information, the result is stark, with only BB surviving, even in a state like ω_M, where each player knows

---

[8]Payoffs are determined by θ, which is not known to player 2, but is known to player 1, so both players' payoffs depend nontrivially on player 1's payoff type (i.e., the value of θ). Since player 2 does not privately know anything that is payoff relevant, his space of payoff types is degenerate.

the state, knows that the other knows the state, and so on, to some large finite order.

Carlsson and van Damme (1993) introduced the term *global games* to emphasize the importance of viewing the benchmark complete information game in a broader (global), i.e., perturbed, context. The term *global game* is now commonly understood to refer to a model that incorporates both a *richness* assumption on the uncertainty (so that each action is dominant for at least one value of the uncertainty) *and small noise* (as illustrated next).

Example 3.5.2 (Global Games). The stage game is as in Example 3.5.1. We change the information structure: We now assume θ is uniformly distributed on the interval $[0, 20]$. For $\theta < 5$, B is strictly dominant, while if $\theta > 16$, A is strictly dominant.

Each player i receives a signal x_i, with x_1 and x_2 independently and uniformly drawn from the interval $[\theta - \varepsilon, \theta + \varepsilon]$ for $\varepsilon > 0$. A pure strategy for player i is a function

$$s_i : [-\varepsilon, \ 20 + \varepsilon] \to \{A, B\}.$$

First observe that, for $x_i \in [\varepsilon, \ 20 - \varepsilon]$, player i's posterior on θ is uniform on $[x_i - \varepsilon, \ x_i + \varepsilon]$. This is most easily seen as follows: Letting g be the density of θ and h be the density of x given θ, we immediately have $g(\theta) = \frac{1}{20}$ for all $\theta \in [0, 20]$ and

$$h(x \mid \theta) = \begin{cases} \frac{1}{2\varepsilon}, & \text{if } x \in [\theta - \varepsilon, \theta + \varepsilon], \\ 0, & \text{otherwise.} \end{cases}$$

Since

$$h(x \mid \theta) = \frac{f(x, \theta)}{g(\theta)},$$

where f is the joint density, we have

$$f(x, \theta) = \begin{cases} \frac{1}{40\varepsilon}, & \text{if } x \in [\theta - \varepsilon, \theta + \varepsilon] \text{ and } \theta \in [0, 20], \\ 0, & \text{otherwise.} \end{cases}$$

The marginal density for $x \in [\varepsilon, \ 20 - \varepsilon]$ is thus simply the constant function $\frac{1}{20}$, and so the density of θ conditional on an $x \in [\varepsilon, \ 20 - \varepsilon]$ is the constant function $\frac{1}{2\varepsilon}$ on the interval $[x - \varepsilon, \ x + \varepsilon]$.

Similar considerations show that for $x_i \in [\varepsilon, \ 20 - \varepsilon]$, player i's posterior on x_j is symmetric around x_i with support $[x_i - 2\varepsilon, \ x_i + 2\varepsilon]$. Hence, for $x_i \in [\varepsilon, \ 20 - \varepsilon]$, we have $\Pr\{x_j > x_i \mid x_i\} = \Pr\{x_j < x_i \mid x_i\} = \frac{1}{2}$ and $E(\theta \mid x_i) = x_i$.

Claim 3.5.2. *For $\varepsilon < \frac{5}{2}$, the game has an essentially unique Nash equilibrium (s_1^*, s_2^*), given by*

$$s_i^*(x_i) = \begin{cases} A, & \text{if } x_i \geq 10\frac{1}{2}, \\ B, & \text{if } x_i < 10\frac{1}{2}. \end{cases}$$

Proof. We again apply iterated deletion of dominated strategies. Suppose $x_i < 5 - \varepsilon$. Then, player i's conditional expected payoff from A is less than that from B *irrespective* of player j's action, and so i plays B for $x_i < 5 - \varepsilon$ (as does j for $x_j < 5 - \varepsilon$). But then at $x_i = 5 - \varepsilon$, since $\varepsilon < 5 - \varepsilon$, player i assigns at least probability $\frac{1}{2}$ to j playing B, and so i strictly prefers B. Let x_i^* be the largest signal for which B is implied by iterated dominance (i.e., $x_i^* = \sup\{x_i' \mid B \text{ is implied by iterated strict dominance for all } x_i < x_i'\}$). By symmetry, $x_1^* = x_2^* = x^*$. At $x_i = x^*$, player i cannot strictly prefer B to A (otherwise, we can expand the set of signals for which iterated dominance implies B), and he assigns at least probability $\frac{1}{2}$ to j playing B. The expected payoff from B is at least 6, while the payoff from A is at most $x^* - 4\frac{1}{2}$, and so, $x^* \geq 10\frac{1}{2}$.

Similarly, for $x_i > 16$, player i's conditional expected payoff from A is greater than that from B irrespective of player j's action, and so i plays A for $x_i > 16$ (as does j for $x_j > 16$). Let x_i^{**} be the smallest signal for which A is implied by iterated dominance (i.e., $x_i^{**} = \inf\{x_i' \mid A \text{ is implied by iterated strict dominance for all } x_i > x_i'\}$). By symmetry, $x_1^{**} = x_2^{**} = x^{**}$. At $x_i = x^{**}$, player i cannot strictly prefer A to B, and he assigns at least probability $\frac{1}{2}$ to j playing A. Hence, $x^{**} \leq 10\frac{1}{2}$.

But then

$$10\frac{1}{2} \leq x^* \leq x^{**} \leq 10\frac{1}{2}.$$

∎

The iterated deletion argument connecting x_i in the dominance regions to values not in the dominance regions is often called an *infection argument*.

This idea is not dependent on the particular distributional assumptions made here. The property that $\Pr\{x_j > x_i \mid x_i\} = \Pr\{x_j < x_i \mid x_i\} = \frac{1}{2}$ is key (and holds under quite weak assumptions, see Morris and Shin, 2003, for details).   ★

Remark 3.5.4 (CAUTION). Some people have interpreted the global games literature as solving the multiplicity problem, at least in some settings. There is in fact a stronger result: Weinstein and Yildiz (2007) show that "almost all" games have a unique rationalizable outcome (which of course implies a unique Nash equilibrium)!

Does this mean that we don't need to worry about multiplicity? Of course not: This is a result about robustness. The uniqueness of the rationalizable outcome is driven by similar ideas to that in Example 3.5.1— "almost all" simply means that all information structures can be approximated by information structures allowing an infection argument. In order for a modeler to be confident that he knows the unique rationalizable outcome, he needs to be confident of the information structure.

Indeed, the stronger result in Weinstein and Yildiz (2007) proves even more: for "almost all" games, the following is true: Fix any rationalizable action a_i'. There are games close by (terms of players' beliefs) for which *that* action is the unique rationalizable action for player i.[9],[10] This means that, in this setting at least, there are no robust refinements of rationalizability: If there were, then some rationalizable action a_i' would have failed the refinement, and yet it must pass the refinement for some games close by (since it is the only rationalizable action on those games). In particular, Nash equilibrium is *not* a robust concept, at least in this belief setting! ◆

Remark 3.5.5. Superficially, purification and global games appear very similar. In both cases, we take a benchmark game, and consider perturbations in which players have small amounts of private information. Yet the conclusions are very different: With Harsanyi purification, given *any* equilibrium (pure or mixed), we obtain essentially strict equilibria of the perturbed game that approximate that equilibrium. With global games, given a benchmark game with multiple equilibria, we obtain a *unique* essentially strict equilibrium of the perturbed game. This difference arises from the private information in purification being independently distributed (one player's type is uninformative about the realized type of the other player), while this is *not* true for global games (one player's type can be very informative about the realized type of the other player). ◆

3.6 Problems

3.6.1. There are two firms, 1 and 2, producing the same good. The inverse demand curve is given by $P = \theta - q_1 - q_2$, where $q_i \in \mathbb{R}_+$ is firm

---

[9] I have not introduced rationalizability for games of incomplete information. For our purposes here, interpret the phrase "an action is rationalizable" to mean that the action is consistent with common knowledge of rationality.

[10] As Example 3.5.1 suggests, this result requires common knowledge of rationality. The result is not true once we admit the possibility that rationality is not common knowledge (see Heifetz and Kets, 2018, and Germano, Weinstein, and Zuazo-Garin, 2017).

	A	B
A	9, 9	0, 5
B	4, 0	7, 7

Figure 3.6.1: The game for Problem 3.6.2.

i's output. (Note that we are allowing negative prices.) There is demand uncertainty with nature determining the value of θ, assigning probability $\alpha \in (0, 1)$ to $\theta = 3$, and complementary probability $1 - \alpha$ to $\theta = 4$. Firm 2 knows (is informed of) the value of θ, while firm 1 is not. Finally, each firm has zero costs of production. As usual, assume this description is common knowledge. Suppose the two firms choose quantities simultaneously. Define a strategy profile for this game. Describe the Nash equilibrium behavior (which may be unique).

3.6.2. Redo Example 3.2.1 for the game in Figure 3.6.1.

3.6.3. Prove that it is optimal for bidder 1 to bid according to the strategy in (3.3.2) when bidder 2 is following that strategy in the sealed bid first price auction of Example 3.3.1. [**Hint:** First show that it is not optimal to bid $b_1 < \underline{v}$ or $b_1 > Ev$. Since $\tilde{\sigma}$ is strictly increasing and continuous, any bid in $[\underline{v}, Ev]$ is the bid of some valuation v. Prove that bidding as if valuation v has valuation v' is suboptimal.]

3.6.4. Consider the following variant of a sealed-bid auction in a setting of independent private values. The highest bidder wins, and pays a price determined as the weighted average of the highest bid and second highest bid, with weight $\alpha \in (0, 1)$ on the highest bid (ties are resolved by a fair coin). Suppose there are two bidders, with bidder i's value v_i randomly drawn from the interval $[\underline{v}_i, \bar{v}_i]$ according to the distribution function F_i, with density f_i.

(a) What are the interim payoffs of player i?

(b) Suppose (σ_1, σ_2) is a Nash equilibrium of the auction, and assume σ_i is a strictly increasing and differentiable function, for $i = 1, 2$. Describe the pair of differential equations the strategies must satisfy.

(c) Suppose v_1 and v_2 are uniformly and independently distributed on $[0, 1]$. Describe the differential equation a symmetric increasing and differentiable equilibrium bidding strategy must satisfy.

(d) Solve the differential equation found in part (c). [**Hint:** Conjecture a functional form.]

(e) For the assumptions under part (c), prove the strategy found in part (d) is a symmetric equilibrium strategy.

3.6.5. Consider the two bidder sealed-bid auction of Example 3.3.1, when the two bidders' valuations are independent uniform draws from the interval $[0, 1]$.

(a) What is the expected revenue of the symmetric equilibrium?

In a first price sealed bid auction with reserve price r, the highest bidder only wins the auction (and pays his bid) if the bid is at least r.

(b) What is the symmetric Nash equilibrium of the first-price sealed-bid auction with reserve price $r \in [0, 1]$? [**Hint:** Since any winning bid must be no less than r, bidders with a value less than r can be assumed to bid 0 (there are multiple symmetric equilibria that only differ in the losing bids of bidders with values less than r). It remains to determine the bidding behavior of those bidders with values $\geq r$. Prove that any bidder with value r must bid r. While the equilibrium is *not* linear in valuation, the recipe from Example 3.3.1 still works.]

(c) What reserve price maximizes expected revenue?

3.6.6. This question asks you to fill in the details of Example 3.3.2.

(a) Prove that in any equilibrium, any bidder with value $\underline{v}$ must bid $\underline{v}$.

(b) Prove that there is no equilibrium in pure strategies.

(c) Prove that in any mixed strategy equilibrium, the minimum of the support of F_2 is given by $\underline{v}$.

(d) Prove that it is not optimal for $\bar{v}$ to bid $\underline{v}$.

(e) Prove that the symmetric profile in which each bidder bids $\underline{v}$ if $v = \underline{v}$, and according to the distribution function $F(b) = (b - \underline{v})/(\bar{v} - b)$ if $v = \bar{v}$, is a Nash equilibrium.

3.6.7. A variant of Example 3.3.2. The two bidders have independent private values drawn independently and identically from the common three point set $\{v_1, v_2, v_3\}$, with $v_1 < v_2 < v_3$. The common distribution assigns probability p_k to value v_k, $k = 1, 2, 3$. There is a symmetric equilibrium in which the low value bidders bid v_1, the mid value bidders randomize over a support $[v_1, \bar{b}_2]$, and the high value bidders randomize over a support $[\bar{b}_2, \bar{b}_3]$. Fully describe the equilibrium, and verify that it is indeed an equilibrium.

3.6.8. Consider again the two bidder sealed-bid auction of Example 3.3.1, but with a different valuation structure. In particular, the two bidders valuations are independent draws from the interval $[1, 2]$, with each bidder's valuation being a uniform draw from that interval with probability $p \in (0, 1)$ and with complementary probability $1 - p$ equaling some $v^* \in [1, 2]$.

(a) Suppose $v^* = 1$. What is the symmetric Nash equilibrium? [**Hint:** What must v^* bid?]

(b) Suppose $v^* = 2$. What is the symmetric Nash equilibrium? [**Hint:** Type v^* must randomize. Why? Can v^* have an atom in his bid distribution?]

(c) Suppose $v^* \in (1, 2)$. What is the symmetric Nash equilibrium?

(d) Compare and contrast the comparative statics with respect to p of the equilibria for different values of v^*.

3.6.9. An asymmetric variant of Example 3.3.2. The two bidders have independent private values drawn independently from the common two point set $\{\underline{v}, \bar{v}\}$, with $\underline{v} < \bar{v}$. The probability that bidder i has value $\bar{v}$ is $p_i \in (0, 1)$, and $p_1 < p_2$. Ties are broken in favor of bidder 2.

(a) As in Problem 3.6.6, in any equilibrium any bidder with value $\underline{v}$ must bid $\underline{v}$. Prove that in any equilibrium, the support of the bids made by bidder 1 of type $\underline{v}$ is the same as that of bidder 2 of type $\underline{v}$, and the minimum of the common support is $\underline{v}$. Denote the common support by $[\underline{v}, \bar{b}]$.

(b) Prove that one of the bidder's behavior strategies has an atom at $\underline{v}$ (the other does not). What is the size of the atom?

(c) Describe the equilibrium.

(d) Suppose now that the high value bidder 1 has value $\bar{v}_1$, while the high value bidder 2 has value $\bar{v}_2$, with the two values not necessarily equal. This game only has an equilibrium if ties are broken in the "right" way. What is that way?

3.6.10. This question asks you to prove Lemma 3.3.2:

(a) Suppose $\bar{s}_2 < \bar{s}_1$, and set $\delta := \Pr\{\bar{s}_2 < \sigma_1(t_1) \leq \bar{s}_1\} > 0$. Prove that there exists $\tilde{s}_1$ satisfying $\Pr\{\sigma_1(t_1) > \tilde{s}_1\} < \delta/2$. [**Hint:** This is trivial if $\bar{s}_1 < \infty$ (why?). The case where $\bar{s}_1 = \infty$ uses a basic continuity property of probability.]

(b) Show that a deviation by type $t_2 > 2\tilde{s}_1$ to a stopping time $s_2 > \tilde{s}_1$ (which implies that t_2 wins the war of attrition with probability of at least $\delta/2$) satisfying $s_2 < t_2/2$ is strictly profitable.

3.6.11. This question builds on Example 3.1.1 to capture the possibility that firm 2 may know that firm 1 has low costs, c_L. This can be done as follows: Firm 1's space of uncertainty (types) is, as before, $\{c_L, c_H\}$, while firm 2's is $\{t_I, t_U\}$. Nature determines the types according to the distribution

$$\Pr(t_1, t_2) = \begin{cases} 1 - p - q, & \text{if } (t_1, t_2) = (c_L, t_I), \\ p, & \text{if } (t_1, t_2) = (c_L, t_U), \\ q, & \text{if } (t_1, t_2) = (c_H, t_U), \end{cases}$$

where $0 < p, q$ and $p + q < 1$. Firm 2's type, t_I or t_U, does not affect his payoffs (in particular, his cost is c_2, as in Example 3.1.1). Firm 1's type is just his cost, c_1.

(a) What is the probability firm 2 assigns to $c_1 = c_L$ when his type is t_I? When his type is t_U?

(b) What is the probability firm 1 assigns to firm 2 knowing firm 1's cost? [This may depend on 1's type.]

(c) Solve for the values of p and q that imply that t_U has the beliefs of player 2 in Example 3.1.1 and c_L assigns probability $1 - \alpha$ to t_I.

(d) For these values of p and q, solve for the Nash equilibrium of this game. Compare your analysis to that of Example 3.1.1.

3.6.12. The linear equilibrium of Example 3.3.6 is not the only equilibrium of the double auction.

(a) Fix a price $p \in (0, 1)$. Show that there is an equilibrium at which, if trade occurs, then it occurs at the price p.

(b) What is the probability of trade?

(c) At what p is the probability of trade maximized?

(d) Compare the expected gains from trade under these "fixed price" equilibria with the linear equilibrium of Example 3.3.6.

3.6.13. In Example 3.5.1, suppose that in the state ω_1, $\theta = 20$, while in states ω_m, $2 \leq m \leq M - 1$, $\theta = 9$ and in state ω_M, $\theta = 4$. Suppose the information partitions are as in the example. In other words, apart from the probability distribution over Ω (which we have not yet specified), the only change is that in state ω_1, $\theta = 20$ rather than 9.

(a) Suppose the probability distribution over Ω is uniform (as in the lecture notes). What is the unique Nash equilibrium, and why? What is the unconditional probability of both players choosing A in the equilibrium?

(b) Suppose now the probability of ω_m is $4^{10-m}\alpha$, for $m = 1, \ldots, 9$, where α is chosen so that $\sum_{m=1}^{9} 4^{10-m}\alpha = 1$. What is the unique Nash equilibrium, and why? What is the unconditional probability of both players choosing A in the equilibrium

3.6.14. This question asks you to fill in some of the details of the calculations in Example 3.5.2.

(a) What is the marginal density of x for $x \in [-\varepsilon, 20 + \varepsilon]$?

(b) What is the expected value of θ conditional on x, for $x \in [-\varepsilon, 20 + \varepsilon]$?

(c) Derive player i's posterior beliefs about player j's signal x_j, conditional on $x_i \in [-\varepsilon, 20 + \varepsilon]$.

Chapter 4

Nash Equilibrium: Existence and Foundations

4.1 Existence

Recall Nash equilibria are fixed points of the best reply correspondence:

$$s^* \in \phi(s^*).$$

When does ϕ have a fixed point?

Theorem 4.1.1 (Kakutani's fixed point theorem). *Suppose $X \subset \mathbb{R}^m$ for some m and $F : X \rightrightarrows X$. Suppose*

1. *X is nonempty, compact, and convex;*

2. *F has nonempty convex-values (i.e., $F(x)$ is a convex set and $F(x) \neq \varnothing \; \forall x \in X$); and*

3. *F has closed graph: $(x^k, \hat{x}^k) \to (x, \hat{x}), \; \hat{x}^k \in F(x^k) \Rightarrow \hat{x} \in F(x)$.*

Then F has a fixed point.

Remark 4.1.1. A correspondence $F : X \rightrightarrows X$ is *upper hemicontinuous* at $x \in X$ if for all open sets $\mathcal{O} \subset \mathbb{R}^m$ satisfying $F(x) \subset \mathcal{O}$, there exists δ such that for all $x' \in X$ satisfying $|x - x'| < \delta$, $F(x') \subset \mathcal{O}$. A correspondence $F : X \rightrightarrows X$ is *upper hemicontinuous* if it is upper hemicontinuous at all

$x \in X$. A singleton-valued correspondence is a function,[1] and the notion of upper hemicontinuity for singleton-valued correspondences is precisely that of continuity of the function.

An upper hemicontinuous correspondence with closed values has a closed graph.[2] Unlike upper hemicontinuity, the closed graph property by itself does not imply continuity of functions: the function $f : \mathbb{R} \to \mathbb{R}$ given by $f(x) = 1/x$ for $x > 0$ and $f(0) = 0$ has a closed graph but is not continuous. *But if F has a closed graph and X is compact, then F is upper hemicontinuous*; for more on this, see Ok (2007, §E.2).

So, the hypotheses of Theorem 4.1.1 could have been written as F is closed- and convex-valued and upper hemicontinuous. See Problem 4.3.1 for an illuminating example.   ♦

Theorem 4.1.2. *Given a normal form game $G = \{(S_i, u_i) : i = 1, \ldots, n\}$, if for all i,*

1. *S_i is a nonempty, convex, and compact subset of $\mathbb{R}^k$ for some k, and*

2. *$u_i : S_1 \times \cdots \times S_n \to \mathbb{R}$ is continuous in $s \in S_1 \times \cdots \times S_n$ and quasiconcave in s_i,*

then G has a Nash equilibrium strategy profile.

Proof. Since u_i is continuous, the Maximum Theorem (Mas-Colell, Whinston, and Green, 1995, Theorem M.K.6) implies that ϕ_i has a closed graph.

The quasiconcavity of u_i implies that ϕ_i is convex-valued: For fixed $s_{-i} \in S_{-i}$, suppose $s_i', s_i'' \in \arg\max u_i(s_i, s_{-i})$. Then, from the quasiconcavity of u_i, for all $\alpha \in [0, 1]$,

$$u_i(\alpha s_i' + (1 - \alpha)s_i'', s_{-i}) \geq \min\{u_i(s_i', s_{-i}), \ u_i(s_i'', s_{-i})\},$$

and so

$$u_i(\alpha s_i' + (1 - \alpha)s_i'', s_{-i}) \geq \max u_i(s_i, s_{-i})$$

so that $\alpha s_i' + (1 - \alpha)s_i'' \in \arg\max u_i(s_i, s_{-i})$.

The theorem then follows from Kakutani's fixed point theorem by taking $X = S_1 \times \cdots \times S_n$ and $F = (\phi_1, \ldots, \phi_n)$.   ■

Theorem 4.1.3 (Nash (1950b, 1951)). *Every finite normal form game, $\{(S_i, u_i)_i\}$, has a mixed strategy Nash equilibrium.*

---

[1]This is a slight abuse of language. Technically, a singleton valued correspondence is not a function, since its value is a subset of X with only one element, i.e., $\{x\}$, while the value of a function is a point in x in X.

[2]The correspondence $F(x) := (0, 1)$ is trivially upper hemicontinuous, does not have closed-values, and does not have a closed graph.

Proof. Define $X_i = \Delta(S_i)$ and $X := \prod_i X_i$. Then $X \subset \mathbb{R}^{\Sigma|S_i|}$ and X is nonempty, convex and compact.

In this case, rather than appealing to the maximum theorem, it is an easy (and worthwhile!) exercise to prove that ϕ_i is convex-valued and has a closed graph.   ∎

In some applications, we need an infinite-dimensional version of Kakutani's fixed point theorem. In the following theorem, a topological space X is Hausdorff if for all $x_1 \neq x_2 \in X$, there exist two disjoint neighborhoods $\mathcal{U}_i$, $i = 1, 2$, with $x_i \in \mathcal{U}_i$. A topological space is *locally convex* if there is a base for the topology consisting of convex sets. In particular, every normed vector space is locally convex Hausdorff.

Theorem 4.1.4 (Fan-Glicksberg fixed point theorem). *Suppose X is a nonempty, compact, and convex subset of a locally convex Hausdorff space, and suppose $F : X \rightrightarrows X$. Suppose*

1. *F has nonempty convex-values (i.e., $F(x)$ is a convex set and $F(x) \neq \varnothing \; \forall x \in X$); and*

2. *F has closed graph: $(x^k, \hat{x}^k) \rightarrow (x, \hat{x})$, $\hat{x}^k \in F(x^k) \Rightarrow \hat{x} \in F(x)$.*

Then, F has a fixed point.

Locally convex Hausdorff spaces generalize many of the nice properties of normed vector spaces. This generalization is needed in the following theorem, since the spaces are typically not normed.

Theorem 4.1.5. *Suppose X is a compact metric space. The space of probability measures on X is a nonempty, compact, and convex subset of a locally convex Hausdorff space. Moreover, if $f : X \rightarrow \mathbb{R}$ is a continuous function, then $\int f d\mu$ is a continuous function of μ.*

Corollary 4.1.1. *Suppose S_i is a compact subset of a finite dimensional Euclidean space $\mathbb{R}^{m_i}$, and suppose $u_i : S \rightarrow \mathbb{R}$ is continuous. The normal form game $\{(S_i, u_i)_i\}$ has a Nash equilibrium in mixed strategies.*

Proof. The proof mimics that of Theorem 4.1.3.   ∎

Example 4.1.1 (Examples of nonexistence). See Problem 4.3.4 for a simple example of nonexistence in a game of complete information.

Nonexistence in games of incomplete information arises more naturally. Consider the following example of a private-value first-price auction with ties broken using a fair coin. Bidder 1 has value 3 and bidder 2 has value 3 with probability $\frac{1}{2}$ and value 4 with probability $\frac{1}{2}$. An intuition for nonexistence can be obtained by observing that since there is positive probability

	$v_2 = 3$	$v_2 = 4$
$v_1 = 3$	$\frac{1}{3}$	$\frac{1}{3}$
$v_1 = 4$	$\frac{1}{3}$	0

Figure 4.1.1: The joint distribution for nonexistence in Example 4.1.1.

that 2 has value 3, neither bidder will bid over 3 when their value is 3 (since he may win). Moreover, neither bidder can randomize on bids below 3 when their value is 3 for standard reasons. But if bidder 1 bids 3, then bidder 2 of value 4 does not have a best reply. Existence is restored if ties are broken in favor of bidder 2. If we discretize the space of bids, then there is an equilibrium in which bidder 2 with value 4 bids $3 + \Delta$, where $\Delta > 0$ is the grid size, and the other bidders bid 3. This equilibrium, for Δ small, effectively breaks ties in favor of the high value bidder.

Suppose now the joint distribution over bidder valuations is given by the table in Figure 4.1.1. The critical feature of this joint distribution is that the bidder with value 4 knows that the other bidder has value 3 (and so the valuations are *not* independent). For any tie breaking rule, the value 4 bidder has no best reply, and so there is no equilibrium (in pure or mixed strategies). Existence can only be restored by using a tie breaking rule that awards the item to the highest value bidder. If we again discretize the space of bids, then there is an equilibrium in which the bidders with value 4 bid $3 + \Delta$, where $\Delta > 0$ is the grid size, and the other bidders bid 3. This equilibrium, for Δ small effectively breaks ties in favor of the high value bidder. For (much) more on this, see Jackson, Simon, Swinkels, and Zame (2002).

★

An accessible reference on fixed point theorems and their application to economics is Border (1985) (though it does not cover the more advanced material briefly mentioned here). A reference that does cover that material is Aliprantis and Border (1999, Sections 14.3 and 16.9).

Existence results that replace continuity assumptions on payoffs with complementarity or supermodularity assumptions (and use Tarski's fixed point theorem) are of increasing importance. For an introduction, see Chapter 7 in Vohra (2005).

4.2   Learning/Evolutionary Foundations

Nash equilibrium requires that players are playing optimally given beliefs about the play of other players (call this the *optimization hypothesis*) and that these beliefs are correct (call this the *consistency hypothesis*).

There is a sense in which the optimization hypothesis is almost true by definition: a player chooses an action he/she believes is optimal. For if he/she believed another action was optimal, (surely) it would have been chosen. This reflects our assumption that a player's payoffs capture the game as perceived by that player (recall Remark 1.1.1). The optimization hypothesis does require that each player is not boundedly rational, in the sense that he or she does not face any computational or other constraints on calculating optimal behavior. In some cases, this is a heroic assumption. For example, under tournament rules,[3] chess is a finite game of perfect information and so has a backward induction solution. But that solution is unknown.

The consistency hypothesis requires that all players have correct beliefs about other players' behavior, including how other players would respond to possible deviations. This is a strong hypothesis. In some settings, the best justification is that it constrains modelers. For example, in the battle of the sexes (Example 2.1.1), the outcome (Opera, Ballet) conforms with the optimization, but not with the consistency, hypothesis. More generally, optimization without constraints on beliefs only implies that players do not play strictly dominated strategies. Unfortunately, most games have large sets of undominated strategies.

At the same time, the assumptions used to constrain modelers need to be themselves disciplined in some way. Considerable energy has been devoted to obtaining other justifications for consistency.

One obvious path is learning. Suppose two players are repeatedly playing the same stage game. Will they eventually learn to play a Nash equilibrium of the stage game? A few minutes of reflection suggests that the answer is not obvious. For example, if players are not myopic, then we should not expect the players to be maximizing stage game payoffs (in which case, a player may have an incentive to manipulate the future player of the other player). But such a scenario seems to be a worse case scenario, since it lacks important context (in particular, if this "repeated play" occurs in a social context, there may relevant history that aids in coordinating play).

In this section, I give a very brief introduction to two more fruitful

---

[3]Strictly speaking, chess is only finite under the assumption that players claim a draw when the rules permit: when either the position is repeated three times, or 50 moves are made without a capture or a pawn move.

approaches. The first considers a scenario of social learning (which has strong connections to evolutionary game theory) and the second of individual learning. Both share the following features:

- the analysis is dynamic, focusing on asymptotic properties,

- the assumed behavior respects myopic optimality, and

- a focus on the interaction of learning with the evolution of the system.

The second bullet point needs some explanation. Since players are learning in some way, it is not necessary to assume that players know (or behave as if they know) the optimal actions given beliefs. Rather, players may learn (discover) the optimal actions through experimentation or observation. It is worth noting that this argument is in the same vein as the arguments economists have given for why we should expect markets to be in equilibrium. Economic actors set prices, buy and sell goods and services, revise prices, proceeding by trial and error until the process settles down at an equilibrium where supply equals demand. Evolutionary game theory can be viewed as a return to the traditional interpretation of economic models (in contrast to the "ultrarational" view of traditional game theory).[4]

In Section 2.5, I argued that not all Nash equilibria are equally deserving of investigation, and commented that equilibrium selection often invokes a broader context, such as learning. Many of the games in this section have multiple equilibria, not all of which can be learned.

A good book length treatment of the general topic covered here is Fudenberg and Levine (1998). For much more on evolution, see Samuelson (1997) and Weibull (1995) (or Mailath, 1998, for a longer nontechnical introduction than here).

4.2.1   Social Learning (Evolutionary Game Theory)

In the social learning approach, the players in a game are drawn from a large population repeatedly over time, learn by observing the history of play, and choose an action. In the basic models, players are not modeled as sophisticated decision makers updating beliefs over their uncertainty over their environment (which includes the behavior of other players). Rather, players adjust their behavior in response to social evidence that suggests that another action may be myopically superior. The analysis provides a foundation for both the consistency and the optimization hypotheses.

---

[4]In his Ph.D. dissertation (Nash, 1950c), Nash gave two interpretations of equilibrium, one corresponding to the rational view, and the other corresponding to evolutionary game theory.

	A	B
A	1	0
B	0	1

Figure 4.2.1: The game for Example 4.2.1.

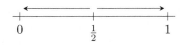

Figure 4.2.2: The dynamics for Example 4.2.1.

Example 4.2.1. There is a large population of players whose members are repeatedly randomly paired to play the game displayed in Figure 4.2.1. Assume there is no *role identification*, so the payoff represents the payoff to a player who chooses the row action, when facing the column action. If α is fraction of population playing A, then average payoff is given by

$$u(A; \alpha) = \alpha,$$
$$\text{and} \quad u(B; \alpha) = 1 - \alpha.$$

Social learning is captured by the assumption that over time, the fraction of the population playing A (B) increases if A (B) has a higher average payoff. Then,

$$\dot{\alpha} = \frac{d\alpha}{dt} > 0 \quad \Longleftrightarrow \quad u(A; \alpha) > u(B; \alpha) \quad \Longleftrightarrow \quad \alpha > \frac{1}{2}$$

and

$$\dot{\alpha} < 0 \quad \Longleftrightarrow \quad u(A; \alpha) < u(B; \alpha) \quad \Longleftrightarrow \quad \alpha < \frac{1}{2}.$$

The dynamics are illustrated in Figure 4.2.2. Clearly, the two symmetric pure strategy equilibria are stable rest points of the dynamics,[5] while the symmetric mixed strategy equilibrium $\frac{1}{2} \circ A + \frac{1}{2} \circ B$ is unstable.

This is an example of the equilibrating process eliminating an equilibrium from the set of equilibria.   ★

Example 4.2.2. Now suppose the players are playing the game in Figure 4.2.3.

---

[5]For now, interpret stable intuitively; this is defined below in Definition 4.2.1.

	A	B
A	1	2
B	2	1

Figure 4.2.3: The game for Example 4.2.2.

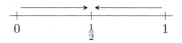

$$0 \qquad \frac{1}{2} \qquad 1$$

Figure 4.2.4: The dynamics for Example 4.2.2.

We again assume no role identification, so that AB and BA are infeasible. If α is the fraction of population playing A, then

$$u(A;\alpha) = \alpha + 2(1-\alpha) = 2 - \alpha,$$
$$\text{and} \quad u(B;\alpha) = 2\alpha + 1 - \alpha = 1 + \alpha.$$

Then,

$$\dot{\alpha} > 0 \quad \Longleftrightarrow \quad 2 - \alpha > 1 + \alpha \quad \Longleftrightarrow \quad \frac{1}{2} > \alpha$$

and

$$\dot{\alpha} < 0 \quad \Longleftrightarrow \quad 2 - \alpha < 1 + \alpha \quad \Longleftrightarrow \quad \frac{1}{2} < \alpha.$$

The symmetric mixed strategy equilibrium is now the only stable equilibrium. The dynamics are illustrated in Figure 4.2.4. ★

Let S denote a finite set of strategies in a symmetric game. In the above examples, $S = \{A, B\}$. The payoff to playing the strategy $s \in S$ against an opponent who plays $r \in S$ is $u(s, r)$.

The state of society is $\sigma \in \Delta(S)$. The expected payoff to s when state of society is σ is

$$u(s, \sigma) = \sum_{r} u(s, r)\sigma(r).$$

The dynamics are described by a function

$$F : \Delta(S) \times \mathbb{R}_+ \to \Delta(S);$$

if the initial state is σ, then under the dynamic F, the state at time t is given by $F(\sigma, t)$. For this interpretation of F to make sense, we clearly must require

$$F(\sigma, t' + t) = F(F(\sigma, t'), t),$$

since if the population strategy profile is σ' at t', then at time $t' + t$ it will be $F(\sigma', t)$.

Definition 4.2.1. *A state σ^* is a* rest *(or* stationary*) point of F if*

$$\sigma^* = F(\sigma^*, t) \qquad \forall t.$$

A rest point *σ^* is* asymptotically stable *under F if there exists $\varepsilon > 0$ such that if $\mid \sigma' - \sigma^* \mid < \varepsilon$, then $\lim_{t \to \infty} F(\sigma', t) = \sigma^*$.*

Assume F is continuously differentiable in all its arguments (on the boundaries, assume the appropriate one-sided derivatives exist and are continuous). Write $\dot\sigma$ for $\partial F(\sigma, t)/\partial t|_{t=0}$.

Note that

$$\sum_s \sigma(s) = 1 \implies \sum_s \dot\sigma(s) = 0.$$

Definition 4.2.2. *The dynamic F is a* myopic adjustment dynamic *if for all σ, $s, r \in S$ satisfying $\sigma(s), \sigma(r) > 0$,*

$$u(s, \sigma) > u(r, \sigma) \implies \dot\sigma(s) > \dot\sigma(r).$$

Theorem 4.2.1. *Suppose F is a myopic adjustment dynamic.*

1. *If σ^* is asymptotically stable under F, then it is a symmetric Nash equilibrium.*

2. *If σ^* is a strict Nash equilibrium, then σ^* is asymptotically stable under F.*

Proof.   1. Left as an exercise.

2. Suppose σ^* is a strict Nash equilibrium. Then, σ^* is a pure strategy s and $u(s, s) > u(r, s)$ for all $r \neq s$. This implies that there exists $\varepsilon > 0$ such that for all σ satisfying $\sigma(s) > 1 - \varepsilon$,

$$u(s, \sigma) > u(r, \sigma), \quad \forall r \neq s.$$

Suppose $1 > \sigma(s) > 1 - \varepsilon$ and $\sigma(r) > 0$ for all r. Myopic adjustment implies $\dot\sigma(s) > \max\{\dot\sigma(r) : r \neq s\}$, and so $\dot\sigma(s) > 0$ (since $\sum_{r \in S} \dot\sigma(r) = 0$).

Consider now σ satisfying $1 > \sigma(s) > 1 - \varepsilon$ with $\sigma(r) = 0$ for some r. Since $\dot\sigma$ is continuous in σ (since F is continuously differentiable, including on the boundaries), $\dot\sigma(s) \geq 0$ and $\dot\sigma(s) \geq \dot\sigma(r)$. Suppose $\dot\sigma(s) = 0$ (so that $\dot\sigma(r) \leq 0$). Then, $\dot\sigma(r') < 0$ for all r' satisfying $\sigma(r') > 0$ and so $\dot\sigma(r) > 0$, a contradiction.

Hence, if $1 > \sigma(s) > 1 - \varepsilon$, then $\dot\sigma(s) > 0$. Defining $\sigma^t := F(\sigma, t) \in \Delta(S)$, this implies $\sigma^t(s) > 1 - \varepsilon$ for all t, and so $\sigma^t(s) \to 1$.

∎

There are examples of myopic adjustment dynamics that do *not* eliminate strategies that are iteratively strictly dominated. Stronger conditions (such as *aggregate monotonicity*) are needed—see Fudenberg and Levine (1998). These conditions are satisfied by the *replicator dynamic*, which I now describe.

This dynamic comes from biology and, more specifically, evolutionary game theory. Payoffs are now interpreted as reproductive fitness (normalize payoffs so $u(s, r) > 0$ for all $s, r \in S$). At the end of each period, each agent is replaced by a group of agents who play the same strategy, with the size of the group given by the payoff (fitness) of the agent. Let $x_s(t)$ be the size of the population playing s in period t. Then,

$$x_s(t + 1) = x_s(t)u(s, \sigma^t),$$

where

$$\sigma^t(s) = \frac{x_s(t)}{\sum_r x_r(t)} =: \frac{x_s(t)}{\bar x(t)}.$$

Then, since $\bar x(t + 1) = \bar x(t)u(\sigma^t, \sigma^t)$,

$$\sigma^{t+1}(s) = \sigma^t(s)\frac{u(s, \sigma^t)}{u(\sigma^t, \sigma^t)},$$

so the difference equation is

$$\sigma^{t+1}(s) - \sigma^t(s) = \sigma^t(s)\frac{u(s, \sigma^t) - u(\sigma^t, \sigma^t)}{u(\sigma^t, \sigma^t)}.$$

Thus, as long as $\sigma^t(s) > 0$, we have $\sigma^{t+1}(s) > (<)\sigma^t(s)$ if, and only if, $u(s, \sigma^t) > (<)u(\sigma^t, \sigma^t)$. In continuous time, this is

$$\dot\sigma(s) = \sigma(s)\frac{u(s, \sigma) - u(\sigma, \sigma)}{u(\sigma, \sigma)}.$$

This dynamic has the same trajectories (paths) as

$$\dot\sigma(s) = \sigma(s)[u(s, \sigma) - u(\sigma, \sigma)].$$

	L	R
T	1, 1	1, 0
B	1, 1	0, 0

Figure 4.2.5: The game for Example 4.2.3.

Note that under the replicator dynamic, every pure strategy profile is a rest point: if $\sigma(s) = 0$ then $\dot{\sigma}(s) = 0$ *even when* $u(s, \sigma) > u(\sigma, \sigma)$.

Idea extends in straightforward fashion to games with role identification. In that case, we have

$$\dot{\sigma}_i(s_i) = \sigma_i(s_i)[u_i(s_i, \sigma_{-i}) - u_i(\sigma_i, \sigma_{-i})].$$

Example 4.2.3 (Domination). The game is displayed in Figure 4.2.5. Let p^t be the fraction of row players choosing T, while q^t is the fraction of column players choosing L. The replicator dynamics are

$$\dot{p} = p(1 - p)(1 - q)$$
$$\text{and} \quad \dot{q} = q(1 - q).$$

The phase diagram is illustrated in Figure 4.2.6.[6] No rest point is asymptotically stable.   ★

Example 4.2.4 (Simplified ultimatum game). In the simplified ultimatum game, the proposer offer either an equal split, or a small payment. The responder only responds to the small payment (he must accept the equal split). The extensive form is given in Figure 4.2.7, and the normal form is in Figure 4.2.8.

Let p be the fraction of row players choosing equal split, while q is the fraction of column players choosing N.

The subgame perfect profile is $(0, 0)$. There is another Nash outcome, given by the row player choosing equal division. The set of Nash equilibrium yielding this outcome is $N = \{(1, q) : 3/8 \le q\}$.

The replicator dynamics are

$$\dot{p} = p(1 - p)(80q - 30)$$
$$\text{and} \quad \dot{q} = -20q(1 - q)(1 - p).$$

Note that $\dot{p} > 0$ if $q > 3/8$ and $\dot{p} < 0$ if $q < 3/8$, while $\dot{q} < 0$ for all (p, q).

The phase diagram is illustrated in Figure 4.2.9. The subgame perfect

---

[6]The phase diagram in Mailath (1998, Figure 11) is incorrect.

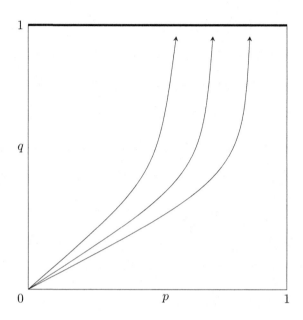

Figure 4.2.6: The phase diagram for Example 4.2.3.

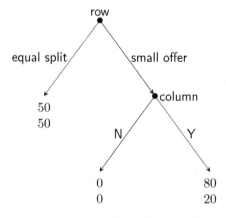

Figure 4.2.7: The extensive form of the simplified ultimatum game.

	N	Y
equal split	50, 50	50, 50
small offer	0, 0	80, 20

Figure 4.2.8: The normal form of the simplified ultimatum game.

equilibrium B is the only asymptotically stable rest point.

In the presence of drift, the dynamics are now given by

$$\dot{p} = p(1 - p)(80q - 30) + \delta_1\left(\frac{1}{2} - p\right)$$

$$\text{and} \quad \dot{q} = -20q(1 - q)(1 - p) + \delta_2\left(\frac{1}{2} - q\right).$$

For small δ_i, with δ_1 sufficiently smaller than δ_2, the dynamics have two asymptotically stable rest points, one near B and one near A (see Samuelson (1997, chapter 5)). ★

4.2.2 Individual learning

Fix an n-player finite normal form game, $G = (S, U)$, $S := S_1 \times \cdots \times S_n$, $U : S \to \mathbb{R}^n$. Players play the infinitely repeated game G^∞ with perfect monitoring, so that players observe in period t the history of play $h^t := (s^0, ..., s^{t-1}) \in H^t := S^t$.

In each period, after observing a history h^t, player i forms a belief over the play of the other players in the next period. Call the function assigning a belief to every history an *assessment*, denoted $\mu_i : H^t \to \Delta(S_{-i})$. Player i 's behavior is described by a *behavior rule* $\phi_i : H^t \to \Delta(S_i)$.

Definition 4.2.3. *A behavior rule ϕ_i is* myopic with respect to *the assessment μ_i if, for all t and h^t, $\phi_i(h^t)$ maximizes $U_i(\sigma_i, \mu_i(h^t))$.*

Definition 4.2.4. *An assessment μ_i is* adaptive *if, for all $\varepsilon > 0$ and t, there exists $T(\varepsilon, t)$ such that for all $t' > T(\varepsilon, t)$ and $h^{t'}$, the belief $\mu_i(h^{t'})$ assigns no more than ε probability to pure strategies not played by $-i$ between t and t' in $h^{t'}$.*

Examples of adaptive assessments are last period's play, the empirical distribution of past play in the history, and exponential weighting of past

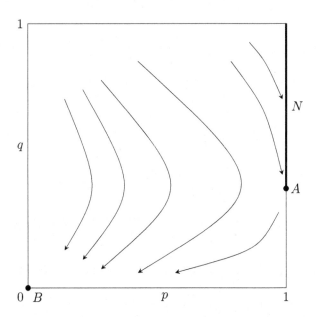

Figure 4.2.9: The phase diagram for the simplified ultimatum example. A is the nonsubgame perfect equilibrium $(1, 3/8)$, and B is the subgame perfect equilibrium.

plays. A myopic behavior rule with the first adaptive assessment yields Cournot dynamics, and with the second is fictitious play.

Behavior rules that are myopic with respect to adaptive assessments do not directly build in rationalizability-type sophisticated analysis of the behavior of the other players, since adaptivity does not impose restrictions on the relative weight on strategies that are not excluded.

Definition 4.2.5. *A history $h := (s^0, s^1, \ldots)$ is compatible with the behavior rules ϕ if s_i^t is in the support of $\phi_i(h^t)$, for all i and t.*

Theorem 4.2.2. *Suppose $(s^0, s^1, \ldots)$ is compatible with behavior that is myopic with respect to an adaptive assessment.*

1. *There exists T such that $s^t \in \bar{S}$ for all $t \geq T$, where $\bar{S}$ is the result of the iterative deletion of all strictly dominated strategies (which is equivalent to the set of rationalizable strategies when $n = 2$).*

2. *If there exists T such that $s^t = s^*$ for all $t > T$, then s^* is a (pure-strategy) Nash equilibrium of G.*

Proof.   1. Let S_i^k denote the set of player i's strategies after k rounds of deletions of strictly dominated strategies. Since S is finite, there exists $K < \infty$ such that $\bar{S} = S^K$. The proof proceeds by induction.

There exists T such that $s^t \in S^1$ for all $t \geq T$: Any $s_i \notin S_i^1$ is not a best reply to any beliefs, and so myopia implies that such a strategy is never chosen.

Suppose there exists T such that $s^t \in S^k$ for all $t \geq T$. If $s_i \notin S_i^{k+1}$, then s_i is not a best reply to any beliefs with support in S_{-i}^k. But then there exists $\varepsilon > 0$ such that s_i is not a best reply to any belief $\mu_i \in \Delta(S_{-i})$ satisfying $\mu_i(S_{-i}^k) > 1 - \varepsilon$. (Exercise: Calculate the bound on ε.) Since assessments are adaptive, there exists $T' > T$ such that $\mu_i(h^t)(S_{-i}^k) > 1 - \varepsilon$ for all $t > T'$. Since behavior is myopic, $s^t \in S^{k+1}$ for all $t \geq T'$.

2. Suppose there exists T such that $s^t = s^*$ for all $t > T$ and s^* is not a Nash equilibrium of G. Then, there exists i and $s_i' \in S_i$ such that $U_i(s_i', s_{-i}^*) > U_i(s^*)$. Further, there exists $\varepsilon > 0$ such that $U_i(s_i', \sigma_{-i}) > U_i(s_i^*, \sigma_{-i})$ if $\sigma_{-i}(s^*) > 1 - \varepsilon$. But then adaptive assessments with myopic behavior implies $s_i^t \neq s_i^*$ for t large, a contradiction.

∎

Stronger results on convergence (such as to mixed strategy equilibria) require more restrictions on assessments. For more, see Fudenberg and Levine (1998).

Convergence of beliefs need *not* imply imply convergence in behavior. For example, in matching pennies, the empirical distribution converges to $(\frac{1}{2}, \frac{1}{2})$, but players always play a pure strategy.

4.3   Problems

4.3.1. Consider the correspondence $F : [-1, 1] \rightrightarrows [-1, 1]$ defined as

$$
F(x) = \begin{cases} (0, 1) & \text{if } x \leq 0, \\ (0, \frac{x}{2}] & \text{if } x > 0. \end{cases}
$$

(a) Prove that F is upper hemicontinuous (i.e., prove that for all $x \in [-1, 1]$, for all open sets $\mathcal{O} \subset R$ satisfying $F(x) \subset \mathcal{O}$, there exists δ such that for all $x' \in (x - \delta, x + \delta) \cap [-1, 1]$, $F(x') \subset \mathcal{O}$).

(b) Does F have a fixed point?

(c) Which hypothesis of Theorem 4.1.1 fails for F?

(d) (A simple exercise to clarify the definition of upper hemicontinuity.) Consider the correspondences $G : [-1, 1] \rightrightarrows [-1, 1]$ and
$H : [-1, 1] \rightrightarrows [-1, 1]$ defined as

$$G(x) = \begin{cases} (0, 1), & \text{if } x \leq 0, \\ \left(0, \frac{x}{2}\right), & \text{if } x > 0, \end{cases}$$

and

$$H(x) = \begin{cases} (0, 1), & \text{if } x \leq 0, \\ \left[-\frac{x}{2}, \frac{x}{2}\right], & \text{if } x > 0. \end{cases}$$

For which values of x is G not upper hemicontinuous? For which
values of x is H not upper hemicontinuous?

4.3.2. (a) Suppose $f : X \times Y \to \mathbb{R}$ is a continuous function and X and Y
are compact subsets of $\mathbb{R}$. Prove that

$$\max_{x \in X} \min_{y \in Y} f(x, y) \leq \min_{y \in Y} \max_{x \in X} f(x, y).$$

Give an example showing that the inequality can hold strictly
(it suffices to do this for X and Y each only containing two
points—recall matching pennies from Section 2.4.1).

(b) von Neumann's celebrated *Minmax Theorem* states the following
equality: Suppose X and Y are finite sets, $f : X \times Y \to \mathbb{R}$, and $f :
\Delta(X) \times \Delta(Y) \to \mathbb{R}$ is given by $f(\alpha, \beta) = \sum_{x,y} f(x, y)\alpha(x)\beta(y)$.
Then,

$$\max_{\alpha \in \Delta(X)} \min_{\beta \in \Delta(Y)} f(\alpha, \beta) = \min_{\beta \in \Delta(Y)} \max_{\alpha \in \Delta(X)} f(\alpha, \beta).$$

Prove this theorem by applying Theorem 4.1.3 to the game G
given by $S_1 = X$, $S_2 = Y$, $u_1(s_1, s_2) = f(s_1, s_2)$, and $u_2(s_1, s_2) =
-u_1(s_1, s_2)$.[7] (Any two-player normal form game satisfying $u_1(s)
= -u_2(s)$ for all $s \in S$ is said to be *zero sum*.)

(c) Prove that (σ_1^*, σ_2^*) is a Nash equilibrium of a zero sum game if
and only if σ_i^* is a security strategy for player i, and that player
i's security level $\underline{v}_i$ is given by i's payoff in any Nash equilibrium.
(Compare with Problem 2.6.12.)

---

[7]von Neumann's original argument (1928), significantly predates Nash's existence
theorem, and the result is true more generally. There are elementary proofs of the
minmax theorem (based on the basic separating hyperplane theorem) that do not rely
on a fixed point theorem. See, for example, Owen (1982, §II.4) for the finite dimensional
case, and Ben-El-Mechaiekh and Dimand (2011) for the general case.

(d) Prove the following generalization of Problem 2.6.12(b): Suppose a two-player normal form game (not necessarily zero sum) has a unique Nash equilibrium, and each player's Nash equilibrium strategy and security strategy are both completely mixed. Prove that each player's security level is given by his/her Nash equilibrium payoff.

4.3.3. (a) Prove that every finite extensive form game has a subgame perfect equilibrium.

(b) (Harris, Reny, and Robson, 1995, Section 2.1) Verify that the following two-stage extensive-form game does not have a subgame perfect equilibrium (it is worth noting that this game has continuous payoffs and compact actions spaces). There are four players. In stage 1, players 1 and 2 simultaneously choose $a_1 \in [-1,1]$ and $a_2 \in \{L, R\}$ respectively. In stage 2, players 3 and 4 are informed of the choices of players 1 and 2, and then simultaneously choose $a_3 \in \{L, R\}$ and $a_4 \in \{L, R\}$ respectively. Payoffs are

$$u_1(a_1, a_2, a_3, a_4) =$$
$$\begin{cases} -|a_1| - \frac{1}{2}|a_1|^2, & \text{if } a_2 = a_3 \text{ and } a_3 = a_4, \\ |a_1| - \frac{1}{2}|a_1|^2, & \text{if } a_2 \neq a_3 \text{ and } a_3 = a_4, \\ -|a_1| - 10 - \frac{1}{2}|a_1|^2, & \text{if } a_2 = a_3 \text{ and } a_3 \neq a_4, \\ |a_1| - 10 - \frac{1}{2}|a_1|^2, & \text{if } a_2 \neq a_3 \text{ and } a_3 \neq a_4, \end{cases}$$

$$u_2(a_1, a_2, a_3, a_4) = \begin{cases} 1, & \text{if } a_2 = a_3 = L, \\ -1, & \text{if } a_2 = L, a_3 = R, \\ 2, & \text{if } a_2 = a_3 = R, \\ -2, & \text{if } a_2 = R, a_3 = L, \end{cases}$$

$$u_3(a_1, a_2, a_3, a_4) = \begin{cases} a_1, & \text{if } a_3 = L, \\ -a_1, & \text{if } a_3 = R, \end{cases}$$

and

$$u_4(a_1, a_2, a_3, a_4) = \begin{cases} a_1, & \text{if } a_4 = L, \\ -a_1, & \text{if } a_4 = R. \end{cases}$$

(c) (Luttmer and Mariotti, 2003) Prove that the following continuous game with perfect information does not have a subgame perfect equilibrium. The game has five stages. In stage 1, player 1 chooses $a_1 \in [0,1]$. In stage 2, player 2, knowing player 1's choice, chooses $a_2 \in [0,1]$. In stage 3, Nature chooses x by randomizing uniformly over the interval $[-2 + a_1 + a_2,\ 2 - a_1 - a_2]$. After observing x, player 3 chooses between U and D. The choice of D ends the game, resulting in payoffs $(a_1, 2a_2, 1, 1)$. After U, player 4, knowing everything that has happened, chooses between u and d, with u yielding payoffs $(2a_1, a_2, 2, 0)$ and d yielding payoffs $(0, 0, 0, x)$.

4.3.4. Two psychologists have to choose locations on a portion of Interstate 5 running through California and Oregon. The relevant portion of Interstate 5 is represented by the interval $[0, 4]$; the California portion is represented by $[0, 3]$ and the Oregon portion by $[3, 4]$. There is a continuum of potential clients, uniformly distributed on the Interstate; each client patronizes the psychologist located closest to him (irrespective of state). If the two psychologists are at the same location, each receives half the market. Finally, assume each psychologist chooses location to maximize his or her number of clients.

   (a) (The classic Hotelling location model.) Suppose both psychologists are licensed to practice in both states, and so can locate anywhere in the interval $[0, 4]$. What is the pure strategy equilibrium? Is it unique?

   (b) Suppose now that one psychologist is only licensed to practice in California (so his/her location is restricted to $[0, 3]$), while the other is licensed to practice in both states. How does this change your answer to part (a)?

   (c) (The next two parts are based on Simon and Zame, 1990.) Suppose that one psychologist is only licensed to practice in California, while the other is only licensed to practice in Oregon. Prove that this game now has no equilibrium (in either pure or mixed strategies).

   (d) Finally, maintain the licensing assumptions of part (c), but suppose that when the two psychologists both locate at 3, the Californian psychologist receives $\frac{3}{4}$ of the market. What is the pure strategy equilibrium?

4.3.5. Fill in the details of Example 4.1.1.

4.3.6. This question is a variation of the Cournot duopoly of Example 1.1.3. The market clearing price is given by $P(Q) = \max\{a - Q, 0\}$, where $Q = q_1 + q_2$ is total quantity and q_i is firm i's quantity. There is a constant marginal cost of production c, with $0 < c < a$. Finally, there is a fixed cost of production κ. Suppose $(a - c)^2/9 < \kappa < (a - c)^2/4$.

(a) Suppose firm i only incurs the fixed cost of production when $q_i > 0$. Then, firm i's profits are given by

$$U_i(q_1, q_2) = \begin{cases} (P(q_1 + q_2) - c)q_i - \kappa, & \text{if } q_i > 0, \\ 0, & \text{if } q_i = 0. \end{cases}$$

This game has two pure strategy Nash equilibria. What are they? Why doesn't this game have a symmetric pure strategy Nash equilibrium? Since firm payoffs are not continuous functions (why not?), the existence of a symmetric Nash equilibrium in mixed strategies is *not* implied by any of the theorems in Section 4.1.

(b) Suppose the fixed cost is an entry cost (and so is sunk). The game is still a simultaneous move game, but now firm i's strategy space is $\{-1\} \cup \mathbb{R}_+$, with the strategy -1 meaning "don't enter", and a number $q_i \in \mathbb{R}_+$ meaning "enter" and choose $q_i \geq 0$ (note the weak inequality). Then, firm i's profits are given by

$$U_i(s_1, s_2) = \begin{cases} (P(q_1 + q_2) - c)q_i - \kappa, & \text{if } (s_1, s_2) = (q_1, q_2), \\ (P(q_i) - c)q_i - \kappa, & \text{if } s_i = q_i \text{ and } s_j = -1, \\ 0, & \text{if } s_i = -1. \end{cases}$$

The existence of a symmetric Nash equilibrium in mixed strategies is implied by Theorems 4.1.4 and 4.1.5 as follows:

i. Argue that we can restrict the strategy space of firm i to $\{-1\} \cup [0, a]$, and so the strategy space is closed and bounded (i.e., compact). Note that the strategy space is not convex. (In contrast, the strategy spaces in part (a) are convex. But see part iii below!)

ii. Prove that U_i is continuous function of the strategy profiles.

iii. We can make the strategy space convex, while maintaining continuity as follows. Define $S_i := [-1, a]$, and extend U_i to S_i be setting

$$U_i(s_1, s_2) = \begin{cases} (P(s_1 + s_2) - c)s_i - \kappa, & \text{if } s_j \geq 0, \\ (P(s_i) - c)s_i - \kappa, & \text{if } s_i \geq 0 \text{ and } s_j < 0, \\ -(1 + s_i)\kappa, & \text{if } s_i < 0. \end{cases}$$

Prove U_i is continuous on S_i. Prove that every strategy in $(-1, 0)$ is strictly dominated (and so the addition of these strategies has no strategic impact). In particular, the set of Nash equilibria (in pure or mixed strategies) is unaffected by this change. Prove that U_i is not quasiconcave.

    iv. As for finite games, since the payoff defined on pure strategies is continuous, by considering mixed strategies, we obtain convex strategy spaces and payoffs that are continuous and quasiconcave. This yields for firm each i, a best reply correspondence (where $j \neq i$)

$$\phi_i : \Delta(S_j) \rightrightarrows \Delta(S_i)$$

    that satisfies all the conditions of Theorem 4.1.4. So, there is nothing to prove here.

    v. Explain why Theorem 4.1.4 implies the existence of a *symmetric* Nash equilibrium in mixed strategies.

(c) The symmetric mixed strategy equilibrium of the game in part (b) can be easily calculated for this parameterization (and coincides with the symmetric mixed strategy equilibrium of the game in part (a)).

    i. In the mixed strategy equilibrium, the support must be $\{-1, q^*\}$, for some $q^* > 0$, since firm payoffs are a strictly concave function of $q > 0$. Explain the role of strict concavity.

    ii. The symmetric mixed strategy equilibrium is thus determined by two numbers, q^* and α, the probability of q^*. Express q^* as a function of α using an appropriate first-order condition.

    iii. Finally, solve for α from the appropriate indifference condition.

4.3.7. Prove that the phase diagram for Example 4.2.3 is as portrayed in Figure 4.2.6. [This essentially asks you to give an expression for dq/dp.]

4.3.8. Prove that if σ^* is asymptotically stable under a myopic adjustment dynamic defined on a game with no role identification, then it is a symmetric Nash equilibrium.

4.3.9. Suppose $F : \Delta(S) \times \mathbb{R}_+ \to \Delta(S)$ is a dynamic on the strategy simplex with F is continuously differentiable (including on the boundaries). Suppose that if

$$\eta < \sigma(s) < 1,$$

for some $\eta \in (0, 1)$, then

$$\dot{\sigma}(s) > 0,$$

	A	B	C
A	1	1	0
B	0	1	1
C	0	0	1

Figure 4.3.1: The game for Problem 4.3.10.

where

$$\dot{\sigma} := \left. \frac{\partial F(\sigma, t)}{\partial t} \right|_{t=0}.$$

Fix σ^0 satisfying $\sigma^0(s) > \eta$. Prove that

$$\sigma^t(s) \to 1,$$

where $\sigma^t := F(\sigma^0, t)$.

4.3.10. Suppose a large population of players are randomly paired to play
the game (where the payoffs are to the row player) displayed in Figure
4.3.1 (such a game is said to have *no role identification*). Let α denote
the fraction of the population playing A, and γ denote the fraction
of the population playing C (so that $1 - \alpha - \gamma$ is the fraction of the
population playing B). Suppose the state of the population adjusts
according to the continuous-time replicator dynamic.

   (a) Give an expression for $\dot{\alpha}$ and for $\dot{\gamma}$.

   (b) Describe all the rest points of the dynamic.

   (c) Describe the phase diagram in the space $\{(\alpha, \gamma) \in \mathbb{R}_+^2 : \alpha + \gamma \leq 1\}$. Which of the rest points are asymptotically stable?

Chapter 5

Nash Equilibrium Refinements in Dynamic Games

5.1 Sequential Rationality

Example 5.1.1 (Selten's (1975) horse). Consider the game illustrated in Figure 5.1.1. Let (p_1, p_2, p_3) denote the mixed strategy profile where

$$\Pr(I \text{ plays } A) = p_1,$$
$$\Pr(II \text{ plays } a) = p_2,$$
$$\text{and} \qquad \Pr(III \text{ plays } L) = p_3.$$

The Nash equilibrium profile $(0, 1, 0)$ (i.e., DaR) is subgame perfect, and yet player II is not playing *sequentially rationally*: Even though player II's information is off the path of play, if player I did deviate and II played d, player III would not know that player I had deviated and so would play R, implying that II has a higher payoff from d.

The profile is also *not* trembling hand perfect (Definition 2.5.1): playing a is not optimal against any mixture close to DR.

The only trembling hand perfect equilibrium outcome is Aa. The set of Nash equilibria with this outcome is $\{(1, 1, p_3) : \frac{3}{4} \leq p_3 \leq 1\}$. In these equilibria, player III's information set is not reached, and so the profile cannot be used to obtain beliefs for III. However, each Nash equilibrium in the set is trembling hand perfect: Fix an equilibrium $(1, 1, p_3)$. Suppose first that $p_3 \in [\frac{3}{4}, 1)$ (so that $p_3 \neq 1!$) and consider the completely mixed

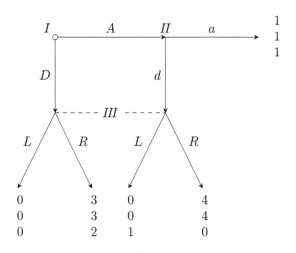

Figure 5.1.1: Selten's horse.

profile

$$p_1^n = 1 - \frac{1}{n},$$

$$p_2^n = 1 - \frac{2}{(n-1)},$$

and $p_3^n = p_3.$

Note that $p_1^n, p_2^n \to 1$ as $n \to \infty$. Suppose $n \geq 4$. It is easy to verify that both I and II are playing optimally against the mixed profile in $(1, 1, p_3)$. What about III? The probability that III is reached is

$$\frac{1}{n} + \frac{(n-1)}{n} \times \frac{2}{(n-1)} = \frac{3}{n},$$

and so the induced beliefs for III at his information set assign probability $\frac{1}{3}$ to the left node and $\frac{2}{3}$ to the right. Player III is therefore indifferent and so willing to randomize.

The same argument shows that $(1, 1, 1)$ is trembling hand perfect, using the trembles

$$p_1^n = 1 - \frac{1}{n},$$

$$p_2^n = 1 - \frac{2}{(n-1)},$$

$$\text{and} \qquad p_3^n = 1 - \frac{1}{n}.$$

Indeed, any sequence of trembles satisfying $p_1^n \to 1$, $p_2^n \to 1$, and $p_3^n \to 1$ will work, providing

$$\limsup_{n \to \infty} \frac{(1 - p_1^n)}{(1 - p_1^n p_2^n)} \le \frac{1}{3}.$$

(It is not even necessary for $(1 - p_1^n)/(1 - p_1^n p_2^n)$ to have a well-defined limit.)

$\bigstar$

Example 5.1.1 illustrates an observation made at the beginning of Section 2.5.1. The Nash equilibrium DaR involves a selection from the multiple best replies a and d for player II, and requiring that the behavior at player II's unreached information sets be optimal (sequentially rational) precludes the selection of a from the multiple best replies. We now explore sequential rationality in more detail.

Recall that $X := T \setminus Z$ is the set of decision nodes (nonterminal nodes) of an extensive form (Definition 1.3.1); we denote a typical element of X by x.

Definition 5.1.1. *A system of beliefs μ in a finite extensive form is a specification of a probability distribution over the decision nodes in every information set, i.e., $\mu : X \to [0, 1]$ such that*

$$\sum_{x \in h} \mu(x) = 1, \qquad \forall h.$$

Note that $\mu \in \prod_{h \in \cup_i H_i} \Delta(h)$, a compact set. We can write μ as a vector $(\mu_h)_h$ specifying $\mu_h \in \Delta(h)$ for each $h \in \cup_i H_i$.

We interpret μ as describing player beliefs. In particular, if h is player i's information set, then $\mu_h \in \Delta(h)$ describes i's beliefs over the nodes in h.

Let $\mathbf{P}^b$ denote the probability distribution on the set of terminal nodes Z implied by the behavior profile b (with b_0 describing nature's moves ρ).

Example 5.1.2. Consider the extensive form displayed in Figure 5.1.2. The label $[p]$ indicates that nature chooses the node t_1 with probability p (so that $\rho(t_1) = p$ and $\rho(t_2) = 1 - p$). As indicated, both nodes are owned by player I, and each is a singleton information set.[1] The induced distribution $\mathbf{P}^b$ on Z by the profile (LR, UD) is $p \circ z_1 + (1 - p) \circ z_8$. $\bigstar$

---

[1] In subsequent extensive forms with an initial move by nature, if all the nodes reached by the move of nature are owned by the same player (typically player I), that player name is typically omitted.

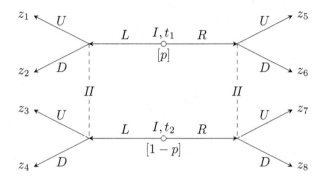

Figure 5.1.2: Game for Example 5.1.2.

In what follows, it is helpful to be more explicit about the determination of $\mathbf{P}^b$. In what follows, to avoid expressions of the form $b_{\iota(\ell)}(h^\ell)(a^\ell)$, where h^ℓ is the ℓ-th information set on the path to a terminal node, label actions so that $A(h) \cap A(h') = \varnothing$ for all $h \neq h' \in H_i$. Under this labeling convention, actions uniquely identify information sets and h^ℓ can be suppressed.[2]

For all noninitial nodes $t \in T = X \cup Z$, there is a unique path from the initial node $x_0 \in X$ to t. Denote the unique sequence of actions on the path from x_0 to t by $a^0, a^1, \ldots, a^L$ (where a^ℓ may be an action of nature), with player $\iota(\ell)$ choosing the action a^ℓ. For any behavior profile b (with b_0 describing nature's moves ρ), we define

$$\mathbf{P}^b(t) := \prod_{\ell=0}^{L} b_{\iota(\ell)}(a^\ell). \tag{5.1.1}$$

In particular, this gives the expression for the probability of reaching any terminal node $z \in Z$, and the expected payoff to player i is (recalling Definition 1.3.1)

$$E^b[u_i] := \sum_{z \in Z} u_i(z)\mathbf{P}^b(z).$$

Exercise 5.1.1. Prove that for any $x \in X$,

$$\mathbf{P}^b(x) = \sum_{\{z \in Z : x \prec z\}} \mathbf{P}^b(z).$$

---

[2]Equivalently, when $A(h) \cap A(h') = \varnothing$ for all $h \neq h' \in H_i$, every behavior strategy b_i can be viewed as a function

$$b_i : \bigcup_{h \in H_i} A(h) \to [0,1] \quad \text{such that} \quad \sum_{a \in A(h)} b_i(a) = 1 \quad \forall h \in H_i.$$

Similarly, for any decision node $x \in X$, we can calculate the probability of reaching a terminal node $z \in Z(x) := \{z \in Z : x \prec z\}$ from x; clearly the probability is zero for any $z \notin Z(x)$: For $z \in Z(x)$, denoting the unique sequence of actions on the path from x to z by $a^0, \ldots, a^L$, we have

$$\mathbf{P}^b(z \mid x) = \prod_{\ell=0}^{L} b_{\iota(\ell)}(a^\ell).$$

Let $Z(h) := \{z \in Z : \exists x \in h, x \prec z\} = \bigcup_{x \in h} Z(x)$. Define $\mathbf{P}^{\mu,b}(\cdot \mid h) \in \Delta(Z(h))$ by, for all $z \in Z(h)$,

$$\mathbf{P}^{\mu,b}(z \mid h) := \mu_h(x')\mathbf{P}^b(z \mid x'),$$

where x' is the *unique* decision node in h that precedes z. The distribution $\mathbf{P}^{\mu,b}(\cdot|h)$ is the probability distribution on $Z(h)$ implied by $\mu_h \in \Delta(h)$ (the beliefs specified by μ over the nodes in h), the behavior profile b (interpreted as describing behavior at the information set h and any that could be reached from h, and nature ρ if there are any moves of nature following h). By setting $\mathbf{P}^{\mu,b}(z|h) = 0$ for all $z \notin Z(h)$, $\mathbf{P}^{\mu,b}(\cdot|h)$ can be interpreted as the distribution on Z, "conditional" on h being reached. Note that $\mathbf{P}^{\mu,b}(\cdot|h)$ only depends on μ through μ_h; it does not depend on $\mu_{h'}$ for any $h' \neq h$. With this in mind, I no longer include the subscript h on μ. Note the different roles μ and b play in the expression for $\mathbf{P}^{\mu,b}$.

Player i's expected payoff conditional on h is

$$E^{\mu,b}[u_i \mid h] := \sum_{z \in Z} \mathbf{P}^{\mu,b}(z|h)u_i(z)$$
$$= \sum_{x \in h} \sum_{\{z:x \prec z\}} \mu(x)\mathbf{P}^b(z \mid x)u_i(z).$$

Definition 5.1.2. *A behavior strategy profile b in a finite extensive form is* sequentially rational at $h \in H_i$, *given a system of beliefs μ, if*

$$E^{\mu,b}[u_i \mid h] \geq E^{\mu,(\hat{b}_i,b_{-i})}[u_i \mid h],$$

for all $\hat{b}_i$.

A behavior strategy profile b in an extensive form is sequentially rational, *given a system of beliefs μ, if for all players i and all information sets $h \in H_i$, b is sequentially rational at h.*

A behavior strategy profile in an extensive form is sequentially rational *if it is sequentially rational given some system of beliefs.*

Definition 5.1.3. *A* one-shot deviation *by player i from b is a strategy b'_i with the property that there exists a unique information set $h' \in H_i$ such that $b_i(h) = b'_i(h)$ for all $h \neq h'$, $h \in H_i$, and $b_i(h') \neq b'_i(h')$.*

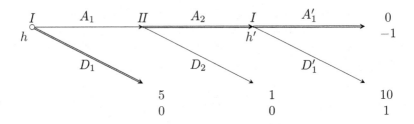

Figure 5.1.3: The game for Example 5.1.3.

A one-shot deviation b'_i (from b, given a system of beliefs μ) is profitable if

$$E^{\mu,(b'_i,b_{-i})}[u_i \mid h'] > E^{\mu,b}[u_i \mid h'],$$

where $h' \in H_i$ is the information set for which $b'_i(h') \neq b_i(h')$.

Example 5.1.3. Consider the profile $((D_1, A'_1), A_2)$ in the game in Figure 5.1.3. Player I is not playing sequentially rationally at his first information set h, but does *not* have a profitable one-shot deviation there. Player I does have a profitable one-shot deviation at his second information set h'. Player II also has a profitable one-shot deviation.

Consider now the profile $((D_1, A'_1), D_2)$. Now, player I is playing sequentially rationally at h, even though he still has a profitable one-shot deviation from the specified play at h'.   ★

The following result is obvious.

Lemma 5.1.1. *If a strategy profile b of a finite extensive form game is sequentially rational given μ, then there are no profitable one-shot deviations from b.*

Without further restrictions on μ (see Theorems 5.1.1 and 5.3.2 for examples), the converse need not hold: Even if a profile has no profitable one-shot deviations, it may fail to be sequentially rational.

Example 5.1.4. Consider the profile $((D_1, A'_1), A_2)$ in the game in Figure 5.1.4. Player I is not playing sequentially rationally at his first information set h, but does *not* have a profitable one-shot deviation at *any* information set, given the system of beliefs indicated.   ★

The following is the first instance of what is often called the *one-shot deviation principle* (it appears again in Theorems 5.3.2 and 7.1.3).

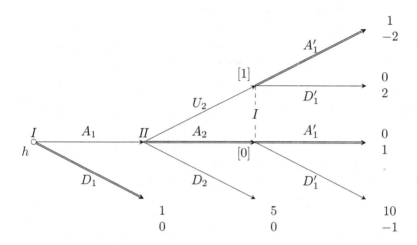

Figure 5.1.4: The game for Example 5.1.4. Player I's beliefs at his nontrivial information set is indicated near the relevant two nodes
in square brackets. Player I is not playing sequentially rationally at h. Nonetheless, player I does *not* have a profitable
one-shot deviation at any information set (given the beliefs
specified).

Theorem 5.1.1. *The following three statements about a strategy profile b
in a finite game of perfect information are equivalent:*

1. *The strategy profile b is subgame perfect.*

2. *The strategy profile b is sequentially rational.*

3. *The strategy profile b has no profitable one-shot deviations.*

Proof. Recall from Definition 1.3.6 that a game of perfect information has
singleton information sets. In such a case, the system of beliefs is trivial,
and sequential rationality is equivalent to subgame perfection.

By Lemma 5.1.1, a sequentially rational strategy profile has no profitable one-shot deviations.

The proof of the remaining direction is left as an exercise (Problem
5.4.2). ∎

5.2   Perfect Bayesian Equilibrium

Without some restrictions connecting beliefs to behavior, even Nash equilibria need *not* be sequentially rational.

Definition 5.2.1. *The information set h in a finite extensive form game is reached with positive probability under b, or is on the path-of-play, if*

$$\mathbf{P}^b(h) = \sum_{x \in h} \mathbf{P}^b(x) > 0.$$

Theorem 5.2.1. *The behavior strategy profile b of a finite extensive form game is Nash if and only if it is sequentially rational at every information set on the path of play, given a system of beliefs μ obtained using Bayes' rule at those information sets, i.e., for all h satisfying $\mathbf{P}^b(h) > 0$,*

$$\mu(x) = \frac{\mathbf{P}^b(x)}{\mathbf{P}^b(h)} \quad \forall x \in h. \tag{5.2.1}$$

The proof of Theorem 5.2.1 is left as an exercise (Problem 5.4.3).

Example 5.2.1. Recall the extensive form from Example 2.3.4, reproduced in Figure 5.2.1. The label $[p]$ indicates that the player owning that information set assigns probability p to the labeled node. The profile RBr (illustrated) is Nash and satisfies the conditions of the theorem.   ★

Theorem 5.2.1 implies the result in Problem 2.6.10.

In Theorem 5.2.1, sequential rationality is only imposed at information sets on the path of play. Strengthening this to all information sets yields:

Definition 5.2.2. *A strategy profile b of a finite extensive form game is a weak perfect Bayesian equilibrium (weak PBE) if there exists a system of beliefs μ such that*

 1. *b is sequentially rational given μ, and*

 2. *for all h on the path of play,*

$$\mu(x) = \frac{\mathbf{P}^b(x)}{\mathbf{P}^b(h)} \quad \forall x \in h.$$

Thus, a strategy profile b is a weak perfect Bayesian equilibrium if, and only if, it is a Nash equilibrium that is sequentially rational. A simple example of a Nash strategy profile that fails to be weak perfect Bayesian

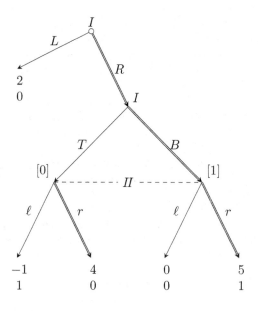

Figure 5.2.1: Game for Example 5.2.1.

is a Nash equilibrium of a perfect information game that is not subgame perfect, such as (Out, Fight) in Example 2.2.1.

While weak perfect Bayesian equilibrium does impose sequential rationality everywhere, it is insufficiently demanding in the following sense: It places *no* restrictions on the system of beliefs on information sets off the path of play. In particular, this means that for games that do not have perfect information, there will be strategy profiles that are weak perfect Bayesian, and yet are not subgame perfect. An example is the profile (LB, ℓ) in Figure 5.2.1, with beliefs μ that assign probability 1 to the node reached by T.

A minimum desiderata for a solution concept to capture sequential rationality with respect to sensible beliefs is that the solution concept imply subgame perfection. This we now do by taking seriously the phrase that we should use Bayes' rule "where possible." The phrase "where possible" is meant to suggest that we apply Bayes' rule in a conditional manner. We need the "follows" relation $\prec^*$ from Problem 1.4.4 (recall also from Problem 1.4.4 that information sets are *not* partially ordered by $\prec^*$). Note that the two information sets h and h' need *not* be owned by the same player.

Definition 5.2.3. *The information set h *-follows h' (written $h' \prec^* h$) if* *there exists $x \in h$ and $x' \in h'$ such that $x' \prec x$. If $h' \prec^* h$, define $h_{h'}$ as the set of nodes in h that can be reached from h', i.e., $h_{h'} := \{x \in h : \exists x' \in h', x' \prec x\}$.*

An information set h (*-following h') is reached with positive probability from h' under (μ, b) if

$$\mathbf{P}^{\mu,b}(h_{h'} \mid h') = \sum_{x \in h_{h'}} \mathbf{P}^{\mu,b}(x \mid h') > 0.$$

Since $\prec^*$ is not antisymmetric,[3] care must be taken in the following definition (see Problem 5.4.4). We must also allow for the possibility that $\sum_{\hat{x} \in h_{h'}} \mu(\hat{x}) = 0$.

Definition 5.2.4. *A strategy profile b of a finite extensive form game is an almost perfect Bayesian equilibrium (almost PBE) if there exists a system of beliefs μ such that*

1. *b is sequentially rational given μ, and*

2. *for any information set h' and *-following information set h reached with positive probability from h' under (μ, b) (i.e., $\mathbf{P}^{\mu,b}(h_{h'} \mid h') > 0$),*

$$\mu(x) = \frac{\mathbf{P}^{\mu,b}(x \mid h')}{\mathbf{P}^{\mu,b}(h_{h'} \mid h')} \sum_{\hat{x} \in h_{h'}} \mu(\hat{x}) \qquad \forall x \in h_{h'}. \qquad (5.2.2)$$

This strengthens weak perfect Bayesian by imposing a consistency requirement (motivated by Bayes' rule) on beliefs *across* multiple information sets related by *-follows, where the "first" (and so subsequent) information set is off the path of play.

Theorem 5.2.2. *Every almost perfect Bayesian equilibrium is subgame perfect.*

The proof is left as an exercise (Problem 5.4.5).

Example 5.2.2. Continuing with the extensive form from Example 2.3.4 displayed in Figure 5.2.2: The profile $LB\ell$ (illustrated) is weak perfect Bayesian, but not almost perfect Bayesian. Note that $LT\ell$ is *not* weak perfect Bayesian. The only subgame perfect equilibrium is RBr. ★

---

[3]That is, there can be two information sets h and h', owned by different players, with h *-following h' *and* h' *-following h; this cannot occur if the information sets are owned by the same player.

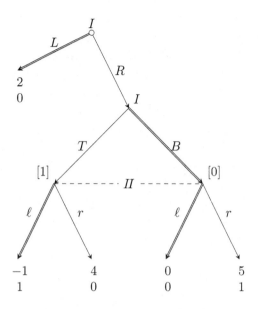

Figure 5.2.2: The profile $LB\ell$ (illustrated) is weak perfect Bayesian, but not almost perfect Bayesian.

Example 5.2.3. In the game displayed in Figure 5.2.3, for $x = -3$, the indicated profile $(C, R, (T_L, T_R))$ is not an almost perfect Bayesian equilibrium: If $(C, R, (T_L, T_R))$ was an almost perfect Bayesian equilibrium, then Definition 5.2.4 requires that in any beliefs supporting this equilibrium, players II and III after R assign the same probability to I's initial deviation being U. But any beliefs that make II's behavior sequentially rational must put significant probability on I's initial deviation being U, while the only belief that makes III's behavior after R sequentially rational puts zero probability on I's initial deviation being U, a contradiction.   ★

While almost perfect Bayesian equilibria are subgame perfect, they still allow "unreasonable" behavior. In particular, players can be interpreted as "signaling what they don't know."

Example 5.2.4. Continuing with the game displayed in Figure 5.2.3, for $x = 0$, the indicated profile $(C, R, (T_L, T_R))$ *is* an almost perfect Bayesian equilibrium. In the beliefs supporting this equilibrium, player III after L must assign at least probability $\frac{1}{2}$ to I's initial deviation having been U, while after R, III must assign *zero* probability to that deviation. For

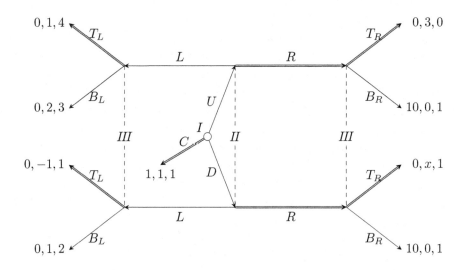

Figure 5.2.3: A game illustrating almost perfect Bayesian equilibrium. The indicated profile is not an almost perfect Bayesian equilibrium for $x = -3$, but is for $x = 0$.

the profile under consideration, Definition 5.2.4 does not impose any joint restriction on the beliefs at II's information set and the information set reached by L. But II cannot distinguish between U and D, so it seems unreasonable that a play by II is interpreted by III as a signal about the nature of I's deviation.   ★

While it is straightforward to directly deal with the issue raised by Example 5.2.4 (a failure of "not signaling what you don't know") in simple examples, the conditions that deal with the general phenomenon are complicated and can be hard to interpret. It is rare for the complicated conditions to be used in practice, though *no signaling what you don't know* is often directly imposed (as in our discussion of incomplete information bargaining in Section 8.2).[4]

The term *perfect Bayesian equilibrium* (or *PBE*) is often used in applications to describe the collections of restrictions on the system of beliefs that "do the right/obvious thing," and as such is one of the more abused notions in the literature. I will similarly abuse the term. See Fudenberg and Tirole (1991) for an early attempt and Watson (2016) for a recent attempt at a general definition.

---

[4]Problem 8.4.6 illustrates the importance of doing so in that application.

5.3   Sequential Equilibrium

A natural way of restricting the system of beliefs without simply adding one seemingly ad hoc restriction after another is to use Bayes' rule on completely mixed profiles as follows:

Definition 5.3.1 (Kreps and Wilson, 1982). *In a finite extensive form game, a system of beliefs μ is consistent with the strategy profile b if there exists a sequence of completely mixed sequence of behavior strategy profiles $\{b^k\}_k$ converging to b such that the associated sequence of system of beliefs $\{\mu^k\}_k$ obtained via Bayes' rule converges to μ.*
A strategy profile b is a sequential equilibrium if, for some consistent system of beliefs μ, b is sequentially rational at every information set.

It is helpful to be a little more explicit here. Since every information set is reached with positive probability, the beliefs obtained by Bayes' rule are, for all information sets h,

$$\mu^k(x) = \mathbf{P}^{b^k}(x \mid h) = \frac{\mathbf{P}^{b^k}(x)}{\mathbf{P}^{b^k}(h)} = \frac{\mathbf{P}^{b^k}(x)}{\sum_{x' \in h} \mathbf{P}^{b^k}(x')} \qquad \forall x \in h.$$

Since $b^k \to b$, from (5.1.1) we have $\mathbf{P}^{b^k}(x) \to \mathbf{P}^b(x)$, and so if h is on the path of play of b (i.e., $\mathbf{P}^b(h) > 0$), we immediately have

$$\mu^k(x) = \frac{\mathbf{P}^{b^k}(x)}{\mathbf{P}^{b^k}(h)} \longrightarrow \frac{\mathbf{P}^b(x)}{\mathbf{P}^b(h)},$$

the expression for $\mu(x)$ in (5.2.1).

Example 5.3.1. We first apply consistency to the profile $(C, R, (T_L, T_R))$ in the game in Figure 5.2.3. The tremble probabilities for I are $b_I^k(U) = \eta^k$, $b_I^k(D) = \zeta^k$, and complementary probability on C, and for II are $b_{II}^k(L) = \alpha^k$, and complementary probability on R. Since the completely mixed behavior strategies converge to C and R, η^k, ζ^k and α^k all converge to 0. Note that player II's trembles must be equal at the nodes of his information set.

Let p^k denote the conditional probability that II's information set is reached by U (i.e., that II is at the top decision node). Similarly let q^k and r^k denote the conditional probability that III's left (right, respectively) information set is reached by U (i.e., that III is at the top decision node after L, R respectively).

We then have

$$p^k = \frac{\eta^k}{\eta^k + \zeta^k}.$$

Any probability can be obtained in the limit through appropriate choices of the rates at which η^k and ζ^k converge to 0 (depending on the limiting behavior of η^k/ζ^k).

In addition, we also have

$$q^k = \frac{\eta^k \alpha^k}{\eta^k \alpha^k + \zeta^k \alpha^k},$$

which equals p^k for all k, since $\alpha^k > 0$. In other words, if players cannot signal what they don't know, $p^k = q^k$, and so the limit beliefs must agree. Finally,

$$r^k = \frac{\eta^k(1 - \alpha^k)}{\eta^k(1 - \alpha^k) + \zeta^k(1 - \alpha^k)},$$

which also equals p^k, and so once again the limit beliefs must agree.

Hence, while the profile $(C, R, (T_L, T_R))$ is almost perfect Bayesian for $x = 0$, it cannot be a sequential equilibrium.   ★

Problem 5.4.10 illustrates further restrictions that consistency places on beliefs.

Theorem 5.3.1. *A sequential equilibrium is almost perfect Bayesian.*

Proof. While this is "obvious," it is again useful to spell out the details.

Suppose b is a sequential equilibrium, with associated system of beliefs μ and sequence (b^k, μ^k). We need to prove that (5.2.2) holds for all $x \in h_{h'}$, where the information set h *-follows the information set h' and is reached with positive probability from h' under (b, μ).

For any $x \in h_{h'}$, there is a unique path from some $x^0 \in h'$. Denote the unique sequence of actions on the path from x^0 to x by $a^1, \ldots, a^L$, with player $\iota(\ell)$ choosing the action a^ℓ. Then, for all $x \in h_{h'}$, from (5.1.1) we have

$$\mathbf{P}^{b^k}(x) = \mathbf{P}^{b^k}(x^0) \prod_{\ell=0}^{L} b^k_{\iota(\ell)}(a^\ell),$$

and so

$$\begin{aligned}
\mathbf{P}^{b^k}(x \mid h') &= \frac{\mathbf{P}^{b^k}(x^0) \prod_{\ell=0}^{L} b^k_{\iota(\ell)}(a^\ell)}{\mathbf{P}^{b^k}(h')} \\
&= \mathbf{P}^{b^k}(x^0 \mid h') \prod_{\ell=0}^{L} b^k_{\iota(\ell)}(a^\ell) \\
&= \mu^k(x^0) \prod_{\ell=0}^{L} b^k_{\iota(\ell)}(a^\ell) \\
&\to \mu(x^0) \prod_{\ell=0}^{L} b_{\iota(\ell)}(a^\ell) = \mathbf{P}^{\mu,b}(x \mid h').
\end{aligned}$$

If $\sum_{\hat{x} \in h_{h'}} \mu(\hat{x}) = 0$, then (5.2.2) trivially holds.

Suppose $\sum_{\hat{x} \in h_{h'}} \mu(\hat{x}) > 0$. Then, for all $x \in h_{h'}$,

$$\frac{\mu^k(x)}{\sum_{\hat{x} \in h_{h'}} \mu^k(\hat{x})} = \frac{\mathbf{P}^{b^k}(x)/\mathbf{P}^{b^k}(h)}{\sum_{\hat{x} \in h_{h'}} \mathbf{P}^{b^k}(\hat{x})/\mathbf{P}^{b^k}(h)}$$

$$= \frac{\mathbf{P}^{b^k}(x)}{\sum_{\hat{x} \in h_{h'}} \mathbf{P}^{b^k}(\hat{x})}$$

$$= \frac{\mathbf{P}^{b^k}(x)}{\mathbf{P}^{b^k}(h_{h'})}$$

$$= \frac{\mathbf{P}^{b^k}(x)/\mathbf{P}^{b^k}(h')}{\mathbf{P}^{b^k}(h_{h'})/\mathbf{P}^{b^k}(h')}$$

$$= \frac{\mathbf{P}^{b^k}(x \mid h')}{\mathbf{P}^{b^k}(h_{h'} \mid h')}.$$

The equality is preserved when taking limits (since the limits of the first and last denominator are both strictly positive), and so (5.2.2) again holds. ∎

Sequential equilibrium satisfies the *one-shot deviation principle*, which we first saw in Theorem 5.1.1:

Theorem 5.3.2. *In a finite extensive form game, suppose μ is consistent with a profile b. The profile b is sequentially rational given μ (and so a sequential equilibrium) if and only if there are no profitable one-shot deviations from b (given μ).*

Proof. Lemma 5.1.1 is the easy direction.

Suppose b is not sequentially rational given μ. Then there is a player, denoted i, with a profitable deviation. Denote the profitable deviation (by player i) by b_i' and the information set h'. Player i's information sets H_i are strictly partially ordered by $\prec^*$. Let $H_i(h')$ denote the finite (since the game is finite) collection of information sets that follow h'. Let K be the length of the longest chain in $H_i(h')$, and say an information set $h \in H_i(h')$ is of level k if the successor chain from h' to h has k links (h' is 0-level and its immediate successors are all 1-level). If i has a profitable deviation from b_i (given μ) at any K-level information set, then that deviation is a profitable one-shot deviation (given μ), and we are done (note that this includes the possibility that $K = 0$).

Suppose i does not have a profitable deviation from b_i at any K-level information set and $K \geq 1$. Define a strategy $b_i^{(K)}$ by

$$b_i^{(K)}(h) = \begin{cases} b_i(h), & \text{if } h \text{ is a } K\text{-level information set or } h \notin H_i(h'), \\ b_i'(h), & \text{if } h \text{ is a } k\text{-level information set, } k = 0, \ldots, K-1. \end{cases}$$

Then $E^{\mu,(b_i^{(K)},b_{-i})}[u_i|h'] \geq E^{\mu,(b_i',b_{-i})}[u_i|h']$. (This requires proof, which is left as an exercise, see Problem 5.4.13. This is where consistency is important.)

But this implies that, like b_i', the strategy $b_i^{(K)}$ is a profitable deviation at h'. We now induct on k. Either there is profitable one-shot deviation from b_i at a $(K-1)$-level information set (in which case we are again done), or we can define a new strategy $b_i^{(K-1)}$ that is a profitable deviation at h' and which agrees with b_i on the $(K-1)$-level as well as the K-level information sets.

Proceeding in this way, we either find a profitable one-shot deviation at some k-level information set, or the action specified at h' by b_i' is a profitable one-shot deviation. $\blacksquare$

Remark 5.3.1. Sequential equilibrium is only defined for finite games. For some informative examples of what can go wrong when considering finite horizon games with infinite action spaces, see Myerson and Reny (2015). ◆

Remark 5.3.2. Sequential equilibrium is very close to (and is implied by) *trembling hand perfect in the extensive form*. Roughly speaking (see Section 14.2.2 for a more precise description), sequential equilibrium requires behavior to be sequentially rational against a limit assessment, while trembling hand perfection in the extensive form requires the behavior to be sequentially rational along the sequence as well. Problem 2.6.25 shows that trembling hand perfect in the normal form does not imply sequential equilibrium. ◆

Remark 5.3.3. Sequential equilibrium does not require that zero beliefs remain at zero. In the extensive form in Figure 5.3.1, the profile $(L_1L_2L_3L_4, r_1\ell_2)$, with the indicated system of beliefs, is a sequential equilibrium. This example demonstrates that sequentiality does not preclude *switching away from probability one beliefs*. In many applications, a further refinement of a *support restriction* (precluding switching away from probability zero beliefs) on sequentiality is imposed to prevent such switching. However, in some cases this can result in nonexistence (Problem 5.4.11 is an example due to Madrigal, Tan, and Werlang, 1987), as well as the elimination of plausible equilibria (Nöldeke and van Damme, 1990). ◆

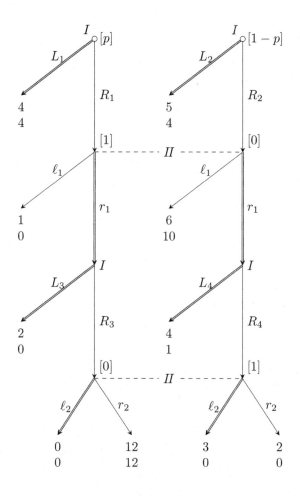

Figure 5.3.1: The game for Remark 5.3.3.

5.4 Problems

5.4.1. This problem concerns the game given in Figure 5.4.1.

(a) Show that $(\mathsf{GoStop_1Stop_2}, \mathsf{StopGo_1})$ is a Nash equilibrium.

(b) Identify all of the profitable one-shot deviations.

(c) Does player I choose Go in any subgame perfect equilibrium?

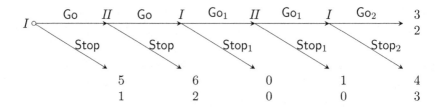

Figure 5.4.1: The game for Problem 5.4.1.

5.4.2. Complete the proof of Theorem 5.1.1.

5.4.3. Prove Theorem 5.2.1 (recall Problem 2.6.10).

5.4.4. This problem illustrates how badly behaved the "*-follows" relation $\prec^*$ on information sets can be, and the reason for the introduction of $h_{hh'}$ into the definition of almost perfect Bayesian equilibria. Consider the game in Figure 5.4.2.

  (a) Verify that the profile $(\frac{1}{2} \circ U + \frac{1}{2} \circ D, B, A)$ is a Nash equilibrium.

  (b) Denote player II's information set by h and player III's information set by h'. Verify that h *-follows h' and that h' *-follows h. Evaluate
  $$\mathbf{P}^{\mu,b}(x \mid h') \text{ and } \mathbf{P}^{\mu,b}(h \mid h')$$
  for all $x \in h$. Conclude that $\mu(x)$ should not equal their ratio.

5.4.5. Prove Theorem 5.2.2. [Hint: If h' is the singleton information at the beginning of a subgame, then for any information set h that *-follows h', $h = h_{h'}$.]

5.4.6. Show that (A, a, L) is a sequential equilibrium of Selten's horse (Figure 5.1.1) by exhibiting the sequence of converging completely mixed strategies and showing that the profile is sequentially rational with respect to the limit beliefs.

5.4.7. Prove by direct verification that the only sequential equilibrium of the first extensive form in Example 2.3.4 is (RB, r), but that (L, ℓ) is a sequential equilibrium of the second extensive form.

5.4.8. We return to the environment of Problem 3.6.1, but with one change. Rather than the two firms choosing quantities simultaneously, firm 1

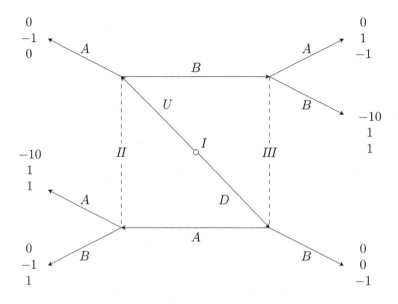

Figure 5.4.2: The game for Problem 5.4.4.

is a Stackelberg leader: Firm 1 chooses its quantity, q_1, first. Firm 2, knowing firm 1's quantity choice then chooses its quantity. Describe a strategy profile for this dynamic game. What is the appropriate equilibrium notion for this game and why? Describe an equilibrium of this game. Is it unique?

5.4.9. Consider the game in Figure 5.4.3.[5]

    (a) Prove that player I plays r in the unique sequential equilibrium.

    (b) Does the analysis of the sequential equilibria of this game change if the two types of player II are modeled as distinct players (receiving the same payoff)?

    (c) Prove that $(\ell d, UD)$ is an almost perfect Bayesian equilibrium.

5.4.10. Consider the game in Figure 5.4.4.

    (a) Suppose $x = 0$. Verify that (L, ℓ, r') is an almost perfect Bayesian equilibrium that is not a sequential equilibrium.

---

[5]This example was motivated by Section 8.2.1.

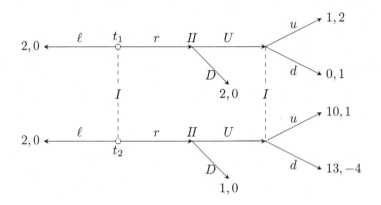

Figure 5.4.3: The game for Problem 5.4.9. The two types of player II are equally likely.

(b) Describe a pure strategy sequential equilibrium for this game for $x = 0$.

(c) Suppose $x = 3$. Verify that (L, ℓ, r') is a sequential equilibrium, and describe the supporting beliefs, showing they are consistent.

5.4.11. (Madrigal, Tan, and Werlang, 1987) Consider the game in Figure 5.4.5.

(a) What is the unique Nash equilibrium of this game?

Suppose this game is played twice. Nature chooses t_1 or t_2 only once at the very beginning of the game, and it remains fixed for both periods. Player I is informed of t_k and chooses an action L or R in period one. Player II observes I's action but not the value of t_k, and chooses an action U or D in period one. In period two, I again moves first and II observes that action and moves again. The payoff to each player in the two period game is the sum of what they get in each period individually. They receive the payoffs only at the end of the second period and do not observe their interim payoffs at the end of the first period.

(b) Prove that the only Nash equilibrium *outcome* of the repeated game is the repetition of the static Nash equilibrium outcome of the one period game.

(c) Prove every sequential equilibrium with this outcome violates the support restriction (see Remark 5.3.3).

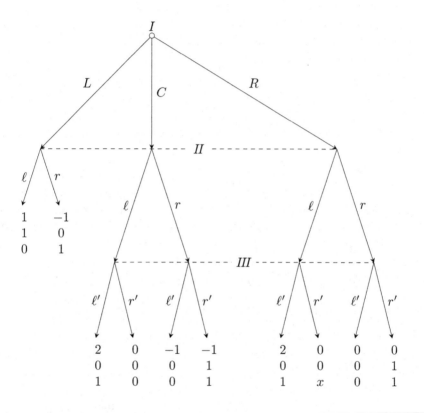

Figure 5.4.4: The game for Problem 5.4.10.

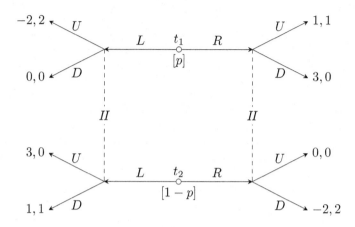

Figure 5.4.5: The game for Problem 5.4.11.

5.4.12. Fix a finite extensive form game. Suppose μ is consistent with b. Suppose for some player i there are two information sets $h, h' \in H_i$ with $h' \prec^* h$ and $\mathbf{P}^{\mu,b}(h|h') = 0$. Prove that if there exists another strategy $\hat{b}_i$ for player i with the property that $\mathbf{P}^{\mu,(\hat{b}_i,b_{-i})}(h|h') > 0$, then

$$\mu(x) = \frac{\mathbf{P}^{\mu,(\hat{b}_i,b_{-i})}(x|h')}{\mathbf{P}^{\mu,(\hat{b}_i,b_{-i})}(h|h')}, \qquad \forall x \in h.$$

5.4.13. Complete the proof of Theorem 5.3.2 by showing that

$$E^{\mu,(b_i^{(K)},b_{-i})}[u_i|h'] \geq E^{\mu,(b_i',b_{-i})}[u_i|h']$$

using the following steps.

(a) Let $Z^\dagger := \{z \in Z : \nexists x \in h, h \text{ a } K\text{-level information set, with } x \prec z\}$ be the set of terminal nodes that cannot be reached from any K-level information set. Prove that for any $\hat{b}_i$,

$$E^{\mu,(\hat{b}_i,b_i)}[u_i \mid h'] =$$

$$\sum_{\{h \text{ a } K\text{-level info set}\}} \sum_{x \in h} \mathbf{P}^{\mu,(\hat{b}_i,b_{-i})}(x \mid h') \sum_{\{z:x\prec z\}} \mathbf{P}^{(\hat{b}_i,b_{-i})}(z \mid x)u_i(z)$$

$$+ \sum_{z \in Z^\dagger} \mathbf{P}^{\mu,(\hat{b}_i,b_{-i})}(z \mid h')u_i(z),$$

where $\mathbf{P}^{(\hat{b}_i, b_{-i})}(z \mid x) := \prod_\ell b_{\iota(\ell)}(a^\ell)$ and (a^ℓ) is the sequence of actions on the path from x to z.

(b) Complete the argument using Problem 5.4.12.

Chapter 6

Signaling

6.1 General Theory

The canonical signaling game has one informed player (the sender) sending a signal (message) to one uninformed player. The sender's type $t \in T \subset \mathbb{R}$ is first drawn by nature according to a probability distribution $\rho \in \Delta(T)$. The sender then chooses a signal $m \in M \subset \mathbb{R}$. After observing the signal, the responder then chooses $r \in R \subset \mathbb{R}$.

Payoffs are $u(m, r, t)$ for the sender and $v(m, r, t)$ for the responder.

A pure strategy for the sender is $\tau : T \to M$.

A pure strategy for the responder is $\sigma : M \to R$.

A strategy profile (equilibrium) (τ, σ) is *separating* if τ is one-to-one (so that different types choose different actions). A strategy profile (equilibrium) (τ, σ) is *pooling* if τ is constant (so that all types choose the same action).

Example 6.1.1. In the game given in Figure 6.1.1, the profile (qq, pf) is a pooling Nash equilibrium,[1] but it is not even weak perfect Bayesian (why?).

The profile (bq, fr) is a separating Nash equilibrium. Since this equilibrium has no information sets off the path of play, this Nash equilibrium satisfies all the refinements of Nash described in Chapter 5 (weak perfect Bayesian, almost perfect Bayesian, sequential). ★

Since the different information sets for the responder are not ordered by $\prec^*$ (recall Problem 1.4.4), consistency places no restrictions on beliefs at different information sets of the responder. This implies the following result (which Problem 6.3.3 asks you to prove).

---

[1]The sender's strategy lists actions in the order of types, and the responder's strategy lists actions by information set, left to right (so pf is p after b and f after q).

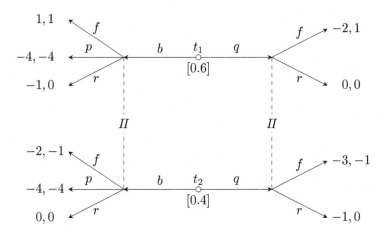

Figure 6.1.1: A signaling game. Footnote 1 on page 123 explains the notation.

Theorem 6.1.1. *Suppose T, M, and R are finite. A profile is a weak
perfect Bayesian equilibrium if, and only if, it is a sequential equilibrium.*

We often wish to study signaling games where T, M, and/or R are
infinite sets (typically continua). If any of these sets are continua, then
consistency (and so sequentiality) is not defined.[2]

Weak perfect Bayesian equilibrium is well defined when T, M, and/or
R are infinite sets. It is Nash equilibrium plus sequential rationality at
all information sets with respect to some set of beliefs. Due to Theorem
6.1.1, we drop the qualifer "weak" and define a notion of perfect Bayesian
equilibrium for signaling games.

Definition 6.1.1. *The pure strategy profile $(\hat{\tau}, \hat{\sigma})$ is a perfect Bayesian
equilibrium of the signaling game if*

   1. for all $t \in T$,
$$\hat{\tau}(t) \in \arg\max_{m \in M} \; u(m, \hat{\sigma}(m), t),$$

   2. for all m, there exists some $\mu_m \in \Delta(T)$ such that
$$\hat{\sigma}(m) \in \arg\max_{r \in R} \; E^{\mu_m}[v(m, r, t)],$$

   where E^{μ_m} denotes expectation with respect to μ_m, and

---

[2]The difficulty is that there is no single "obvious" notion of convergence.

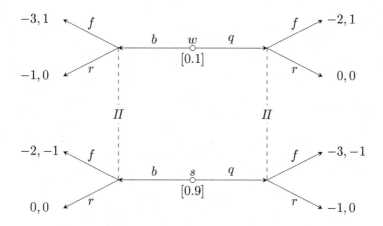

Figure 6.1.2: The beer-quiche game.

3. *for* $m \in \hat{\tau}(T)$, μ_m *in part 2 is given by*

$$\mu_m(t) = \rho\{t \mid m = \hat{\tau}(t)\}.$$

We have already seen (in Chapter 5) that sequential rationality can eliminate (refine) Nash equilibria by restricting the selection from the multiple best replies at unreached information sets; Example 6.1.1 is another illustration. However, in many games, not all sequential equilibria are plausible (self-enforcing). Example 6.1.2 gives an example of an intuitive argument that further restricts sequential rationality by imposing restrictions on the beliefs a player can have at unreached information sets. This argument is formalized after the example.

Example 6.1.2 (Beer-quiche). In the game in Figure 6.1.2, (bb, rf) and (qq, fr) are both pooling equilibria.[3]

The equilibrium in which the types pool on q is often argued to be unintuitive: Would the w type ever "rationally" deviate to b? In this pooling equilibrium, w receives 0, and this is strictly larger than his payoff from b *no matter* how II responds. On the other hand, if by deviating to b, s can "signal" that he is indeed s, he is strictly better off, since II's best response is r, yielding payoff of 0. This is an example of the *intuitive criterion*.   ★

---

[3]This is the beer-quiche game (Cho and Kreps, 1987, Section II): the two types of sender are wimpy (w) and surly (s), q is quiche while b is beer for breakfast, f is fight, and r is runaway (player II is a bully). The labels are an allusion to a popular 1982 book, "Real Men Don't Eat Quiche" by Bruce Feirstein.

Let $BR(T', m)$ be the set of best replies to m of the responder, when the beliefs have support in T', i.e.,

$$BR(T', m) = \{r \in R : \exists \mu \in \Delta(T'), r \in \arg\max_{r' \in R} E^\mu[v(m, r', t)]\}$$

$$= \bigcup_{\mu \in \Delta(T')} \arg\max_{r' \in R} E^\mu[v(m, r', t)].$$

Definition 6.1.2 (Cho and Kreps, 1987). *Fix a perfect Bayesian equilibrium $(\hat{\tau}, \hat{\sigma})$, and let $\hat{u}(t) = u(\hat{\tau}(t), \hat{\sigma}(\hat{\tau}(t)), t)$. Define $D(m) \subset T$ as the set of types satisfying*

$$\hat{u}(t) > \max_{r \in BR(T, m)} u(m, r, t).$$

The equilibrium $(\hat{\tau}, \hat{\sigma})$ fails the intuitive criterion if there exists m' (necessarily not in $\hat{\tau}(T)$, i.e., an unsent message) and a type t' (necessarily not in $D(m')$) such that

$$\hat{u}(t') < \min_{r \in BR(T \setminus D(m'), m')} u(m', r, t').$$

Remark 6.1.1. The test concerns equilibrium outcomes, and not the specification of behavior after out-of-equilibrium messages. In particular, if an equilibrium $(\hat{\tau}, \hat{\sigma})$ fails the intuitive criterion, then *every* equilibrium with the *same* outcome as that implied by $(\hat{\tau}, \hat{\sigma})$ fails the intuitive criterion. (Such equilibria differ from $(\hat{\tau}, \hat{\sigma})$ in the specification of the responses to out-of-equilibrium messages.)

For messages m' that satisfy $\varnothing \neq D(m') \subsetneq T$, it is in the spirit of the test to require responses r to out-of-equilibrium messages m' satisfy $r \in BR(T \setminus D(m'), m')$.   ♦

Remark 6.1.2. The intuitive criterion is an example of the *forward induction* notion discussed in Example 2.5.1, which requires responses to deviations from equilibrium be, if possible, consistent with some rationalization of the deviation (much as the choice of b in Example 6.1.2 is rationalized as coming from s). It is worth rereading Example 2.5.1 and the following discussion; in particular, the argument underlying the elimination of the qq equilibrium in Example 6.1.2 is very similar to that which eliminates (LB, r) in Example 2.5.1.

The intuitive criterion is implied by Kohlberg and Mertens's (1986) implementation of forward induction.   ♦

6.2   Job Market Signaling

There is a worker with private ability $\theta \in \Theta \subset \mathbb{R}$. The worker can signal ability through choice of level of education, $e \in \mathbb{R}_+$. The worker's utility is

$$w - c(e, \theta),$$

where w is the wage, and c is the disutility of education. Assume c is $\mathcal{C}^2$ and satisfies the *single-crossing condition*:

$$\frac{\partial^2 c(e, \theta)}{\partial e \partial \theta} < 0.$$

Also assume $c(e, \theta) \geq 0$, $c_e(e, \theta) := \partial c(e, \theta)/\partial e \geq 0$, $c_e(0, \theta) = 0$, $c_{ee}(e, \theta) := \partial c(e, \theta)/\partial e^2 > 0$, and $\lim_{e \to \infty} c_e(e, \theta) = \infty$.

Two identical firms compete for the worker by simultaneously posting a wage. Each firm values a worker of type θ with education e at $f(e, \theta)$. Neither firm knows the worker's ability and has prior $\rho \in \Delta(\Theta)$ on θ.

Since the notions of almost perfect Bayesian and sequential equilibrium are defined only for finite games, we will occasionally consider discretizations of the game. In any discretization of the game, in any almost perfect Bayesian equilibrium, after any e, firms have identical beliefs $\mu \in \Delta(\Theta)$ about worker ability (see Problem 6.3.6). Consequently, the two firms are effectively playing a sealed-bid common-value first-price auction, and so both firms bid their value $E_\mu f(e, \theta)$. To model as a signaling game, replace the two firms with a single uninformed receiver (the "market") with payoff

$$-(f(e, \theta) - w)^2.$$

Strategy for worker, $e : \Theta \to \mathbb{R}_+$.

Strategy for "market", $w : \mathbb{R}_+ \to \mathbb{R}_+$.

Assume f is $\mathcal{C}^2$. Assume $f(e, \theta) \geq 0$, $f_e(e, \theta) := \partial f(e, \theta)/\partial e \geq 0$, $f_\theta(e, \theta) > 0$, $f_{ee}(e, \theta) := \partial^2 f(e, \theta)/\partial e^2 \leq 0$, and $f_{e\theta}(e, \theta) := \partial^2 f(e, \theta)/\partial e \partial \theta \geq 0$.

1. **Unproductive education.** When f is independent of e, we can interpret θ as the productivity of the worker, and so assume $f(e, \theta) = \theta$.

2. **Productive education.** $f_e(e, \theta) > 0$.

If the market believes the worker has ability $\hat\theta$, it pays a wage of $f(e, \hat\theta)$. The result is a signaling game as described in Section 6.1, and so we can apply the equilibrium notion of perfect Bayesian as defined there.

This scenario is often called Spence job-market signaling, since Spence (1973) introduced it.

6.2.1   Full Information

If firms *know* the worker has ability θ, the worker chooses e to maximize

$$f(e, \theta) - c(e, \theta). \tag{6.2.1}$$

For each θ there is a unique e^* maximizing (6.2.1) (this is the efficient level of education), that is,

$$e^*(\theta) := \arg\max_{e \geq 0} f(e, \theta) - c(e, \theta).$$

Assuming $f_e(0, \theta) > 0$ (together with the assumption on c above) is sufficient to imply that $e^*(\theta)$ is interior for all θ and so

$$\frac{de^*}{d\theta} = -\frac{f_{e\theta}(e, \theta) - c_{e\theta}(e, \theta)}{f_{ee}(e, \theta) - c_{ee}(e, \theta)} > 0.$$

6.2.2   Incomplete Information

Define

$$U(\theta, \hat{\theta}, e) := f(e, \hat{\theta}) - c(e, \theta).$$

Note that

$$e^*(\theta) = \arg\max_{e \geq 0} U(\theta, \theta, e). \tag{6.2.2}$$

We are first interested in separating perfect Bayesian equilibria, i.e., an equilibrum in which education separates workers of different abilities. The profile $(\hat{e}, \hat{w})$ is *separating* if $\hat{e}$ is one-to-one, so that after observing $e \in \hat{e}(\Theta)$ an uninformed firm infers the worker has ability $\hat{e}^{-1}(e)$.

The outcome associated with a profile $(\hat{e}, \hat{w})$ is

$$(\hat{e}(\theta), \hat{w}(\hat{e}(\theta)))_{\theta \in \Theta}.$$

If $e' = \hat{e}(\theta')$ for some $\theta' \in \Theta$, then $\hat{w}(e') = f(e', (\hat{e})^{-1}(e')) = f(e', \theta')$, and so the payoff to the worker of type θ is

$$\hat{w}(e') - c(e', \theta) = f(e', \theta') - c(e', \theta) = U(\theta, \theta', e').$$

A separating strategy $\hat{e}$ is *incentive compatible* if no type strictly benefits from mimicking another type, i.e.,

$$U(\theta', \theta', \hat{e}(\theta')) \geq U(\theta', \theta'', \hat{e}(\theta'')), \qquad \forall \theta', \theta'' \in \Theta. \tag{6.2.3}$$

Figure 6.2.1 illustrates the case $\Theta = \{\theta', \theta''\}$.

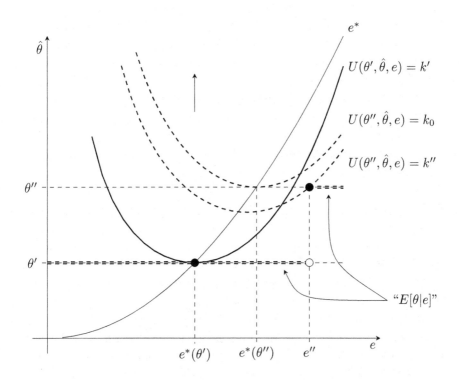

Figure 6.2.1: Indifference curves in $\hat{\theta} - e$ space. Space of types is $\Theta = \{\theta', \theta''\}$. The arrow indicates increasing worker preference. Note that $k' = U(\theta', \theta', e^*(\theta')) = \max_e U(\theta', \theta', e)$, $k'' = U(\theta'', \theta'', e'')$, and $k_0 = U(\theta'', \theta'', e^*(\theta'')) = \max_e U(\theta'', \theta'', e)$, and incentive compatibility is satisfied at the indicated points: $U(\theta'', \theta'', e'') \geq U(\theta'', \theta', e^*(\theta'))$ and $U(\theta', \theta', e^*(\theta')) \geq U(\theta', \theta'', e'')$. For any $e < e''$, firms believe $\theta = \theta'$, and for any $e \geq e''$, firms believe $\theta = \theta''$.

The figures in this subsection are drawn using the production function $f(e, \theta) = e\theta$ and cost function $c(e, \theta) = e^5/(5\theta)$, so that $U(\theta, \hat{\theta}, e) = e\hat{\theta} - (e^5)/(5\theta)$. The set of possible θ's is $\{1, 2\}$, with full information optimal educations of 1 and $\sqrt{2}$.

Definition 6.2.1. *The separating strategy profile* $(\hat{e}, \hat{w})$ *is a* perfect Bayesian equilibrium *if*

1. $\hat{e}$ *satisfies* (6.2.3), *i.e.,* $\hat{e}$ *is incentive compatible,*

2. $\hat{w}(e) = f(e, \hat{e}^{-1}(e))$ *for all* $e \in \hat{e}(\Theta)$,

3. *for all* $e \notin \hat{e}(\Theta)$ *and all* $\theta \in \Theta$,

$$U(\theta, \theta, \hat{e}(\theta)) \geq \hat{w}(e) - c(c, \theta),$$

   and

4. $\hat{w}$ *is sequentially rational, i.e., for all* $e \in \mathbb{R}_+$, *there is some* $\mu \in \Delta(\Theta)$ *such that*
$$\hat{w}(e) = E_\mu f(e, \theta).$$

For $e \in \hat{e}(\Theta)$, μ is of course given by the belief that assigns probability one to the type for which $e = \hat{e}(\theta)$, i.e. $\hat{\theta} = \hat{e}^{-1}(e)$). Sequential rationality restricts wages for $w \notin \hat{e}(\Theta)$.

The Intermediate Value Theorem implies that for all $e \in \mathbb{R}_+$ and all $\mu \in \Delta(\Theta)$, there exists a unique $\hat{\theta} \in \operatorname{conv}(\Theta)$ such that

$$f(e, \hat{\theta}) = E_\mu f(e, \theta). \tag{6.2.4}$$

(This extends to "canonical" signaling games, see Problem 6.3.5(a).) Thus, condition 4 in Definition 6.2.1 can be replaced by: for all $e \in \mathbb{R}_+$, there is some $\hat{\theta} \in \operatorname{conv}(\Theta) = [\min \Theta, \ \max \Theta]$ such that

$$\hat{w}(e) = f(e, \hat{\theta}).$$

In particular, if $\Theta = \{\theta', \theta''\}$, then $\hat{\theta} \in [\theta', \theta'']$. The importance of the replacement of beliefs μ by types θ is that it is easier dealing with a one-dimensional object (type) than with a multi-dimensional object (beliefs, which live in a $|\Theta - 1|$-dimensional simplex). As one illustration, this allows us to interpreting the vertical axes in the figures in this section as capturing both on-path and off-path beliefs.

Sequential rationality implies that the worker cannot be treated worse than the lowest ability worker after any deviation. This implies that in any separating equilibrium, the lowest ability worker's education choice cannot be distorted from the full information optimum.

Lemma 6.2.1. *Let* $\underline{\theta} = \min \Theta$. *In any separating perfect Bayesian equilibrium,*

$$\hat{e}(\underline{\theta}) = e^*(\underline{\theta}). \tag{6.2.5}$$

Proof. This is a straightforward implication of sequential rationality: Suppose (6.2.5) does not hold, and let $\hat{\theta}$ solve $\hat{w}(e^*(\underline{\theta})) = f(e^*(\underline{\theta}), \hat{\theta})$. Recall that $\hat{\theta} \geq \underline{\theta}$. The payoff to $\underline{\theta}$ from deviating to $e^*(\underline{\theta})$ is $U(\underline{\theta}, \hat{\theta}, e^*(\underline{\theta}))$, and we have

$$U(\underline{\theta}, \hat{\theta}, e^*(\underline{\theta})) \geq U(\underline{\theta}, \underline{\theta}, e^*(\underline{\theta})) > U(\underline{\theta}, \underline{\theta}, \hat{e}(\underline{\theta})),$$

where the first inequality is an implication of $f_\theta > 0$ and the second from the strict concavity of U with respect to e (which is implied by $c_{ee} > 0$ and $f_{ee} \leq 0$).   ∎

It is important to understand the role of sequential rationality in Lemma 6.2.1. Equation (6.2.5) typically need not hold in separating Nash equilibria: Suppose $f(e^*(\underline{\theta}), \underline{\theta}) > 0$. Specifying a wage of 0 after $e^*(\underline{\theta})$ (which is inconsistent with sequential rationality) gives the type-$\underline{\theta}$ worker negative (or zero if $e^*(\underline{\theta}) = 0$) payoffs from choosing that level of education.

The set of separating perfect Bayesian equilibrium outcomes is illustrated in Figure 6.2.2. Note that in this figure, the full information choices are not incentive compatible (i.e., they violate (6.2.3)). This means that in any separating equilibrium, the choice of θ'' is distorted from the full information choice of $e^*(\theta'')$.

The *Riley outcome* (Riley, 1979) is the separating outcome that minimizes the distortion (as measured by $|\hat{e}(\theta) - e^*(\theta)|$). If the full information choices are consistent with (6.2.3), then no distortion is necessary.

Suppose there are two types (as in Figure 6.2.2). Suppose moreover, as in the figures, that the full information choices are not consistent with (6.2.3). The Riley outcome is

$$((e^*(\theta'), f(e^*(\theta'), \theta')), (e_1'', f(e_1'', \theta''))).$$

It is the separating outcome that minimizes the distortion, since θ' is indifferent between $((e^*(\theta'), f(e^*(\theta'), \theta'))$ and $(e_1'', f(e_1'', \theta'')))$. Any lower education level for θ'' violates (6.2.3).

Finally, worth noting that inequality (6.2.3) can be rewritten as

$$U(\theta', \theta', \hat{e}(\theta')) \geq U(\theta', (\hat{e})^{-1}(e), e) \qquad \forall e \in \hat{e}(\Theta).$$

That is, the function $\hat{e} : \Theta \to \mathbb{R}_+$ satisfies the functional equation

$$\hat{e}(\theta') \in \underset{e \in \hat{e}(\Theta)}{\arg\max}\ U(\theta', (\hat{e})^{-1}(e), e), \qquad \forall \theta' \in \Theta. \tag{6.2.6}$$

Note that (6.2.2) and (6.2.6) differ in two ways: the set of possible maximizers and how e enters into the objective function.

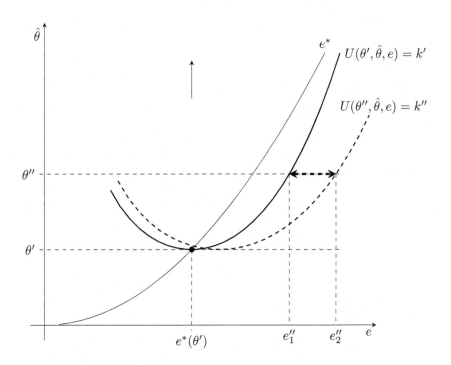

Figure 6.2.2: Separating equilibria when space of types is $\Theta = \{\theta', \theta''\}$.
The separating perfect Bayesian equilibrium outcomes set is
$\{((e^*(\theta'), f(e^*(\theta'), \theta')), (e'', f(e'', \theta''))) : e'' \in [e_1'', e_2'']\}$. Note
that $k' = \max_e U(\theta', \theta', e)$ and θ'' cannot receive a lower payoff than $\max_e U(\theta'', \theta', e) = k''$.

6.2.3 Refining to Separation

In addition to the plethora of separating equilibria, signaling games typically have nonseparating equilibria as well. A pooling equilibrium is illustrated in Figure 6.2.3, for the case of two types, θ' and θ''.

Suppose f is *affine in* θ, so that $Ef(e,\theta) = f(e, E\theta)$ (this is unnecessary, see (6.2.4) and Problem 6.3.5(a), but simplifies the discussion).

The pooling outcome in Figure 6.2.3 is a perfect Bayesian outcome, but is ruled out by the intuitive criterion. To see this, consider the out-of-equilibrium message $\tilde{e}$ in the figure. Note first that (using the notation from Definition 6.1.2) $D(\tilde{e}) = \{\theta'\}$. Then, the market wage after $\tilde{e}$ must be θ'' (since the market puts zero probability on θ'). Since

$$U(\theta'', \theta'', \tilde{e}) > k_p'' = U(\theta'', E[\theta], e^p),$$

the equilibrium fails the intuitive criterion.

Indeed, for two types, the intuitive criterion selects the *Riley* separating outcome, i.e., the separating outcome that minimizes the signaling distortion. Consider a separating outcome that is not the Riley outcome. Then, $e(\theta'') > e_1''$ (see Figure 6.2.2). Consider an out-of-equilibrium message $\tilde{e} \in (e_1'', e(\theta''))$. Since $\tilde{e} > e_1''$, $D(\tilde{e}) = \{\theta'\}$. Then, as above, the market wage after $\tilde{e}$ must be θ'', and so since $\tilde{e} < e(\theta'')$,

$$U(\theta'', \theta'', \tilde{e}) > U(\theta'', \theta'', e(\theta'')),$$

and non-Riley separating equilibria also fail the intuitive criterion.

Similarly, every pooling outcome fails the intuitive criterion (see Figure 6.2.3).

With three or more types, many equilibrium outcomes in addition to the Riley equilibrium outcome survive the intuitive criterion. Suppose we add another type θ''' to the situation illustrated in Figure 6.2.3 (see Figure 6.2.4). Then, for any education that signals that the worker is not of type θ' (i.e., $e > \bar{e}$), there is still an inference (belief) putting zero probability on θ' that makes the deviation unprofitable for the worker of type θ''.

There are stronger refinements (also implied by the Kohlberg and Mertens (1986) version of forward induction) that do select the Riley outcome (Cho and Kreps, 1987; Cho and Sobel, 1990).

6.2.4 Continuum of Types

Suppose $\Theta = [\underline{\theta}, \bar{\theta}]$ (so that there is a continuum of types), and suppose $\hat{e}$ is differentiable (this can be justified, see Mailath, 1987, and Mailath and von Thadden, 2013).

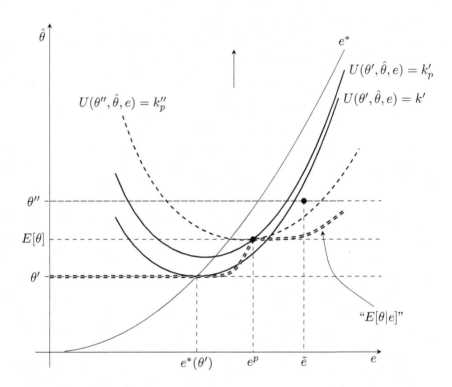

Figure 6.2.3: A pooling outcome at $e = e^p$ when the space of types is $\Theta = \{\theta', \theta''\}$. The constants are $k'_p = U(\theta', E\theta, e^p)$, $k''_p = U(\theta'', E\theta, e^p)$. Note that the firms' beliefs after potential deviating e's (denoted $E[\theta \mid e]$) must lie below the θ' and θ'' indifference curves indexed by k'_p and k''_p, respectively. This outcome fails the intuitive criterion: consider a deviation by θ'' to $\tilde{e}$. Finally, e^p need not maximize $U(\theta'', E\theta, e)$ (indeed, in the figure e^p is marginally larger than the optimal e) for θ''.

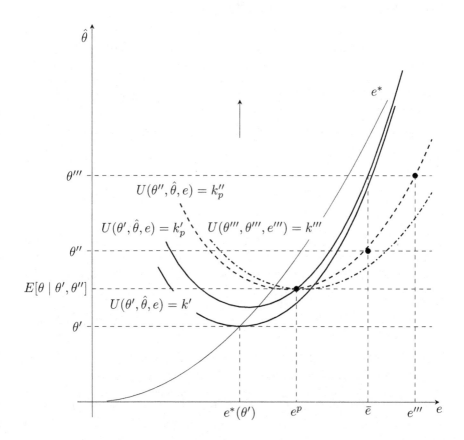

Figure 6.2.4: A semi-pooling outcome, with θ' and θ'' pooling on $e = e^p$ (as in Figure 6.2.3). Type θ''' separates out at e'''. For any education level $e \leq \bar{e}$, $D(e) = \varnothing$, and so this equilibrium passes the intuitive criterion (a semi-pooling equilibrium where θ''' chooses an education level strictly larger than e''' fails the intuitive criterion-why?).

Then the first derivative of the objective function in (6.2.6) with respect to e is

$$U_{\hat{\theta}}(\theta, (\hat{e})^{-1}(e), e)\frac{d(\hat{e})^{-1}(e)}{de} + U_e(\theta, (\hat{e})^{-1}(e), e)$$

$$= U_{\hat{\theta}}(\theta, (\hat{e})^{-1}(e), e)\left(\frac{d\hat{e}(\theta)}{d\theta}\bigg|_{\theta=(\hat{e})^{-1}(e)}\right)^{-1} + U_e(\theta, (\hat{e})^{-1}(e), e).$$

The first-order condition is obtained by evaluating this derivative at $e = \hat{e}(\theta)$ (so that $(\hat{e})^{-1}(e) = \theta$) and setting the result equal to 0:

$$U_{\hat{\theta}}(\theta, \theta, e)\left(\frac{d\hat{e}(\theta)}{d\theta}\right)^{-1} + U_e(\theta, \theta, e) = 0.$$

The result is a differential equation characterizing $\hat{e}$,

$$\frac{d\hat{e}}{d\theta} = -\frac{U_{\hat{\theta}}(\theta, \theta, \hat{e})}{U_e(\theta, \theta, e)} = -\frac{f_\theta(\hat{e}, \theta)}{f_e(\hat{e}, \theta) - c_e(\hat{e}, \theta)}. \qquad (6.2.7)$$

Together with (6.2.5), we have an initial value problem that characterizes the unique separating perfect Bayesian equilibrium strategy for the worker. In contrast to the situation with a finite set of types, there is only one separating perfect Bayesian equilibrium, and so its outcome is the Riley outcome.[4]

Note that because of (6.2.5), as $\theta \to \underline{\theta}$, $d\hat{e}(\theta)/d\theta \to +\infty$, and that for $\theta > \underline{\theta}$, $\hat{e}(\theta) > e^*(\theta)$, that is, there is necessarily a signalling distortion (see Figure 6.2.5).

Remark 6.2.1. The above characterization of separating strategies works for any signaling game for which the payoff to the informed player of type θ, when the uninformed player best responds to a belief that the type is $\hat{\theta}$, and e is chosen by the informed player can be written as a function $U(\theta, \hat{\theta}, e)$. See Problem 6.3.5 for a description of the canonical signaling model. ◆

6.3 Problems

6.3.1. Show that for the game in Figure 6.3.1, for all values of x, the outcome in which both types of player I play L is sequential by explicitly describing the converging sequence of completely mixed behavior strategy profiles and the associated system of beliefs. For what values of x does this equilibrium pass the intuitive criterion?

---

[4]As the set of types becomes dense in the interval, all separating perfect Bayesian outcomes converge to the unique separating outcome of the model with a continuum of types, (Mailath, 1988).

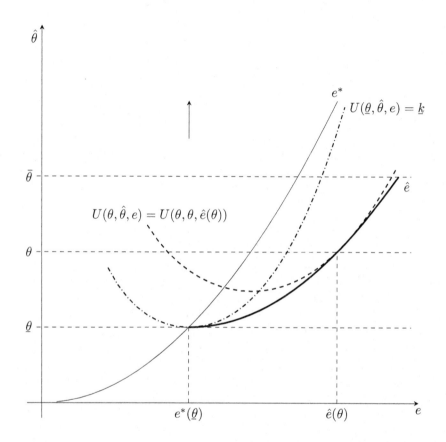

Figure 6.2.5: The separating equilibrium strategy $\hat{e}$ for the continuum types case. As usual, $\underline{k} = \max_e U(\underline{\theta}, \underline{\theta}, e)$. The separating strategy is the lower envelope of the indifference curves $U(\theta, \hat{\theta}, e) = U(\theta, \theta, \hat{e}(\theta))$, since $\hat{e}$ must be tangential to each indifference curve (by (6.2.7)) at $(\hat{e}(\theta), \theta)$ and lie below all the indifference curves.

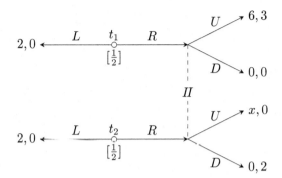

Figure 6.3.1: Game for Problem 6.3.1. The probability that player I is type t_1 is 1/2 and the probability that he is type t_2 is 1/2. The first payoff is player I's payoff, and the second is player II's.

6.3.2. (a) Verify that the signaling game illustrated in Figure 6.3.2 has no Nash equilibrium in pure strategies.

 (b) Definition 6.1.1 gives the definition of a pure strategy perfect Bayesian equilibrium for signaling games. How should this definition be changed to cover behavior (mixed) strategies? More precisely, give the definition of a perfect Bayesian equilibrium for the case of finite signaling games.

 (c) For the game in Figure 6.3.2, suppose $p = \frac{1}{2}$. Describe a perfect Bayesian equilibrium.

 (d) How does your answer to part (c) change if $p = 0.1$?

6.3.3. Prove Theorem 6.1.1.

6.3.4. Fill in the details of the proof of Lemma 6.2.1.

6.3.5. The canonical signaling game has a sender with private information, denoted $\theta \in \Theta \subset \mathbb{R}$ choosing a message $m \in \mathbb{R}$, where Θ is compact. A receiver, observing m, but not knowing θ then chooses a response $r \in \mathbb{R}$. The payoff to the sender is $u(m, r, \theta)$ while the payoff to the receiver is $v(m, r, \theta)$. Assume both u and v are C^2. Assume v is strictly concave in r, so that $v(m, r, \theta)$ has a unique maximizer in r for all (m, θ), denoted $\xi(m, \theta)$. Define

$$U(\theta, \hat{\theta}, m) = u(m, \xi(m, \hat{\theta}), \theta).$$

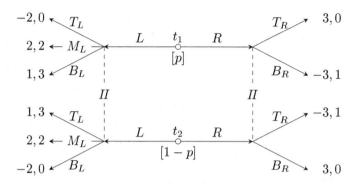

Figure 6.3.2: Figure for Problem 6.3.2.

Assume u is strictly increasing in r, and ξ is strictly increasing in $\hat{\theta}$, so that U is also strictly increasing in $\hat{\theta}$. Finally, assume that for all $(\theta, \hat{\theta})$, $U(\theta, \hat{\theta}, m)$ is bounded above (and so has a well-defined maximum).

(a) Given a message m^* and a belief F over θ, suppose r^* maximizes the receiver's expected payoff. Prove there exists $\hat{\theta}$ such that $r^* = \xi(m^*, \hat{\theta})$. Moreover, if the support of F is a continuum, conv Θ, prove that $\hat{\theta}$ is in the support of F.

Assume u satisfies the *single-crossing condition*:

If $\theta < \theta'$ and $m < m'$, then $u(m, r, \theta) \leq u(m', r', \theta)$ implies $u(m, r, \theta') < u(m', r', \theta')$.

(Draw the indifference curves for different types in $m - r$ space to see that they can only cross once.)

(b) Provide restrictions on the productive education case covered in Section 6.2 so that the sender's payoff satisfies the single-crossing condition as defined here.

(c) Prove that U satisfies an analogous version of the single-crossing condition: If $\theta < \theta'$ and $m < m'$, then $U(\theta, \hat{\theta}, m) \leq U(\theta, \hat{\theta}', m')$ implies $U(\theta', \hat{\theta}, m) < U(\theta', \hat{\theta}', m')$.

(d) Prove that the messages sent by the sender in any separating Nash equilibrium are strictly increasing in type.

(e) Prove that in any separating perfect Bayesian equilibrium, type $\underline{\theta} := \min \Theta$ chooses the action $\underline{m}$ maximizing $u(m, \xi(m, \underline{\theta}), \underline{\theta})$ (recall (6.2.5)). How is this implication of separating perfect Bayesian equilibrium changed if u is strictly decreasing in r? If ξ is strictly decreasing in $\hat{\theta}$?

6.3.6. Prove that in any discretization of the job market signaling game, in any almost perfect Bayesian equilibrium, after any e, firms have identical beliefs about worker ability.

6.3.7. Suppose that, in the incomplete information model of Section 6.2, the payoff to a firm from hiring a worker of type θ with education e at wage w is

$$f(e, \theta) - w = 3e\theta - w.$$

The utility of a worker of type θ with education e receiving a wage w is

$$w - c(e, \theta) = w - \frac{e^3}{\theta}.$$

Suppose the support of the firms' prior beliefs ρ on θ is $\Theta = \{1, 3\}$.

(a) Describe a perfect Bayesian equilibrium in which both types of worker choose their full information eduction level. Be sure to verify that all the incentive constraints are satisfied.

(b) Are there other separating perfect Bayesian equilibria? What are they? Do they depend on the prior distribution ρ?

Now suppose the support of the firms' prior beliefs on θ is $\Theta = \{1, 2, 3\}$.

(c) Why is it no longer consistent with a separating perfect Bayesian equilibrium to have $\theta = 3$ choose his full information eduction level $e^*(3)$? Describe the Riley outcome (the separating equilibrium outcome that minimizes the distortion), and verify that it is indeed the outcome of a perfect Bayesian equilibrium.

(d) What is the largest education level for $\theta = 2$ consistent with separating perfect Bayesian equilibrium? Prove that any separating equilibrium in which $\theta = 2$ chooses that level of education fails the intuitive criterion. [Hint: consider the out-of-equilibrium education level $e = 3$.]

(e) Describe the separating perfect Bayesian equilibria in which $\theta = 2$ chooses $e = 2.5$. Some of these equilibria fail the intuitive criterion and some do not. Give an example of one of each (i.e., an equilibrium that fails the intuitive criterion, and an equilibrium that does not fail).

6.3.8. Verify the claims in the caption of Figure 6.2.4.

6.3.9. The owner of a small firm is contemplating selling all or part of his firm to outside investors. The profits from the firm are risky and the owner is risk averse. The owner's preferences over x, the fraction of the firm the owner retains, and p, the price "per share" paid by the outside investors, are given by

$$u(x, \theta, p) = \theta x - x^2 + p(1 - x),$$

where $\theta > 1$ is the value of the firm (i.e., expected profits). The quadratic term reflects the owner's risk aversion. The outside investors are risk neutral, and so the payoff to an outside investor of paying p per share for $1 - x$ of the firm is then

$$\theta(1 - x) - p(1 - x).$$

There are at least two outside investors, and the price is determined by a first price sealed bid auction: The owner first chooses the fraction of the firm to sell, $1 - x$; the outside investors then bid, with the $1 - x$ fraction going to the highest bidder (ties are broken with a coin flip).

(a) Suppose θ is public information. What fraction of the firm will the owner sell, and how much will he receive for it?

(b) Suppose now θ is privately known by the owner. The outside investors have common beliefs, assigning probability $\alpha \in (0, 1)$ to $\theta = \theta_1 > 0$ and probability $1 - \alpha$ to $\theta = \theta_2 > \theta_1$. Characterize the separating perfect Bayesian equilibria. Are there any other perfect Bayesian equilibria?

(c) Maintaining the assumption that θ is privately known by the owner, suppose now that the outside investors' beliefs over θ have support $[\theta_1, \theta_2]$, so that there a continuum of possible values for θ. What is the initial value problem (differential equation plus initial condition) characterizing separating perfect Bayesian equilibria?

6.3.10. Firm 1 is an incumbent firm selling widgets in a market in two periods. In the first period, firm 1 is a monopolist, facing a demand curve $P^1 = A - q_1^1$, where $q_1^1 \in \mathbb{R}_+$ is firm 1's output in period 1 and P^1 is the first period price. In the second period, a second firm, firm 2, will enter the market, observing the first period quantity choice of firm 1. In the second period, the two firms choose quantities simultaneously. The inverse demand curve in period 2 is given by $P^2 = A - q_1^2 - q_2^2$, where $q_i^2 \in \mathbb{R}_+$ is firm i's output in period 2 and P^2 is the second period price. Negative prices are possible (and will arise if quantities exceed A). Firm i has a constant marginal cost of production $c_i > 0$. Firm 1's overall payoff is given by

$$(P^1 - c_1)q_1^1 + (P^2 - c_1)q_1^2,$$

while firm 2's payoff is given by

$$(P^2 - c_2)q_2^2.$$

Firm 2's marginal cost, c_2, is common knowledge (i.e., each firm knows the marginal cost of firm 2), and satisfies $c_2 < A/2$.

(a) Suppose c_1 is also common knowledge (i.e., each firm knows the marginal cost of the other firm), and also satisfies $c_1 < A/2$. What are the subgame perfect equilibria and why?

(b) Suppose now that firm 1's costs are private to firm 1. Firm 2 does not know firm 1's costs, assigning prior probability $p \in (0, 1)$ to cost c_1^L and complementary probability $1 - p$ to cost c_1^H, where $c_1^L < c_1^H < A/2$.

   i. Define a pure strategy almost perfect Bayesian equilibrium for *this* game of incomplete information . What restrictions on second period quantities must be satisfied in any pure strategy almost perfect Bayesian equilibrium? [Make the game finite by considering discretizations of the action spaces. Strictly speaking, this is not a signaling game, since firm 1 is choosing actions in both periods, so the notion from Section 6.1 does not apply.]

  ii. What do the equilibrium conditions specialize to for *separating* pure strategy almost perfect Bayesian equilibria?

(c) Suppose now that firm 2's beliefs about firm 1's costs have support $[c_1^L, c_1^H]$; i.e., the support is now an interval and not two points. What is the direction of the signaling distortion in the separating pure strategy almost perfect Bayesian equilibrium? What differential equation does the function describing first period quantities in that equilibrium satisfy?

6.3.11. Suppose that in the setting of Problem 3.6.1, firm 2 is a Stackelberg leader, i.e., we are reversing the order of moves from Problem 5.4.8.

  (a) Illustrate the preferences of firm 2 in q_2-$\hat{\theta}$ space, where q_2 is firm 2's quantity choice, and $\hat{\theta}$ is firm 1's belief about θ.

  (b) There is a separating perfect Bayesian equilibrium in which firm 2 chooses $q_2 = \frac{1}{2}$ when $\theta = 3$. Describe it, and prove it is a separating perfect Bayesian equilibrium (the diagram from part (a) may help).

  (c) Does the equilibrium from part (b) pass the intuitive criterion? Why or why not? If not, describe a separating perfect Bayesian equilibrium that does.

6.3.12. We continue with the setting of Problem 3.6.1, but now suppose that firm 2 is a Stackelberg leader who has the option of *not* choosing before firm 1: Firm 2 either chooses its quantity, q_2, first, or the action W (for wait). If firm 2 chooses W, then the two firms simultaneously choose quantities, *knowing* that they are doing so. If firm 2 chooses its quantity first (so that it did not choose W), then firm 1, knowing firm 2's quantity choice then chooses its quantity.

  (a) Describe a strategy profile for this dynamic game. Following the practice in signaling games, say a strategy profile is *perfect Bayesian* if it satisfies the conditions implied by sequential equilibrium in discretized versions of the game. (In the current context, a discretized version of the game restricts quantities to some finite subset.) What conditions must a perfect Bayesian equilibrium satisfy, and why?

(b) For which parameter values is there an equilibrium in which firm 2 waits for all values of θ.

(c) Prove that the outcome in which firm 2 does not wait for any θ, and firms behave as in the separating outcome of question (b) is *not* an equilibrium outcome of this game.

Chapter 7

Repeated Games

7.1 Perfect Monitoring

It is obvious that opportunistic behavior can be deterred using future rewards and punishments. However, care needs to be taken to make sure that the future rewards/punishments are self-enforcing. The cleanest setting to explore such questions is one in which the same game (the stage game) is played repeatedly.

7.1.1 The Stage Game

The stage game is a simultaneous move game, $G := \{(A_i, u_i)\}$. The action space for player i is A_i, with typical *action* $a_i \in A_i$. An *action profile* is denoted $a = (a_1, \ldots, a_n)$. Player i's *stage* or *flow payoffs* are given by $u_i : A \to \mathbb{R}$.

7.1.2 The Repeated Game

The game G is played at each date $t = 0, 1, \ldots$. At the end of each period, all players observe the action profile a chosen. Actions of every player are *perfectly monitored* by all other players, i.e., the actions chosen in a period become public information (are observed by all players) at the end of that period.

The history up to date t is denoted by

$$h^t := (a^0, \ldots, a^{t-1}) \in \underbrace{A \times A \times \cdots \times A}_{t \text{ times}} = A^t =: H^t,$$

where $H^0 := \{\varnothing\}$.

The set of all possible histories: $H := \bigcup_{t=0}^{\infty} H^t$.

A *strategy* for player i is denoted $s_i : H \to A_i$. A strategy is often written $s_i = (s_i^0, s_i^1, s_i^2, \ldots)$, where $s_i^t : H^t \to A_i$. Since $H^0 = \{\varnothing\}$, we have $s^0 \in A$, and so can write a^0 for s^0. The set of all strategies for player i is S_i.

Note the distinction between

- actions $a_i \in A_i$ (describing behavior in the stage game) and

- strategies $s_i : H \to A_i$ (describing behavior in the repeated game).

Given a strategy profile $s := (s_1, s_2, \ldots, s_n)$, the outcome path induced by the strategy profile s is $a(s) = (a^0(s), a^1(s), a^2(s), \ldots)$, where

$$
\begin{aligned}
a^0(s) &= (s_1(\varnothing), s_2(\varnothing), \ldots, s_n(\varnothing)), \\
a^1(s) &= (s_1(a^0(s)), s_2(a^0(s)), \ldots, s_n(a^0(s))), \\
a^2(s) &= (s_1(a^0(s), a^1(s)), s_2(a^0(s), a^1(s)), \ldots, s_n(a^0(s), a^1(s))), \\
&\ \ \vdots
\end{aligned}
$$

Each player discounts future payoffs using the same discount factor $\delta \in (0, 1)$. Player i payoffs in $G^\delta(\infty)$ are the *average* discounted sum of the infinite sequence of stage game payoffs $(u_i(a^0(s)), u_i(a^1(s)), \ldots)$ given by

$$
U_i^\delta(s) = (1 - \delta) \sum_{t=0}^{\infty} \delta^t u_i(a^t(s)).
$$

We have now described a normal form game: $G^\delta(\infty) = \{(S_i, U_i^\delta)_{i=1}^n\}$. Note that the payoffs of this normal form game are convex combinations of the stage game payoffs, and so the set of payoffs in $\{(S_i, u_i)\}$ trivially contains $u(A) := \{v \in \mathbb{R}^n : \exists a \in A, v = u(a)\}$ and is a subset of the convex hull of $u(A)$, conv $u(A)$, i.e.,

$$
u(A) \subset U^\delta(S) \subset \text{conv}\, u(A).
$$

Moreover, if A is finite, the first inclusion is *strict*. Finally, for δ sufficiently close to 1, every payoff in conv $u(A)$ can be achieved as a payoff in $G^\delta(\infty)$. More specifically, if $\delta > 1 - 1/|A|$, then for all $v \in \text{conv}\, u(A)$ there exists $s \in S$ such that $v = U^\delta(s)$ (for a proof, see Lemma 3.7.1 in Mailath and Samuelson, 2006).

Definition 7.1.1. *The set of* feasible payoffs *for the infinitely repeated game is given by*

$$
\text{conv}\, u(A) = \text{conv}\{v \in \mathbb{R}^n : \exists a \in S, v = u(a)\}.
$$

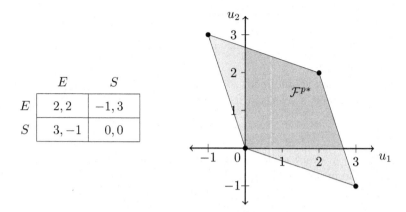

	E	S
E	2, 2	−1, 3
S	3, −1	0, 0

Figure 7.1.1: A prisoner's dilemma on the left, and the set of payoffs on the right. Both players' minmax payoff is 0. The set of feasible payoffs is the union of the two lightly shaded regions (with their borders) and $\mathcal{F}^{p*}$, the darkly shaded region plus its northeast borders.

While the superscript δ is often omitted from $G^\delta(\infty)$ and from U_i^δ, this should not cause confusion, as long as the role of the discount factor in determining payoffs is not forgotten.

Definition 7.1.2. *Player i's pure strategy minmax utility is*

$$\underline{v}_i^p := \min_{a_{-i}} \max_{a_i} u_i(a_i, a_{-i}).$$

The profile $\hat{a}_{-i} \in \arg\min_{a_{-i}} \max_{a_i} u_i(a_i, a_{-i})$ *minmaxes* player i. The set of *(pure strategy) strictly individually rational payoffs* in $\{(S_i, u_i)\}$ is $\{v \in \mathbb{R}^n : v_i > \underline{v}_i^p\}$. Define $\mathcal{F}^{p*} := \{v \in \mathbb{R}^n : v_i > \underline{v}_i^p\} \cap \mathrm{conv}\{v \in \mathbb{R}^n : \exists a \in S, v = u(a)\}$. The set is illustrated in Figure 7.1.1 for the prisoner's dilemma. Note that the relevant parts of the u_1- and u_2-axes belong to the lightly shaded regions, and not to $\mathcal{F}^{p*}$.

Definition 7.1.3. *The strategy profile $\hat{s}$ is a (pure strategy) Nash equilibrium of $G^\delta(\infty)$ if, for all i and all $\tilde{s}_i : H \to A_i$,*

$$U_i^\delta(\hat{s}_i, \hat{s}_{-i}) \geq U_i^\delta(\tilde{s}_i, \hat{s}_{-i}).$$

Theorem 7.1.1. *Suppose s^* is a pure strategy Nash equilibrium. Then,*

$$U_i^\delta(s^*) \geq \underline{v}_i^p.$$

Proof. Let $\hat{s}_i$ be a strategy satisfying

$$\hat{s}_i(h^t) \in \arg\max_{a_i} u_i(a_i, s^*_{-i}(h^t)), \qquad \forall h^t \in H^t$$

(if the arg max is unique for some history h^t, $\hat{s}_i(h^t)$ is uniquely determined, otherwise make a selection from the argmax). Since

$$U_i^\delta(s^*) \geq U_i^\delta(\hat{s}_i, s^*_{-i}),$$

and since in every period $\underline{v}_i^p$ is a lower bound for the flow payoff received under the profile $(\hat{s}_i, s^*_{-i})$, we have

$$U_i^\delta(s^*) \geq U_i^\delta(\hat{s}_i, s^*_{-i}) \geq (1-\delta) \sum_{t=0}^{\infty} \delta^t \underline{v}_i^p = \underline{v}_i^p.$$

∎

In some settings it is necessary to allow players to randomize. For example, in matching pennies, the set of pure strategy feasible and individually rational payoffs is empty.

Definition 7.1.4. *Player i's mixed strategy minmax utility is*

$$\underline{v}_i := \min_{\alpha_{-i} \in \prod j \neq i \Delta(A_j)} \max_{\alpha_i \in \Delta(A_i)} u_i(\alpha_i, \alpha_{-i}).$$

The profile $\hat{\alpha}_{-i} \in \arg\min_{\alpha_{-i}} \max_{\alpha_i} u_i(\alpha_i, \alpha_{-i})$ minmaxes player i. The set of (mixed strategy) strictly individually rational payoffs in $\{(S_i, u_i)\}$ is $\{v \in \mathbb{R}^n : v_i > \underline{v}_i\}$. Define $\mathcal{F}^* := \{v \in \mathbb{R}^n : v_i > \underline{v}_i\} \cap \text{conv}\{v \in \mathbb{R}^n : \exists a \in S, v = u(a)\}$.

The Minmax Theorem (Problem 4.3.2) implies that $\underline{v}_i$ is i's security level (Definition 2.4.2).

An essentially identical proof to that of Theorem 7.1.1 (applied to the behavior strategy profile realization equivalent to σ^*) gives:

Theorem 7.1.2. *Suppose σ^* is a (possibly mixed) Nash equilibrium. Then,*

$$U_i^\delta(\sigma^*) \geq \underline{v}_i.$$

Since $\underline{v}_i \leq \underline{v}_i^p$ (with a strict inequality in some games, such as matching pennies), lower payoffs often can be enforced using mixed strategies. The possibility of enforcing lower payoffs raises the possibility of using these lower payoffs to enforce higher payoffs.

7.1.3   Subgame Perfection

Given $h^t = (a^0, \ldots, a^{t-1}) \in H^t$ and $\bar{h}^\tau = (\bar{a}^0, \ldots, \bar{a}^{\tau-1}) \in H^\tau$, the history $(a^0, \ldots, a^{t-1}, \bar{a}^0, \ldots, \bar{a}^{\tau-1}) \in H^{t'+t}$ is the *concatenation* of h^t followed by $\bar{h}^\tau$, denoted by $(h^t, \bar{h}^\tau)$. Given s_i, define $s_i|_{h^t} \colon H \to A_i$ as

$$s_i|_{h^t}(\bar{h}^\tau) = s_i(h^t, \bar{h}^\tau) \quad \forall \bar{h}^\tau, \tau.$$

Note that for all histories h^t,

$$s_i|_{h^t} \in S_i.$$

Remark 7.1.1. There is a natural isomorphism for each player between histories and that player's information sets. Every history, h^t, reaches a continuation game that is strategically identical to the original repeated game, and for every strategy s_i in the original game, h^t induces a well-defined continuation strategy $s_i|_{h^t}$. In particular, the subgame reached by any history h^t is an infinitely repeated game that is strategically equivalent to the original infinitely repeated game, $G^\delta(\infty)$.   ♦

Definition 7.1.5. *The strategy profile $\hat{s}$ is a* subgame perfect equilibrium *of $G^\delta(\infty)$ if, for all histories, $h^t \in H^t$, $\hat{s}|_{h^t} = (\hat{s}_i|_{h^t}, \ldots, \hat{s}_n|_{h^t})$ is a Nash equilibrium of $G^\delta(\infty)$.*

Example 7.1.1 (Grim trigger in the repeated prisoner's dilemma). Consider the prisoner's dilemma in Figure 7.1.1.

A *grim trigger* strategy profile is a profile where a deviation triggers Nash reversion (hence *trigger*) and the Nash equilibrium minmaxes the players (hence *grim*). For the prisoner's dilemma, grim trigger can be described as follows: Player i's strategy is given by

$$\hat{s}_i(\varnothing) = E,$$

and for $t \geq 1$,

$$\hat{s}_i(a^0, \ldots, a^{t-1}) = \begin{cases} E, & \text{if } a^{t'} = EE \text{ for all } t' = 0, 1, \ldots, t-1, \\ S, & \text{otherwise.} \end{cases}$$

The payoff to player 1 from $(\hat{s}_1, \hat{s}_2)$ is $(1-\delta)\sum 2 \times \delta^t = 2$.

Consider a deviation by player 1 to another strategy $\tilde{s}_1$. In response to the first play of S, player 2 responds with S in every subsequent period, so player 1 can do no better than always play S after the first play of S.

The maximum payoff from deviating in period $t = 0$ (the most profitable deviation) is $(1 - \delta)3$. The profile is Nash if

$$2 \geq 3(1 - \delta) \iff \delta \geq \frac{1}{3}.$$

Strategy profile is subgame perfect: Note first that the profile only induces two different strategy profiles after any history. Denote by $s^\dagger = (s_1^\dagger, s_2^\dagger)$ the profile in which each player plays S for all histories, $s_i^\dagger(h^t) = S$ for all $h^t \in H$. Then,[1]

$$\hat{s}_i|_{(a^0, \ldots, a^{t-1})} = \begin{cases} \hat{s}_i, & \text{if } a^{t'} = EE \text{ for all } t' = 0, 1, \ldots, t-1, \\ s_i^\dagger, & \text{otherwise.} \end{cases}$$

We have already verified that $\hat{s}$ is a Nash equilibrium of $G(\infty)$, and it is immediate that $s^\dagger$ is Nash.   ★

Remark 7.1.2. The finitely-repeated prisoner's dilemma has a unique subgame perfect (indeed, Nash) equilibrium outcome, given by SS in every period (see Remark 2.3.1).[2] This conclusion is driven by *end-game effects*: play in the final period determines play in earlier periods. There is thus a discontinuity in equilibrium outcomes with respect to the horizon: EE in every period can be supported as a subgame-perfect equilibrium outcome in the infinitely-repeated game (when δ is sufficiently large), yet only SS in every period can be supported as a subgame perfect equilibrium outcome in the finitely-repeated game (no matter how large the number of repetitions). Which is the better description? If players believe end-game effects are important, then it is the finitely-repeated game. If players do not believe end-game effects are important, then it is the infinitely repeated game (even though the agents themselves are finitely lived).   ◆

Grim trigger is an example of a strongly symmetric strategy profile (the deviator is not treated differently than the other player(s)):

Definition 7.1.6. *Suppose $A_i = A_j$ for all i and j. A strategy profile is strongly symmetric if*

$$s_i(h^t) = s_j(h^t) \qquad \forall h^t \in H, \forall i, j.$$

---

[1]This is a statement about the strategies as functions, i.e., for all $\bar{h}^\tau \in H$,

$$\hat{s}_i|_{(a^0, \ldots, a^{t-1})}(\bar{h}^\tau) = \begin{cases} \hat{s}_i(\bar{h}^\tau), & \text{if } a^{t'} = EE \text{ for all } t' = 0, 1, \ldots, t-1, \\ s_i^\dagger(\bar{h}^\tau), & \text{otherwise.} \end{cases}$$

[2]Discounting payoffs does not alter the analysis of the finitely repeated game.

7.1.4 Automata

Represent strategy profiles by *automata*, $(\mathcal{W}, w^0, f, \tau)$, where

- $\mathcal{W}$ is set of states,

- w^0 is initial state,

- $f : \mathcal{W} \to A$ is output function (decision rule),[3] and

- $\tau : \mathcal{W} \times A \to \mathcal{W}$ is transition function.

Any automaton $(\mathcal{W}, w^0, f, \tau)$ induces a pure strategy profile as follows: First, extend the transition function from the domain $\mathcal{W} \times A$ to the domain $\mathcal{W} \times (H \backslash \{\varnothing\})$ by recursively defining

$$\tau(w, h^t) = \tau\left(\tau(w, h^{t-1}), a^{t-1}\right) \qquad t \geq 2, \tag{7.1.1}$$

where $\tau(w, h^1) := \tau(w, a^0)$ (recall that $h^1 = (a^0)$). With this definition, the strategy s induced by the automaton is given by $s(\varnothing) = f(w^0)$ and

$$s(h^t) = f(\tau(w^0, h^t)), \forall h^t \in H \backslash \{\varnothing\}.$$

Conversely, it is straightforward that any strategy profile can be represented by an automaton. Take the set of histories H as the set of states, the null history $\varnothing$ as the initial state, $f(h^t) = s(h^t)$, and $\tau(h^t, a) = h^{t+1}$, where $h^{t+1} := (h^t, a)$ is the concatenation of the history h^t with the action profile a.

The representation in the previous paragraph leaves us in the position of working with the full set of histories H, a countably infinite set. However, strategy profiles can often be represented by automata with finite sets $\mathcal{W}$. The set $\mathcal{W}$ is then a partition of H, grouping together those histories that lead to identical continuation strategies.

The advantage of the automaton representation is most obvious when $\mathcal{W}$ can be chosen finite, but the automaton representation also has conceptual advantages (Theorem 7.1.3 is one illustration).

An automaton is *strongly symmetric* if $f_i(w) = f_j(w)$ for all w and all i and j.[4] While many of the examples are strongly symmetric, there is no requirement that the automaton be strongly symmetric (for an example of an asymmetric automaton, see Figure 7.1.5) nor that the game be symmetric (see Example 7.2.1).

---

[3]A profile of behavior strategies $(b_1, \ldots, b_n)$, $b_i : H \to \Delta(A_i)$, can also be represented by an automaton. The output function now maps into profiles of mixtures over action profiles, i.e., $f : \mathcal{W} \to \prod_i \Delta(A_i)$.

[4]The induced pure strategy profile of a strongly symmetric automaton is strongly symmetric. Moreover, a strongly symmetric pure strategy profile can be represented by a strongly symmetric automaton.

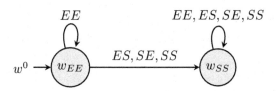

Figure 7.1.2: The automaton for grim trigger in Example 7.1.2. See that example for the notational convention.

Example 7.1.2 (Grim trigger in the repeated prisoner's dilemma, cont.). Grim trigger profile has as its automata representation, $(\mathcal{W}, w^0, f, \tau)$, with $\mathcal{W} = \{w_{EE}, w_{SS}\}$, $w^0 = w_{EE}$, $f(w_{EE}) = EE$ and $f(w_{SS}) = SS$, and

$$\tau(w, a) = \begin{cases} w_{EE}, & \text{if } w = w_{EE} \text{ and } a = EE, \\ w_{SS}, & \text{otherwise.} \end{cases}$$

The automaton is illustrated in Figure 7.1.2. Observe the notational convention: The subscript on the state indicates the action profile specified by the output function (i.e., $f(w_a) = a$; if the same action profile a is specified at two distinct states, the distinct states could be denoted by $\hat{w}_a$ and $\tilde{w}_a$, so that $f(\hat{w}_a) = a$ and $f(\tilde{w}_a) = a$); it is distinct from the transition function.

⋆

If s is represented by $(\mathcal{W}, w^0, f, \tau)$, the continuation strategy profile after a history h^t, $s|_{h^t}$, is represented by the automaton $(\mathcal{W}, \tau(w^0, h^t), f, \tau)$, where $\tau(w^0, h^t)$ is given by (7.1.1).

Definition 7.1.7. *The state $w \in \mathcal{W}$ of an automaton $(\mathcal{W}, w^0, f, \tau)$ is reachable from w^0 if there exists a history $h^t \in H$ such that*

$$w = \tau(w^0, h^t).$$

Denote the set of states reachable from w^0 by $\mathcal{W}(w^0)$.

Lemma 7.1.1. *The strategy profile represented by $(\mathcal{W}, w^0, f, \tau)$ is a subgame perfect equilibrium if, and only if, for all states $w \in \mathcal{W}(w^0)$, the strategy profile represented by $(\mathcal{W}, w, f, \tau)$ is a Nash equilibrium of the repeated game.*

Given an automaton $(\mathcal{W}, w^0, f, \tau)$, let $V_i(w)$ be i's value from being in the state $w \in \mathcal{W}$, i.e.,

$$V_i(w) = (1 - \delta)u_i(f(w)) + \delta V_i(\tau(w, f(w))).$$

Note that if $\mathcal{W}$ is finite, V_i solves a finite set of linear equations (see Problem 7.6.3).

Theorem 7.1.1 and Lemma 7.1.1 imply that if $(\mathcal{W}, w^0, f, \tau)$ represents a pure strategy subgame perfect equilibrium, then for all states $w \in \mathcal{W}$, and all i,

$$V_i(w) \geq \underline{v}_i^p.$$

Compare the following definition with Definition 5.1.3, and the proofs of Theorem 7.1.3 with that of Theorem 5.3.2.

Definition 7.1.8. *Player i has a* profitable one-shot deviation *from the strategy profile (induced by the automaton) $(\mathcal{W}, w^0, f, \tau)$, if there is some state $w \in \mathcal{W}(w^0)$ and some action $a_i \in A_i$ such that*

$$V_i(w) < (1 - \delta)u_i(a_i, f_{-i}(w)) + \delta V_i(\tau(w, (a_i, f_{-i}(w)))).$$

Another instance of the *one-shot deviation principle* (recall Theorems 5.1.1 and 5.3.2):

Theorem 7.1.3. *A strategy profile is subgame perfect if, and only if, there are no profitable one-shot deviations.*

Proof. Clearly, if a strategy profile is subgame perfect, then there are no profitable deviations.

We need to argue that if a profile is not subgame perfect, then there is a profitable one-shot deviation.

Suppose first that $\mathcal{W}$ and A_i are finite. Let $\widetilde{V}_i(w)$ be player i's payoff from the best response to $(\mathcal{W}, w, f_{-i}, \tau)$ (i.e., the strategy profile for the other players specified by the automaton with initial state w). From the principle of optimality in dynamic programming,[5] the optimal value of a sequential problem is the solution to the Bellman equation:

$$\widetilde{V}_i(w) = \max_{a_i \in A_i} \left\{ (1 - \delta)u_i(a_i, f_{-i}(w)) + \delta \widetilde{V}_i\left(\tau\left(w, (a_i, f_{-i}(w))\right)\right) \right\}. \tag{7.1.2}$$

Note that $\widetilde{V}_i(w) \geq V_i(w)$ for all w. Denote by $\bar{w}_i$, the state that maximizes $\widetilde{V}_i(w) - V_i(w)$ (if there is more than one, choose one arbitrarily).

Let a_i^w be the action solving the maximization in (7.1.2), and define

$$V_i^{\dagger}(w) := (1 - \delta)u_i(a_i^w, f_{-i}(w)) + \delta V_i(\tau(w, (a_i^w, f_{-i}(w)))).$$

Observe that $V_i^{\dagger}(w)$ is the value of the one-shot deviation from $(\mathcal{W}, w^0, f, \tau)$ to a_i^w at w.

---

[5]See Problem 7.6.4 or Stokey and Lucas (1989, Theorems 4.2 and 4.3).

If the automaton does not represent a subgame perfect equilibrium, then there exists a player i for which

$$\widetilde{V}_i(\bar{w}_i) > V_i(\bar{w}_i). \tag{7.1.3}$$

Then, for that player i,

$$\widetilde{V}_i(\bar{w}_i) - V_i(\bar{w}_i) > \delta[\widetilde{V}_i(w) - V_i(w)]$$

for all w (the strict inequality is key here, and only holds because of (7.1.3)), and so

$$\widetilde{V}_i(\bar{w}_i) - V_i(\bar{w}_i)$$
$$> \delta[\widetilde{V}_i(\tau(\bar{w}_i, (a_i^{\bar{w}_i}, f_{-i}(\bar{w}_i)))) - V_i(\tau(\bar{w}_i, (a_i^{\bar{w}_i}, f_{-i}(\bar{w}_i))))]$$
$$= \widetilde{V}_i(\bar{w}_i) - V_i^\dagger(\bar{w}_i).$$

Thus,

$$V_i^\dagger(\bar{w}_i) > V_i(\bar{w}_i),$$

which is the assertion that player i has a profitable one-shot deviation at $\bar{w}_i$.

The argument extends in a straightforward manner to infinite $\mathcal{W}$, compact A_i, and continuous u_i once we reinterpret $\widetilde{V}_i$ as the supremum of player i's payoff when the other players play according to $(\mathcal{W}, w, f_{-i}, \tau)$. ∎

Note that a strategy profile can have no profitable one-shot deviations *on* the path of play, and yet *not* be Nash, see Example 7.1.5/Problem 7.6.6 for a simple example.

See Problem 7.6.5 for an alternative proof that can be extended to arbitrary infinite horizon games without a recursive structure (and does not appeal to the principle of optimality).

Corollary 7.1.1. *The strategy profile represented by $(\mathcal{W}, w^0, f, \tau)$ is subgame perfect if, and only if, for all $w \in \mathcal{W}(w^0)$, $f(w)$ is a Nash equilibrium of the normal form game with payoff function $g^w : A \to \mathbb{R}^n$, where*

$$g_i^w(a) = (1 - \delta)u_i(a) + \delta V_i(\tau(w, a)).$$

Example 7.1.3 (Continuation of grim trigger). We clearly have $V_1(w_{EE}) = 2$ and $V_1(w_{SS}) = 0$, so that $g^{w_{EE}}$, the normal form associated with w_{EE}, is given in Figure 7.1.3, while $g^{w_{SS}}$, the normal form for w_{SS}, is given in Figure 7.1.4.

As required, EE is a (but not the only!) Nash equilibrium of the w_{EE} normal form for $\delta \geq \frac{1}{3}$, while SS is a Nash equilibrium of the w_{SS} normal form.

	E	S
E	$2,2$	$-(1-\delta),3(1-\delta)$
S	$3(1-\delta),-(1-\delta)$	$0,0$

Figure 7.1.3: The normal form $g^{w_{EE}}$ associated with the automaton state w_{EE} of the automaton in Figure 7.1.2.

	E	S
E	$2(1-\delta),2(1-\delta)$	$-(1-\delta),3(1-\delta)$
S	$3(1-\delta),-(1-\delta)$	$0,0$

Figure 7.1.4: The normal form $g^{w_{SS}}$ associated with the automaton state w_{EE} of the automaton in Figure 7.1.2.

★

Example 7.1.4 (Repeated prisoner's dilemma, cont.). A different strategy profile to support mutual effort in the repeated prisoner's dilemma is $(\mathcal{W}, w^0, f, \tau)$, where $\mathcal{W} = \{w_{EE}, w_{ES}, w_{SE}\}$, $w^0 = w_{EE}$, $f_i(w_a) = a_i$, and

$$\tau(w,a) = \begin{cases} w_{EE}, & \text{if } w = w_{EE} \text{ and } a = EE \text{ or } SS, w = w_{ES} \text{ and } a_1 = E, \\ & \quad \text{or } w = w_{SE} \text{ and } a_2 = E, \\ w_{ES}, & \text{if } w = w_{EE} \text{ and } a = SE, \text{ or } w = w_{ES} \text{ and } a_1 = S, \\ w_{SE}, & \text{if } w = w_{EE} \text{ and } a = ES, \text{ or } w = w_{SE} \text{ and } a_2 = S. \end{cases}$$

The automaton is displayed in Figure 7.1.5. For that automaton,

$$V_1(w_{SE}) = 3(1-\delta) + 2\delta = 3 - \delta > 2,$$
$$V_2(w_{SE}) = -(1-\delta) + 2\delta = -1 + 3\delta < 2,$$
$$\text{and} \quad V_1(w_{EE}) = V_2(w_{EE}) = 2.$$

The payoffs are illustrated in Figure 7.1.6.

In particular, state w_{ES} punishes player 1. The normal form associated with w_{EE} is given in Figure 7.1.7, and EE is Nash equilibrium if and only if

$$2 \geq 3 - 4\delta + 3\delta^2 \iff 0 \geq (1 - 3\delta)(1 - \delta) \iff \delta \geq \tfrac{1}{3}.$$

The normal form for w_{ES} is given in Figure 7.1.8, and ES is a Nash equi-

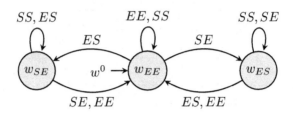

Figure 7.1.5: The automaton for Example 7.1.4.

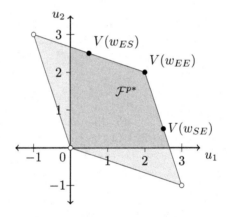

Figure 7.1.6: The payoffs for the automaton in Figure 7.1.5, for $\delta = \frac{1}{2}$.

	E	S
E	$2, 2$	$-1 + 4\delta - \delta^2, 3 - 4\delta + 3\delta^2$
S	$3 - 4\delta + 3\delta^2, -1 + 4\delta - \delta^2$	$2\delta, 2\delta$

Figure 7.1.7: The normal form associated with the automaton state w_{EE} for the automaton displayed in Figure 7.1.5.

	E	S
E	$2,2$	$-1+3\delta, 3-\delta$
S	$3-4\delta+3\delta^2, -1+4\delta-\delta^2$	$-\delta+3\delta^2, 3\delta-\delta^2$

Figure 7.1.8: The normal form associated with the automaton state w_{ES} for the automaton displayed in Figure 7.1.5.

	A	B	C
A	$4,4$	$3,2$	$1,1$
B	$2,3$	$2,2$	$1,1$
C	$1,1$	$1,1$	$-1,-1$

Figure 7.1.9: The stage game for Example 7.1.5.

librium if player 1 does not wish deviate, i.e.,

$$-1+3\delta \geq -\delta+3\delta^2 \iff 0 \geq (1-3\delta)(1-\delta) \iff \delta \geq \tfrac{1}{3},$$

and player 2 does not wish to deviate, i.e.,

$$3-\delta \geq 2 \iff 1 \geq \delta.$$

This last inequality is worth comment: The state transition from state w_{ES} does not depend on player 2's action, and so player 2 faces no intertemporal consequences from his action choice at w_{ES}. Maximizing for 2 then reduces to maximizing stage game payoffs (for any δ), and S maximizes stage game payoffs.

The normal form for w_{SE} switches the players, and so there is no need to separately consider that game. Thus the automaton is a subgame perfect equilibrium for $\delta \geq \tfrac{1}{3}$.

This profile's successful deterrence of deviations using a one period punishment (when $\delta \geq \tfrac{1}{3}$) is due to the specific payoffs, see Problem 7.6.8. ★

Example 7.1.5. The stage game in Figure 7.1.9 has a unique Nash equilibrium: AA. We will argue that, for $\delta \geq 2/3$, there is a subgame perfect equilibrium of $G(\infty)$ with outcome path $(BB)^\infty$: $(\mathcal{W}, w^0, f, \tau)$, where

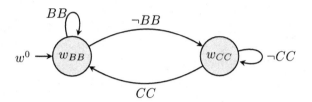

Figure 7.1.10: The automaton for Example 7.1.5.

	A	B	C
A	$4(1-\delta)+\delta(3\delta-1)$	$3(1-\delta)+\delta(3\delta-1)$	$1-\delta+\delta(3\delta-1)$
B	$2(1-\delta)+\delta(3\delta-1)$	2	$1-\delta+\delta(3\delta-1)$
C	$1-\delta+\delta(3\delta-1)$	$1-\delta+\delta(3\delta-1)$	$-(1-\delta)+\delta(3\delta-1)$

Figure 7.1.11: The payoffs for player 1 in the normal form associated with the automaton state w_{BB} for the automaton displayed in Figure 7.1.10.

$\mathcal{W}=\{w_{BB},w_{CC}\}$, $w^0=w_{BB}$, $f_i(w_a)=a_i$, and

$$\tau(w,a)=\begin{cases} w_{BB}, & \text{if } w=w_{BB} \text{ and } a=BB, \text{ or } w=w_{CC} \text{ and } a=CC, \\ w_{CC}, & \text{otherwise.} \end{cases}$$

The automaton is illustrated in Figure 7.1.10.

Values of the states are

$$V_i(w_{BB}) =(1-\delta)2+\delta V_i(w_{BB}),$$
$$\text{and} \qquad V_i(w_{CC}) =(1-\delta)\times(-1)+\delta V_i(w_{BB}).$$

Solving,

$$V_i(w_{BB}) =2,$$
$$\text{and} \qquad V_i(w_{CC}) =3\delta-1.$$

Player 1's payoffs in the normal form associated with w_{BB} are given in Figure 7.1.11, and since the game is symmetric, BB is a Nash equilibrium of this normal form only if

$$2 \geq 3(1-\delta)+\delta(3\delta-1),$$

	A	B	C
A	$4(1-\delta)+\delta(3\delta-1)$	$3(1-\delta)+\delta(3\delta-1)$	$1-\delta+\delta(3\delta-1)$
B	$2(1-\delta)+\delta(3\delta-1)$	$2(1-\delta)+\delta(3\delta-1)$	$1-\delta+\delta(3\delta-1)$
C	$1-\delta+\delta(3\delta-1)$	$1-\delta+\delta(3\delta-1)$	$-(1-\delta)+\delta2$

Figure 7.1.12: The payoffs for player 1 in the normal form associated with the automaton state w_{CC} for the automaton displayed in Figure 7.1.10.

i.e.,

$$0 \geq 1 - 4\delta + 3\delta^2 \Leftrightarrow 0 \geq (1-\delta)(1-3\delta),$$

or $\delta \geq 1/3$.

Player 1's payoffs in the normal form associated with w_{CC} are given in Figure 7.1.12, and since the game is symmetric, CC is a Nash equilibrium of this normal form only if

$$-(1-\delta) + \delta2 \geq 1 - \delta + \delta(3\delta-1),$$

i.e.,

$$0 \geq 2 - 5\delta + 3\delta^2 \Leftrightarrow 0 \geq (1-\delta)(2-3\delta),$$

or $\delta \geq 2/3$. This completes our verification that the profile represented by the automaton in Figure 7.1.10 is subgame perfect.

Note that even though there is no profitable deviation at w_{BB} if $\delta \geq 1/3$, the profile is only subgame perfect if $\delta \geq 2/3$. Moreover, the profile is not Nash if $\delta \in [1/3,\ 1/2)$ (Problem 7.6.6). ★

7.1.5 Renegotiation-Proof Equilibria

Are Pareto inefficient equilibria plausible threats?

Definition 7.1.9 (Farrell and Maskin, 1989). *The subgame perfect automaton $(\mathcal{W}, w^0, f, \tau)$ is* weakly renegotiation proof *if for all $w, w' \in \mathcal{W}(w_0)$ and i,*

$$V_i(w) > V_i(w') \implies V_j(w) \leq V_j(w') \text{ for some } j.$$

This is a notion of *internal dominance*, since it does not preclude the existence of a Pareto dominating equilibrium played by an unrelated automaton.

	c	s
H	3, 3	0, 2
L	4, 0	2, 1

Figure 7.2.1: The product-choice game.

While grim trigger is *not* weakly renegotiation proof, Example 7.1.4 describes a weakly renegotiation proof equilibrium supporting mutual effort in the repeated PD.

Stronger notions of renegotiation proofness easily lead to nonexistence problems in infinitely repeated games. For finitely repeated games, renegotiation proofness has a nice characterization (Benoit and Krishna, 1993).

7.2 Short-Lived Players and Modeling Competitive Agents

The analysis of the previous section easily carries over to the case where not all the players are infinitely lived. A short-lived player lives only one period, and in the infinite-horizon model, is replaced with a new short-lived player at the end of the period. For example, a firm may be interacting with a sequence of short-lived customers.

Example 7.2.1. The *product-choice game* is given by the normal form in Figure 7.2.1. Player I (row player, a firm) is long-lived, choosing high (H) or low (L) effort, player II (the column player, a customer) is short lived, choosing the customized (c) or standardized (s) product. The set of feasible payoffs is illustrated in Figure 7.2.2.

The two state automaton given in Figure 7.2.3 describes a subgame perfect equilibrium for large δ: The action profile Ls is a static Nash equilibrium, and since w_{Ls} is an *absorbing state* (i.e., the automaton never leaves the state once entered), we trivially have that Ls is a Nash equilibrium of the associated one-shot game, $g^{w_{Ls}}$.

Note that $V_1(w_{Hc}) = 3$ and $V_1(w_{Ls}) = 2$. Since player 2 is short-lived, he must myopically optimize in each period. The one-shot game from Corollary 7.1.1 has only one player. The one-shot game $g^{w_{Hc}}$ associated with w_{Hc} is given in Figure 7.2.4, and player I finds H optimal if $3 \geq 4 - 2\delta$, i.e., if $\delta \geq 1/2$.

Thus, the profile is a subgame perfect equilibrium if, and only if, $\delta \geq 1/2$.   ★

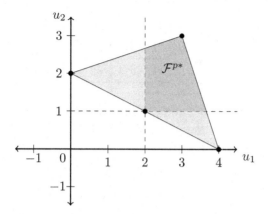

Figure 7.2.2: The set of feasible and individually rational payoffs for the product-choice game, Example 7.2.1.

Figure 7.2.3: The automaton for the equilibrium in Example 7.2.1.

Example 7.2.2. Consider now the modification of the product-choice game in Figure 7.2.5. The action profile Ls is no longer a static Nash equilibrium, and so Nash reversion cannot be used to discipline player I's behavior.

The two-state automaton in Figure 7.2.6 describes a subgame perfect equilibrium. Since player 2 is short-lived, he must myopically optimize in each period, and he is.

Note that $V_1(w_{Hc}) = 3$ and $V_1(w_{Ls}) = (1-\delta)0 + \delta 3 = 3\delta$. There are two one-shot games we need to consider. The one-shot game $g^{w_{Hc}}$ associated with w_{Hc} is given in Figure 7.2.7, and player I finds H optimal if

$$3 \geq 4 - 4\delta + 3\delta^2 \iff 0 \geq (1-\delta)(1-3\delta) \iff \delta \geq 1/3.$$

The one-shot game $g^{w_{Ls}}$ associated with w_{Ls} is given in Figure 7.2.8, and player I finds L optimal if

$$3\delta \geq 2 - 2\delta + 3\delta^2 \iff 0 \geq (1-\delta)(2-3\delta) \iff \delta \geq 2/3.$$

$$c$$

	c
H	$(1-\delta)3 + \delta3$
L	$(1-\delta)4 + \delta2$

Figure 7.2.4: The one-shot game $g^{w_{Hc}}$ associated with w_{Hc} in Figure 7.2.3.

	c	s
H	$3,3$	$2,2$
L	$4,0$	$0,1$

Figure 7.2.5: A modification of the product-choice game.

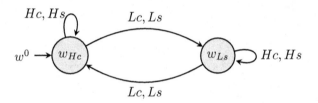

Figure 7.2.6: The automaton for the equilibrium in Example 7.2.2.

	c
H	$(1-\delta)3 + \delta3$
L	$(1-\delta)4 + 3\delta^2$

Figure 7.2.7: The one-shot game $g^{w_{Hc}}$ associated with w_{Hc} in the automaton in Figure 7.2.6.

	s
H	$(1-\delta)2 + 3\delta^2$
L	$(1-\delta)0 + 3\delta$

Figure 7.2.8: The one-shot game $g^{w_{Ls}}$ associated with w_{Hc} in the automaton in Figure 7.2.6.

Thus, the profile is a subgame perfect equilibrium if, and only if, $\delta \geq 2/3$.   ★

The next example concerns a seller facing a population of small consumers (so that each consumer is a price taker). Because each consumer is a price taker, he/she effectively behaves as a short-lived agent.

Example 7.2.3. In the stage game, the seller chooses quality, "H" or "L", and announces a price. The cost of producing H quality is $c_H = 2$, and the cost of producing L quality is $c_L = 1$.

Demand for the good is given by

$$x(p) = \begin{cases} 10 - p, & \text{if } H, \text{ and} \\ 4 - p, & \text{if } L. \end{cases}$$

Monopoly pricing: If consumers know the seller is seller low quality, the monopoly price solves $\max_p (4 - p)(p - c_L)$, giving a monopoly price $p = \frac{5}{2}$ with monopoly quantity $x = \frac{3}{2}$ and profits $\pi^L = \frac{9}{4}$.

If consumers know the seller is seller high quality, the monopoly price solves $\max_p (10 - p)(p - c_H)$, giving a monopoly price $p = 6$ with monopoly quantity $x = 4$ and profits $\pi^H = 16$.

Suppose now the stage game is repeated and quality is only observed after purchase. To model as a game, we assume the seller publicly chooses the price first, and then chooses quality when consumers make their purchase decisions. (The stage game is effectively a version of the product-choice game preceded by a public pricing stage.) The action space for the seller is $\{(p, Q) : p \in \mathbb{R}_+, Q : \mathbb{R}_+ \to \{L, H\}\}$.

There is a continuum of (long-lived) consumers of mass 10, each consumer buying zero or one unit of the good in each period. Consumer $i \in [0, 10]$ values one unit of good as follows

$$v_i = \begin{cases} i, & \text{if } H, \text{ and} \\ \max\{0, i - 6\}, & \text{if } L. \end{cases}$$

The action space for consumer i is $\{s : \mathbb{R}_+ \to \{0, 1\}\}$, where 1 is buy and 0 is not buy and the consumer's choice is a function of the price chosen by the seller.

The action profile is $((p, Q), \xi)$, where $\xi(i)$ is consumer i's strategy. Write ξ_i for $\xi(i)$. Consumer i's payoff function is

$$u_i((p, Q), \xi) = \begin{cases} i - p, & \text{if } Q(p) = H \text{ and } \xi_i(p) = 1, \\ \max\{0, i - 6\} - p, & \text{if } Q(p) = L \text{ and } \xi_i(p) = 1, \text{ and} \\ 0, & \text{if } \xi_i(p) = 0. \end{cases}$$

Firm's payoff function is

$$\pi((p, Q), \xi) = (p - c_{Q(p)})\hat{x}(p, \xi)$$

$$:= (p - c_{Q(p)}) \int_0^{10} \xi_i(p)\,di$$

$$= (p - c_{Q(p)})\lambda\{i \in [0, 10] : \xi_i(p) = 1\},$$

where λ is Lebesgue measure. (Note that we need to assume that ξ is measurable.)

Assume the seller only observes $\hat{x}(p, \xi)$ at the end of the period, so that consumers are *anonymous*.

Note that $\hat{x}(p, \xi)$ is independent of Q, and that for any p, the choice L strictly dominates H (in the stage game) whenever $\hat{x}(p, \xi) \neq 0$.

If consumer i believes the firm has chosen Q, then i's best response to p is $\xi_i(p) = 1$ only if $u_i((p, Q), \xi) \geq 0$. Let $\xi_i^{Q(p)}(p)$ denote the maximizing choice of consumer i when the consumer observes price p and believes the firm also chose quality $Q(p)$. Then,

$$\xi_i^H(p) = \begin{cases} 1, & \text{if } i \geq p, \text{ and} \\ 0, & \text{if } i < p, \end{cases}$$

so $x(p, \xi^H) = \int_p^{10} di = \max\{0, 10 - p\}$. Also,

$$\xi_i^L(p) = \begin{cases} 1, & \text{if } i \geq p + 6, \text{ and} \\ 0, & \text{if } i < p + 6, \end{cases}$$

so $\hat{x}(p, \xi^L) = \int_{p+6}^{10} di = \max\{0, 10 - (p + 6)\} = \max\{0, 4 - p\}$.

The outcome of the *unique* subgame perfect equilibrium of the stage game is $((\frac{5}{2}, L), \xi^L)$.

Why isn't the *outcome path* $((6, H), \xi^H(6))$ consistent with subgame perfection in the stage game? Note that there are two distinct deviations by the firm to consider: an *unobserved* deviation to $(6, L)$, and an *observed* deviation involving a price different from 6. In order to deter an observed deviation, we specify that consumers' believe that, in response to any price different from 6, the firm had chosen $Q = L$, leading to the best response $\tilde{\xi}_i$ given by

$$\tilde{\xi}_i(p) = \begin{cases} 1, & \text{if } p = 6 \text{ and } i \geq p, \text{ or } p \neq 6 \text{ and } i \geq p + 6, \\ 0, & \text{otherwise}, \end{cases}$$

implying aggregate demand

$$\hat{x}(p, \tilde{\xi}) = \begin{cases} 4, & \text{if } p = 6, \\ \max\{0, 4 - p\}, & p \neq 6. \end{cases}$$

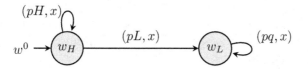

Figure 7.2.9: Grim trigger in the quality game. Note that the transitions are only a function of observed quality, $q \in \{L, H\}$.

Clearly, this implies that observable deviations by the firm are not profitable. Consider then the profile $((6, H), \tilde{\xi})$: the unobserved deviation to $(6, L)$ is profitable, since profits in this case are $(10 - 6)(6 - 1) = 20 > 16$. Note that for the deviation to be profitable, firm must still charge 6 (not the best response to ξ^H).

In the repeated game, there is a subgame perfect equilibrium with high quality: buyers believe H will be produced as long as H has been produced in the past. If ever L is produced, then L is expected to always be produced in future. See Figure 7.2.9.

It only remains to specify the decision rules:

$$f_1(w) = \begin{cases} (6, H), & \text{if } w = w_H, \text{ and} \\ (\frac{5}{2}, L), & \text{if } w = w_L. \end{cases}$$

and

$$f_2(w) = \begin{cases} \tilde{\xi}, & \text{if } w = w_H, \text{ and} \\ \xi^L, & \text{if } w = w_L. \end{cases}$$

Since the transitions are independent of price, the firm's price is myopically optimal in each state.

Since the consumers are small and myopically optimizing, in order to show that the profile is subgame perfect, it remains to verify that the firm is behaving optimally in each state. The firm value in each state is $V_1(w_q) = \pi^q$, $q \in \{L, H\}$. Trivially, L is optimal in w_L. Turning to w_H, we have

$$(1 - \delta)20 + \delta \frac{9}{4} \leq 16 \Leftrightarrow \delta \geq \frac{16}{71}.$$

There are many other equilibria.   ★

Remark 7.2.1 (Short-lived players). We can model the above as a game between one long-lived player and a sequence of short-lived players. In the stage game, the firm chooses p, and then the firm and consumer simultaneously choose quality $q \in \{L, H\}$, and quantity $x \in [0, 10]$, respectively. If the good is high quality, the consumer receives a utility of $10x - x^2/2$ from consuming x units. If the good is of low quality, his utility is reduced by 6 per unit, giving a utility of $4x - x^2/2$.[6] The consumer's utility is linear in money, so his payoffs are

$$u_c(Q, p) = \begin{cases} (4 - p)x - \frac{x^2}{2}, & \text{if } q = L, \text{ and} \\ (10 - p)x - \frac{x^2}{2}, & \text{if } q = H. \end{cases}$$

Since the period t consumer is *short-lived* (a new consumer replaces him next period), if he expects L in period t, then his best reply is to choose $x = x^L(p) := \max\{4 - p, 0\}$, while if he expects H, his best reply is choose $x = x^H(p) := \max\{10 - p, 0\}$. In other words, his behavior is just like the aggregate behavior of the continuum of consumers.

This is in general true: A short-lived player can typically represent a continuum of long-lived anonymous players.   ♦

7.3   Applications

7.3.1   Efficiency Wages I

Consider an employment relationship between a worker and a firm. Within the relationship, (i.e., in the stage game), the worker (player I) decides whether to exert effort (E) or shirk (S) for the firm (player II). Effort yields output y for sure, while shirking yields output 0 for sure. The firm chooses a wage w that period. At the end of the period, the firm observes output, equivalently effort, and the worker observes the wage. The payoffs in the stage game are given by

$$u_I(a_I, a_{II}) = \begin{cases} \text{w} - e, & \text{if } a_I = E \text{ and } a_{II} = \text{w}, \\ \text{w}, & \text{if } a_I = S \text{ and } a_{II} = \text{w}, \end{cases}$$

and

$$u_{II}(a_I, a_{II}) = \begin{cases} y - \text{w}, & \text{if } a_I = E \text{ and } a_{II} = \text{w}, \\ -\text{w}, & \text{if } a_I = S \text{ and } a_{II} = \text{w}. \end{cases}$$

Suppose

$$y > e.$$

---

[6]For $x > 4$, utility is declining in consumption. This can be avoided by setting his utility equal to $4x - x^2/2$ for $x \le 4$, and equal to 8 for all $x > 4$. This does not affect any of the relevant calculations.

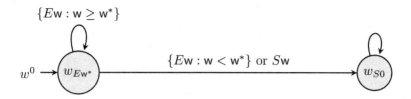

Figure 7.3.1: Grim Trigger for the employment relationship in Section 7.3.1. The transition from w_{Ew^*} labelled $\{Ew : w \geq w^*\}$ means any action profile in which the worker exerted effort and the firm paid at least w^*; the other transition from w_{Ew^*} occurs if either the firm underpaid ($w < w^*$), or the worker shirked (S).

Note that the stage game has $(S, 0)$ as the unique Nash equilibrium, with payoffs $(0, 0)$. This can also be interpreted as the payoffs from terminating this relationship (when both players receive a zero outside option).

Grim trigger at the wage w^* is illustrated in Figure 7.3.1.

Grim trigger is an equilibrium if

$$w^* - e \geq (1 - \delta)w^* \iff \delta w^* \geq e \tag{7.3.1}$$

$$\text{and} \quad y - w^* \geq (1 - \delta)y \iff \delta y \geq w^*. \tag{7.3.2}$$

Combining the worker (7.3.1) and firm (7.3.2) incentive constraints yields bounds on the equilibrium wage w^*:

$$\frac{e}{\delta} \leq w^* \leq \delta y, \tag{7.3.3}$$

Note that both firm and worker must receive a positive surplus from the relationship:

$$w^* - e \geq \frac{(1 - \delta)}{\delta} e > 0 \tag{7.3.4}$$

$$\text{and} \quad y - w^* \geq (1 - \delta)y > 0. \tag{7.3.5}$$

Inequality (7.3.1) (equivalently, (7.3.4)) can be interpreted as indicating that w^* is an *efficiency wage*: the wage is strictly higher than the disutility of effort (if workers are in excess supply, a naive market clearing model would suggest a wage of e). We return to this idea in Section 7.5.1.

Suppose now that there is a labor market where firms and workers who terminate one relationship can *costlessly* form a employment relationship

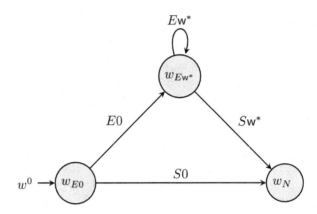

Figure 7.3.2: A symmetric profile for the employment relationship in Section 7.3.1, where the firm is committed to paying w* in every period except the initial period while the relationship lasts (and so transitions are not specified for irrelevant wages). The state w_N means start a new relationship; there are no transitions from this state in *this* relationship (w_N corresponds to w_{E0} in a new relationship).

with a new partner (perhaps there is a pool of unmatched firms and workers who costlessly match).

In particular, new matches are anonymous: it is not possible to treat partners differently on the basis of behavior of past behavior (since that is unobserved).

A specification of behavior is *symmetric* if all firms follow the same strategy and all workers follow the same strategy in an employment relationship. To simplify things, suppose also that firms commit to wage strategy (sequence) at the beginning of each employment relationship. Grim trigger at a constant wage w* satisfying (7.3.3) is not a symmetric equilibrium: After a deviation by the worker, the worker has an incentive to terminate the relationship and start a new one, obtaining the surplus (7.3.4) (as if no deviation had occurred).

Consider an alternative profile illustrated in Figure 7.3.2. Note that this profile has the flavor of being "renegotiation-proof." The firm is willing to commit at the beginning of the relationship to paying w* in every period (after the initial period, when no wage is paid) as long as effort is exerted if

$$y - \delta w^* \geq 0.$$

The worker has two incentive constraints. In state w_{E0}, the value to the worker is

$$-(1 - \delta)e + \delta(w^* - e) = \delta w^* - e =: V_I(w_{E0}).$$

The worker is clearly willing to exert effort in w_{E0} if

$$V_I(w_{E0}) \geq 0 \times (1 - \delta) + \delta V_I(w_{E0}),$$

that is

$$V_I(w_{E0}) \geq 0.$$

The worker is willing to exert effort in w_{Ew^*} if

$$w^* - e \geq (1 - \delta)w^* + \delta V_I(w_{E0})$$
$$= (1 - \delta + \delta^2)w^* - \delta e,$$

which is equivalent to

$$\delta w^* \geq e \iff V_I(w_{E0}) \geq 0.$$

A critical feature of this profile is that the worker must "invest" in the relationship: Before the worker can receive the ongoing surplus of the employment relationship, he/she must pay an upfront cost so the worker does not have an incentive to shirk in the current relationship and then restart with a new firm.

If firms must pay the same wage in every period including the initial period (for example, for legal reasons or social norms), then some other mechanism is needed to provide the necessary disincentive to separate. Frictions in the matching process (involuntary unemployment) is one mechanism. For more on this and related ideas, see Shapiro and Stiglitz (1984); MacLeod and Malcomson (1989); Carmichael and MacLeod (1997).

7.3.2 Collusion Under Demand Uncertainty

This example is of an oligopoly selling perfect substitutes. The demand curve is given by

$$Q = \omega - \min\{p_1, p_2, \ldots, p_n\},$$

where ω is the state of demand and p_i is firm i's price. Demand is evenly divided among the lowest pricing firms. Firms have zero constant marginal cost of production.

The stage game has a unique Nash equilibrium, in which firms price at 0, yielding each firm profits of 0, which is their minmax payoff.

In each period, the state ω is an independent and identical draw from the finite set $\Omega \subset \mathbb{R}_+$, according to the distribution $q \in \Delta(\Omega)$.

The monopoly price is given by $p^m(\omega) := \omega/2$, with associated monopoly profits of $\omega^2/4$.

We are interested in the strongly symmetric equilibrium in which firms jointly maximize expected profits. A profile $(s_1, s_2, \ldots, s_n)$ is *strongly symmetric* if for all histories h^t, $s_i(h^t) = s_j(h^t)$, i.e., after all histories (even asymmetric ones where firms have behaved differently), firms choose the same action.

Along the equilibrium path, firms set a common price $p(\omega)$, and any deviations are punished by perpetual minmax, i.e., grim trigger.[7]

Let v^* be the common expected payoff from such a strategy profile. The necessary and sufficient conditions for grim trigger to be an equilibrium are, for each state ω,

$$\frac{1}{n}(1-\delta)p(\omega)(\omega - p(\omega)) + \delta v^* \geq \sup_{p < p(\omega)} (1-\delta)p(\omega - p)$$
$$= (1-\delta)p(\omega)(\omega - p(\omega)), \qquad (7.3.6)$$

where

$$v^* = \frac{1}{n}\sum_{\omega' \in \Omega} p(\omega')(\omega' - p(\omega'))q(\omega').$$

Inequality (7.3.6) can be written as

$$p(\omega)(\omega - p(\omega)) \leq \frac{\delta n v^*}{(n-1)(1-\delta)}$$
$$= \frac{\delta}{(n-1)(1-\delta)} \sum_{\omega' \in \Omega} p(\omega')(\omega' - p(\omega'))q(\omega').$$

If there is no uncertainty over states (i.e, there exists ω' such that $q(\omega') = 1$), this inequality is independent of states (and the price $p(\omega)$), becoming

$$1 \leq \frac{\delta}{(n-1)(1-\delta)} \iff \frac{n-1}{n} \leq \delta. \qquad (7.3.7)$$

Suppose there are two equally likely states, $L < H$. In order to support collusion at the monopoly price in *each* state, we need:

$$\frac{L^2}{4} \leq \frac{\delta}{(n-1)(1-\delta)2}\left\{\frac{L^2}{4} + \frac{H^2}{4}\right\} \qquad (7.3.8)$$

---

[7]Strictly speaking, this example is not a repeated game with perfect monitoring. An action for firm i in the stage game is a vector $(p_i(\omega))_\omega$. At the end of the period, firms only observe the pricing choices of all firms at the *realized* ω. Nonetheless, the same theory applies. Subgame perfection is equivalent to one-shot optimality: a profile is subgame perfect if, conditional on each information set (in particular, conditional on the realized ω), it is optimal to choose the specified price, given the specified continuation play.

$$\text{and} \quad \frac{H^2}{4} \le \frac{\delta}{(n-1)(1-\delta)2} \left\{ \frac{L^2}{4} + \frac{H^2}{4} \right\}. \tag{7.3.9}$$

Since $H^2 > L^2$, the constraint in the high state (7.3.9) is the relevant one, and is equivalent to

$$\frac{2(n-1)H^2}{L^2 + (2n-1)H^2} \le \delta,$$

a tighter bound than (7.3.7), since the incentive to deviate is unchanged, but the threat of loss of future collusion is smaller.

Suppose

$$\frac{n-1}{n} < \delta < \frac{2(n-1)H^2}{L^2 + (2n-1)H^2},$$

so that colluding on the monopoly price in each state is inconsistent with equilibrium.

Since the high state is the state in which the firms have the strongest incentive to deviate, the most collusive equilibrium sets $p(L) = L/2$ and $p(H)$ to solve the following incentive constraint with equality:

$$p(H)(H - p(H)) \le \frac{\delta}{(n-1)(1-\delta)2} \left\{ \frac{L^2}{4} + p(H)(H - p(H)) \right\}.$$

In order to fix ideas, suppose $n(1-\delta) > 1$. Then, this inequality implies

$$p(H)(H - p(H)) \le \frac{1}{2} \left\{ \frac{L^2}{4} + p(H)(H - p(H)) \right\},$$

that is,

$$p(H)(H - p(H)) \le \frac{L^2}{4} = \frac{L}{2} \left(L - \frac{L}{2} \right).$$

Note that setting $p(H) = \frac{L}{2}$ violates this inequality (since $H > L$). Since profits are strictly concave, this inequality thus requires a lower price, that is,

$$p(H) < \frac{L}{2}.$$

In other words, if there are enough firms colluding ($n(1-\delta) > 1$), collusive pricing is counter-cyclical! This counter-cyclicality of prices also arises with more states and a less severe requirement on the number of firms (Mailath and Samuelson, 2006, §6.1.1).

7.4 Enforceability, Decomposability, and a Folk Theorem

This section gives a quick introduction to the Abreu, Pearce, and Stacchetti (1990) (often abbreviated to APS) method for characterizing equilibrium payoffs.

Definition 7.4.1. *An action profile* $a' \in A$ *is* enforced *by the continuation promises* $\gamma : A \to \mathbb{R}^n$ *if* a' *is a Nash equilibrium of the normal form game with payoff function* $g^\gamma : A \to \mathbb{R}^n$, *where*

$$g_i^\gamma(a) = (1 - \delta)u_i(a) + \delta\gamma_i(a).$$

A payoff v *is* decomposable *on a set of payoffs* $\mathcal{V} \subset \mathbb{R}^n$ *if there exists an action profile* a' *enforced by some continuation promises* $\gamma : A \to \mathcal{V}$ *satisfying, for all* i,

$$v_i = (1 - \delta)u_i(a') + \delta\gamma_i(a').$$

The notions of enforceability and decomposition play a central role in the construction of subgame perfect equilibria. When the other players play their part of an enforceable profile a'_{-i}, the continuation promises γ_i make (enforce) the choice of a'_i optimal (incentive compatible) for i. If a payoff vector v is decomposable on $\mathcal{V}$, then it is "one-period credible" with respect to promises in $\mathcal{V}$. If these promises are themselves decomposable on $\mathcal{V}$, then the payoff vector v is "two-period credible." If a set of payoffs is decomposable on itself, then payoff vector in the set is "infinite-period credible." It turns out that subgame perfect equilibrium payoffs characterize this property.

Fix a stage game $G = \{(A_i, u_i)\}$ and discount factor δ. Let $\mathcal{E}^p(\delta) \subset \mathcal{F}^{p*}$ be the set of pure strategy subgame perfect equilibrium payoffs. Problem 7.6.10 asks you to prove the following result.

Theorem 7.4.1. *A payoff* $v \in \mathbb{R}^n$ *is decomposable on* $\mathcal{E}^p(\delta)$ *if, and only if,* $v \in \mathcal{E}^p(\delta)$.

Example 7.4.1 (Repeated prisoner's dilemma, cont.). From Examples 7.1.1 and 7.1.4, we know that

$$\{(0,0), (3\delta - 1, 3 - \delta), (3 - \delta, 3\delta - 1), (2,2)\} \subset \mathcal{E}^p(\delta)$$

for $\delta \geq \frac{1}{3}$. Then $v = (2 - 3\delta + 3\delta^2, 2 + \delta - \delta^2)$ can be decomposed on $\mathcal{E}^p(\delta)$ by EE and γ given by

$$\gamma(a) = \begin{cases} (3\delta - 1, 3 - \delta), & a = EE, \\ (0,0), & a \neq EE, \end{cases}$$

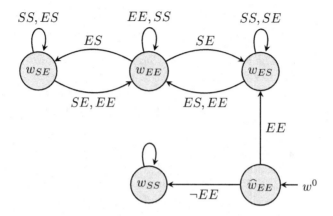

Figure 7.4.1: An automaton consistent with the decomposition of v in Example 7.4.1.

for $\delta \geq 1/\sqrt{3}$ (there are two incentive constraints to check; verify that player 1's is the binding one, and that this yields the indicated lower bound). Hence, v is a subgame perfect equilibrium payoff for $\delta \geq 1/\sqrt{3}$.

There are several automata consistent with this decomposition. Figure 7.4.1 displays one that builds on the automaton from Example 7.1.4. There are simpler ones (it is a good exercise to identify one). ★

There is a well-known "converse" to Theorem 7.4.1:

Theorem 7.4.2. *Suppose every payoff v in some set $\mathcal{V} \subset \mathbb{R}^n$ is decomposable with respect to $\mathcal{V}$. Then, $\mathcal{V} \subset \mathcal{E}^p(\delta)$.*

The proof of this result can be found in Mailath and Samuelson (2006) (Theorem 7.4.2 is Proposition 2.5.1). The essential steps appear in the proof of Lemma 7.4.2 below.

Any set of payoffs with the property described in Theorem 7.4.2 is said to be *self-generating*. Thus, every set of self-generating payoff is a set of subgame perfect equilibrium payoffs. Moreover, the set of subgame perfect equilibrium payoffs is the largest set of self-generating payoffs.

The notions of enforceability and decomposability, as well as Theorem 7.4.2, apply also to games with long and short-lived players.

The following famous result has a long history in game theory (its name comes from it being part of the folklore,[8] this form of the result was first

---

[8]The term "folk theorem" is now used to describe any result asserting that every

proved in Fudenberg and Maskin, 1986). For a proof, see Proposition 3.8.1 in Mailath and Samuelson (2006). Example 7.4.3 illustrates the key ideas.

Theorem 7.4.3 (The Folk Theorem). *Suppose A_i is finite for all i and $\mathcal{F}^*$ has nonempty interior in $\mathbb{R}^n$. For all $v \in \{\tilde{v} \in \mathcal{F}^* : \exists v' \in \mathcal{F}^*, v_i' < \tilde{v}_i \forall i\}$, there exists a sufficiently large discount factor δ', such that for all $\delta \geq \delta'$, there is a subgame perfect equilibrium of the infinitely repeated game whose average discounted value is v.*

Example 7.4.2 (Example 7.2.1, the product-choice game, cont.). Suppose both players 1 and 2 are long lived. For any $\varepsilon > 0$, there is a subgame perfect equilibrium in which player 1 has a payoff of $3\frac{2}{3} - \varepsilon$. But, if player 2 is short lived then the maximum payoff that player 1 can receive in any equilibrium is 3 (because player 2 cannot be induced to play c when player 1 plays L).[9]

                                                                    ★

Example 7.4.3 (Strongly symmetric folk theorem for the repeated prisoner's dilemma, Figure 7.1.1). We restrict attention to *strongly symmetric* strategies, i.e., for all $w \in \mathcal{W}$, $f_1(w) = f_2(w)$. When is $\{(v, v) : v \in [0, 2]\}$ a set of equilibrium payoffs? Since we have restricted attention to strongly symmetric equilibria, we can drop player subscripts. Note that the set of strongly symmetric equilibrium payoffs cannot be any larger than $[0, 2]$, since $[0, 2]$ is the largest set of feasible symmetric payoffs.

Two preliminary calculations (it is important to note that these preliminary calculations make *no* assumptions about $[0, 2]$ being a set of equilibrium payoffs):

1. Let $\mathcal{W}^{EE}$ be the set of player 1 payoffs that could be decomposed on $[0, 2]$ using EE (i.e., $\mathcal{W}^{EE}$ is the set of player 1 payoffs that could be enforceably achieved by EE followed by appropriate symmetric continuations in $[0, 2]$). Then $v \in \mathcal{W}^{EE}$ iff

$$v = 2(1 - \delta) + \delta\gamma(EE)$$
$$\geq 3(1 - \delta) + \delta\gamma(SE),$$

---

feasible and individually rational payoff is the payoff of some equilibrium. At times, it is also used to describe a result asserting that that every payoff in some "large" subset of feasible payoffs is the payoff of some equilibrium.

[9]The upper bound on pure strategy subgame perfect equilibrium payoffs follows immediately from the observation that in any such equilibrium, only the profiles Hc and Ls are played. If the equilibrium involves mixing, observe that in any period in which player 1 randomizes between H and L, 1 is indifferent between H and L, and 3 is the greatest payoff that 1 can get from H in that period.

for some $\gamma(EE), \gamma(SE) \in [0,2]$. The largest value for $\gamma(EE)$ is 2 and the smallest value for $\gamma(SE)$ is 0, so the incentive constraint implies the smallest value for $\gamma(EE)$ is $(1-\delta)/\delta$, so that $\mathcal{W}^{EE} = [3(1-\delta),\ 2]$. See Figure 7.4.2 for an illustration.

2. Let $\mathcal{W}^{SS}$ be the set of player 1 payoffs that could be decomposed on $[0,2]$ using SS. Then $v \in \mathcal{W}^{SS}$ iff

$$v = 0 \times (1 - \delta) + \delta\gamma(SS)$$
$$\geq (-1)(1 - \delta) + \delta\gamma(ES),$$

for some $\gamma(SS), \gamma(ES) \in [0,2]$. Since the inequality is satisfied by setting $\gamma(SS) = \gamma(ES)$, the largest value for $\gamma(SS)$ is 2, while the smallest is 0, and so $\mathcal{W}^{SS} = [0,\ 2\delta]$.

Observe that

$$[0,2] \supset \mathcal{W}^{SS} \cup \mathcal{W}^{EE} = [0, 2\delta] \cup [3(1-\delta), 2].$$

Lemma 7.4.1 (Necessity). *Suppose $[0,2]$ is the set of strongly symmetric strategy equilibrium payoffs. Then,*

$$[0,2] \subset \mathcal{W}^{SS} \cup \mathcal{W}^{EE}.$$

Proof. Suppose v is the payoff of some strongly symmetric strategy equilibrium s. Then either $s^0 = EE$ or SS. Since the continuation equilibrium payoffs must lie in $[0,2]$, we immediately have that if $s^0 = EE$, then $v \in \mathcal{W}^{EE}$, while if $s^0 = SS$, then $v \in \mathcal{W}^{SS}$. But this implies $v \in \mathcal{W}^{SS} \cup \mathcal{W}^{EE}$. So, if $[0,2]$ is the set of strongly symmetric strategy equilibrium payoffs, we must have

$$[0,2] \subset \mathcal{W}^{SS} \cup \mathcal{W}^{EE}.$$

∎

So, when is
$$[0,2] \subset \mathcal{W}^{SS} \cup \mathcal{W}^{EE}?$$

This holds if, and only if, $2\delta \geq 3(1 - \delta)$ (i.e., $\delta \geq \frac{3}{5}$).

Lemma 7.4.2 (Sufficiency). *If*

$$[0,2] = \mathcal{W}^{SS} \cup \mathcal{W}^{EE},$$

then $[0,2]$ is the set of strongly symmetric strategy equilibrium payoffs.

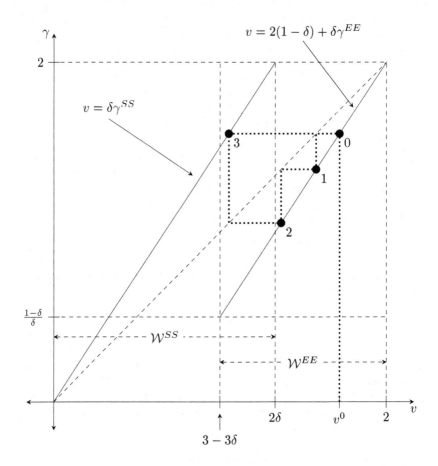

Figure 7.4.2: An illustration of the folk theorem. The continuations that enforce EE are labelled γ^{EE}, while those that enforce SS are labelled γ^{SS}. The value v^0 is the average discounted value of the equilibrium whose current value/continuation value is described by one period of EE, followed by the cycle $1 - 2 - 3 - 1$. In this cycle, play follows $EE, (EE, EE, SS)^\infty$. The Figure was drawn for $\delta = \frac{2}{3}$; $v^0 = \frac{98}{57}$.

Many choices of v^0 will *not* lead to a cycle.

Proof. Fix $v \in [0, 2]$, and define a recursion as follows: Set $\gamma^0 = v$, and

$$\gamma^{t+1} = \begin{cases} \gamma^t/\delta & \text{if } \gamma^t \in \mathcal{W}^{SS} = [0,\, 2\delta], \text{ and} \\ (\gamma^t - 2(1-\delta))/\delta & \text{if } \gamma^t \in \mathcal{W}^{EE} \setminus \mathcal{W}^{SS} = (2\delta,\, 2]. \end{cases}$$

Since $[0, 2] \subset \mathcal{W}^{SS} \cup \mathcal{W}^{EE}$, this recursive definition is well defined for all t. Moreover, since $\delta \geq \frac{3}{5}$, $\gamma^t \in [0, 2]$ for all t. The recursion thus yields a *bounded* sequence of continuations $\{\gamma^t\}_t$. Associated with this sequence of continuations is the outcome path $\{\tilde{a}^t\}_t$:

$$\tilde{a}^t = \begin{cases} EE & \text{if } \gamma^t \in \mathcal{W}^{EE} \setminus \mathcal{W}^{SS}, \text{ and} \\ SS & \text{if } \gamma^t \in \mathcal{W}^{SS}. \end{cases}$$

Observe that, by construction,

$$\gamma^t = (1-\delta)u_i(\tilde{a}^t) + \delta\gamma^{t+1}.$$

Consider the automaton $(\mathcal{W}, w^0, f, \tau)$ where

- $\mathcal{W} = [0, 2]$;

- $w^0 = v$;

- the output function is

$$f(w) = \begin{cases} EE & \text{if } w \in \mathcal{W}^{EE} \setminus \mathcal{W}^{SS}, \text{ and} \\ SS & \text{if } w \in \mathcal{W}^{SS}, \text{ and} \end{cases}$$

- the transition function is

$$\tau(w, a) = \begin{cases} (w - 2(1-\delta))/\delta & \text{if } w \in \mathcal{W}^{EE} \setminus \mathcal{W}^{SS} \text{ and } a = f(w), \\ w/\delta & \text{if } w \in \mathcal{W}^{SS} \text{ and } a = f(w), \text{ and} \\ 0, & \text{if } a \neq f(w). \end{cases}$$

The outcome path implied by this strategy profile is $\{\tilde{a}^t\}_t$. Moreover,

$$\begin{aligned} v = \gamma^0 &= (1-\delta)u_i(\tilde{a}^0) + \delta\gamma^1 \\ &= (1-\delta)u_i(\tilde{a}^0) + \delta\left\{(1-\delta)u_i(\tilde{a}^1) + \delta\gamma^2\right\} \\ &= (1-\delta)\sum_{t=0}^{T-1} \delta^t u_i(\tilde{a}^t) + \delta^T \gamma^T \\ &= (1-\delta)\sum_{t=0}^{\infty} \delta^t u_i(\tilde{a}^t) \end{aligned}$$

(where the last equality is an implication of $\delta < 1$ and the sequence $\{\gamma^T\}_T$ being bounded). Thus, the payoff of this outcome path is exactly v, that is, v is the payoff of the strategy profile described by the automaton $(\mathcal{W}, w^0, f, \tau)$ with initial state $w^0 = v$.

Thus, there is no profitable one-deviation from this automaton (this is guaranteed by the constructions of $\mathcal{W}^{SS}$ and $\mathcal{W}^{EE}$). Consequently the associated strategy profile is subgame perfect.   ■

See Mailath and Samuelson (2006, §2.5) for much more on this.

7.5   Imperfect Public Monitoring

7.5.1   Efficiency Wages II

A slight modification of the example from Section 7.3.1.[10] As before, in the stage game, the worker decides whether to exert effort (E) or to shirk (S) for the firm (player II). Effort has a disutility of e and yields output y for sure, while shirking yields output y with probability p, and output 0 with probability $1 - p$. The firm chooses a wage $\mathsf{w} \in \mathbb{R}_+$. At the end of the period, the firm does *not* observe effort, but does observe output.

Suppose

$$y - e > py$$

so it is efficient for the worker to exert effort.

The payoffs are described in Figure 7.5.1.

Consider the profile described by the automaton illustrated in Figure 7.5.2.

The value functions are

$$V_1(w_{S0}) = 0, \quad V_1(w_{E\mathsf{w}^*}) = \mathsf{w}^* - e,$$

$$V_2(w_{S0}) = py, \quad V_2(w_{E\mathsf{w}^*}) = y - \mathsf{w}^*,$$

In the absorbing state w_{S0}, play is the unique equilibrium of the stage game, and so incentives are trivially satisfied.

The worker does not wish to deviate in H if

$$V_1(w_{E\mathsf{w}^*}) \geq (1 - \delta)\mathsf{w}^* + \delta\{pV_1(w_{E\mathsf{w}^*}) + (1 - p) \times 0\},$$

i.e.,

$$\delta(1 - p)\mathsf{w}^* \geq (1 - \delta p)e$$

---

[10]This is also similar to example in Gibbons (1992, Section 2.3.D), but with the firm also facing an intertemporal trade-off.

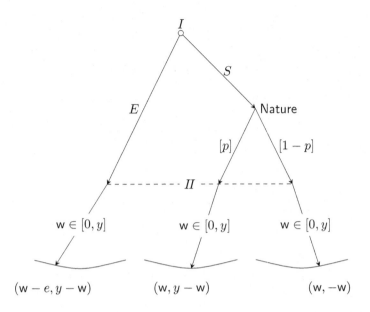

Figure 7.5.1: The extensive form and payoffs of the stage game for the game in Section 7.5.1.

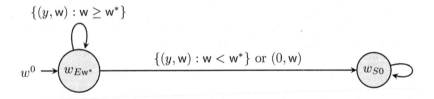

Figure 7.5.2: The automaton for the strategy profile in the repeated game in Section 7.5.1. The transition from the state $w_{E\text{w}^*}$ labelled $\{(y, \text{w}) : \text{w} \geq \text{w}^*\}$ means any signal profile in which output is observed and the firm paid at least w^*; the other transition from $w_{E\text{w}^*}$ occurs if either the firm underpaid ($\text{w} < \text{w}^*$), or no output is observed (0).

or

$$\mathsf{w}^* \geq \frac{1 - \delta p}{\delta(1 - p)}e = \frac{e}{\delta} + \frac{p(1 - \delta)}{\delta(1 - p)}e.$$

To understand the role of the imperfect monitoring, compare this with the analogous constraint when the monitoring is perfect (7.3.1), which requires $\mathsf{w}^* \geq e/\delta$.

The firm does not wish to deviate in $w_{E\mathsf{w}^*}$ if

$$V_2(w_{E\mathsf{w}^*}) \geq (1 - \delta)y + \delta py,$$

i.e.,

$$y - \mathsf{w}^* \geq (1 - \delta)y + \delta py \Leftrightarrow \delta(1 - p)y \geq \mathsf{w}^*.$$

So, the profile is an "equilibrium" if

$$\delta(1 - p)y \geq \mathsf{w}^* \geq \frac{1 - \delta p}{\delta(1 - p)}e.$$

In fact, it is an implication of the next section that the profile is a perfect public equilibrium.

7.5.2   Public Perfect Equilibria

As before, the action space for player i is A_i, with typical action $a_i \in A_i$. An action profile is denoted $a = (a_1, \ldots, a_n)$. At the end of each period, rather than observing the action profile a, all players observe a *public* signal y taking values in some space Y according to the distribution $\Pr\{y|\ (a_1, \ldots, a_n)\} := \rho(y|\ a)$.

Since the signal y is a possibly noisy signal of the action profile a in that period, the actions are *imperfectly monitored* by the other players. Since the signal is public (and so observed by all players), the game is said to have *public monitoring*.

Assume Y is finite.

Player i's *ex post* or realized payoff is given by $u_i^* : A_i \times Y \to \mathbb{R}$. The assumption that each player's payoff only depends upon other players' action through the realized value of the public signal guarantees that the public signal contains all relevant information about the play of other players.

Player i's payoffs do depend upon the behavior of the other players, since the distribution of the public signal depends upon the chosen action profile. Player i's stage game (*ex ante*) payoff is given by

$$u_i(a) := \sum_{y \in Y} u_i^*(a_i, y)\rho(y|\ a).$$

Since players have private information (their own past action choices), a player's information sets are naturally isomorphic to the set of their own private histories. Unlike for games with perfect monitoring (see Remark 7.1.1), there is no continuation game induced by any player history. In order to recover a recursive structure, we focus on behavior that only depends on public histories.

The set of *public histories* is

$$H := \cup_{t=0}^{\infty} Y^t,$$

with $h^t := (y^0, \ldots, y^{t-1})$ being a t period history of public signals ($Y^0 := \{\varnothing\}$).

A strategy is *public* if it specifies the action choice as a function of only public histories,

$$s_i : H \to A_i.$$

Definition 7.5.1. *A* perfect public equilibrium (PPE) *is a profile of public strategies* s *that, after observing any public history* h^t, *specifies a Nash equilibrium for the repeated game, i.e., for all* t *and all* $h^t \in Y^t$, $s|_{h^t}$ *is a Nash equilibrium.*

It is worth emphasizing that in Definition 7.5.1, we do not restrict deviations to public strategies. Nonetheless, if a player does have a profitable deviation from a public profile, there is a profitable deviation to a public strategy. Moreover, the restriction to public strategies is without loss of generality for pure strategies, which are the only strategies we will consider (see Problems 7.6.15 and 7.6.16).[11]

If $\rho(y|a) > 0$ for all y and a, every public history arises with positive probability, and so every Nash equilibrium in public strategies is a perfect public equilibrium.

7.5.3   Automata

Automaton representation of public strategies: $(\mathcal{W}, w^0, f, \tau)$, where

- $\mathcal{W}$ is set of states,

- w^0 is initial state,

- $f : \mathcal{W} \to A$ is output function (decision rule), and

- $\tau : \mathcal{W} \times Y \to \mathcal{W}$ is transition function.

---

[11]There are examples of mixed equilibria in repeated games of public monitoring whose behavior cannot be described by public strategies (Mailath and Samuelson, 2006, Section 10.3).

	$\bar{y}$	y
E	$\dfrac{(3-p-2q)}{(p-q)}$	$-\dfrac{(p+2q)}{(p-q)}$
S	$\dfrac{3(1-r)}{(q-r)}$	$-\dfrac{3r}{(q-r)}$

Figure 7.5.3: The ex post payoffs for the imperfect public monitoring version of the prisoner's dilemma from Figure 7.1.1.

As before, $V_i(w)$ is i's value of being in state w.

Lemma 7.5.1. *Suppose the strategy profile s is represented by* $(\mathcal{W}, w^0, f, \tau)$. *Then s is a PPE if, and only if, for all* $w \in \mathcal{W}$ *(satisfying* $w = \tau(w^0, h^t)$ *for some* $h^t \in H$), $f(w)$ *is a Nash equilibrium of the normal form game with payoff function* $g^w : A \to \mathbb{R}^n$, *where*

$$g_i^w(a) = (1 - \delta)u_i(a) + \delta \sum_y V_i(\tau(w, y))\rho(y \mid a).$$

See Problem 7.6.17 for the proof (another instance of the one-shot deviation principle).

Example 7.5.1 (noisy monitoring version of repeated prisoner's dilemma). Effort determines output $y \in \{\underline{y}, \bar{y}\}$ stochastically according to the distribution

$$\Pr\{\bar{y} \mid a\} := \rho(\bar{y} \mid a) = \begin{cases} p, & \text{if } a = EE, \\ q, & \text{if } a = SE \text{ or } ES, \\ r, & \text{if } a = SS, \end{cases}$$

where $0 < q < p < 1$ and $0 < r < p$.

The ex post payoffs (u_i^*) implying the ex ante payoffs (u_i) given in Figure 7.1.1 are given in Figure 7.5.3.   ★

Example 7.5.2 (One period memory). Consider the following two state automaton: $\mathcal{W} = \{w_{EE}, w_{SS}\}$, $w^0 = w_{EE}$, $f(w_{EE}) = EE$, $f(w_{SS}) = SS$, and

$$\tau(w, y) = \begin{cases} w_{EE}, & \text{if } y = \bar{y}, \\ w_{SS}, & \text{if } y = \underline{y}. \end{cases}$$

The automaton is presented in Figure 7.5.4.

Value functions (I can drop player subscripts by symmetry):

$$V(w_{EE}) = (1 - \delta) \cdot 2 + \delta\{pV(w_{EE}) + (1 - p)V(w_{SS})\}$$

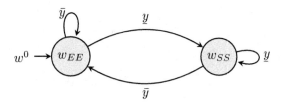

Figure 7.5.4: The automaton for Example 7.5.2.

and
$$V(w_{SS}) = (1 - \delta) \cdot 0 + \delta\{rV(w_{EE}) + (1 - r)V(w_{SS})\}.$$
The automaton describes a PPE if
$$V(w_{EE}) \geq (1 - \delta) \cdot 3 + \delta\{qV(w_{EE}) + (1 - q)V(w_{SS})\}$$
and
$$V(w_{SS}) \geq (1 - \delta) \cdot (-1) + \delta\{qV(w_{EE}) + (1 - q)V(w_{SS})\}.$$
Rewriting the incentive constraint at w_{EE},

$$(1 - \delta) \cdot 2 + \delta\{pV(w_{EE}) + (1 - p)V(w_{SS})\}$$
$$\geq (1 - \delta) \cdot 3 + \delta\{qV(w_{EE}) + (1 - q)V(w_{SS})\}$$

or
$$\delta(p - q)\{V(w_{EE}) - V(w_{SS})\} \geq (1 - \delta).$$

We can obtain an expression for $V(w_{EE}) - V(w_{SS})$ without solving for the value functions separately by differencing the value recursion equations, yielding

$$V(w_{EE}) - V(w_{SS}) = (1 - \delta) \cdot 2 + \delta\{pV(w_{EE}) + (1 - p)V(w_{SS})\}$$
$$- \delta\{rV(w_{EE}) + (1 - r)V(w_{SS})\}$$
$$= (1 - \delta) \cdot 2 + \delta(p - r)\{V(w_{EE}) - V(w_{SS})\},$$

so that
$$V(w_{EE}) - V(w_{SS}) = \frac{2(1 - \delta)}{1 - \delta(p - r)},$$

and so
$$\delta \geq \frac{1}{3p - 2q - r}. \tag{7.5.1}$$

Turning to w_{SS}, we have

$$\delta\{rV(w_{EE}) + (1-r)V(w_{SS})\}$$
$$\geq (1-\delta)\cdot(-1) + \delta\{qV(w_{EE}) + (1-q)V(w_{SS})\}$$

or

$$(1-\delta) \geq \delta(q-r)\{V(w_{EE}) - V(w_{SS})\},$$

requiring

$$\delta \leq \frac{1}{p+2q-3r}. \tag{7.5.2}$$

Note that (7.5.2) is trivially satisfied if $r \geq q$ (make sure you understand why this is intuitive).

The two bounds (7.5.1) and (7.5.2) on δ are consistent if

$$p \geq 2q - r.$$

The constraint $\delta \in (0,1)$ in addition requires

$$3r - p < 2q < 3p - r - 1. \tag{7.5.3}$$

Solving for the value functions,

$$\begin{bmatrix} V(w_{EE}) \\ V(w_{SS}) \end{bmatrix} = (1-\delta) \begin{bmatrix} 1-\delta p & -\delta(1-p) \\ -\delta r & 1-\delta(1-r) \end{bmatrix}^{-1} \begin{bmatrix} 2 \\ 0 \end{bmatrix}$$

$$= \frac{(1-\delta)}{(1-\delta p)(1-\delta(1-r)) - \delta^2(1-p)r} \times$$
$$\begin{bmatrix} 1-\delta(1-r) & \delta(1-p) \\ \delta r & 1-\delta p \end{bmatrix} \begin{bmatrix} 2 \\ 0 \end{bmatrix}$$

$$= \frac{(1-\delta)}{(1-\delta)(1-\delta(p-r))} \begin{bmatrix} 2(1-\delta(1-r)) \\ 2\delta r \end{bmatrix}$$

$$= \frac{1}{1-\delta(p-r)} \begin{bmatrix} 2(1-\delta(1-r)) \\ 2\delta r \end{bmatrix}. \qquad \bigstar$$

Example 7.5.3 (Bounds on PPE payoffs). For the one-period memory automaton from Example 7.5.2, we have for fixed p and r,

$$\lim_{\delta \to 1} V(w_{EE}) = \lim_{\delta \to 1} V(w_{SS}) = \frac{2r}{1-p+r}, \tag{7.5.4}$$

and, for $r > 0$,

$$\lim_{p \to 1} \lim_{\delta \to 1} V(w_{EE}) = 2.$$

In contrast, grim trigger (where one realization of y results in permanent SS) has a limiting (as $\delta \to 1$) payoff of 0 (see Problem 7.6.18(a)). Higher payoffs can be achieved by considering more forgiving versions, such as in Problem 7.6.18(b). Intuitively, as δ gets large, the degree of forgiveness should also increase.

This raises the question of what is the best payoff that can be achieved in *any* strongly symmetric pure strategy public perfect equilibria. Let $\bar{v}^\delta$ be the maximum of strongly symmetric pure strategy public perfect equilibrium payoffs.[12] Then, in any PPE with payoffs $\bar{v}^\delta$, EE must be played in the first period and neither player should have an incentive to play S.[13] Letting $\underline{v}^\delta$ be the continuation value after y in the equilibrium, we have

$$\bar{v}^\delta = 3(1 - \delta) + \delta\{p\bar{v}^\delta + (1 - p)\underline{v}^\delta\} \qquad (7.5.5)$$
$$\geq 3(1 - \delta) + \delta\{q\bar{v}^\delta + (1 - q)\underline{v}^\delta\}. \qquad (7.5.6)$$

The inequality (7.5.6) implies

$$\underline{v}^\delta \leq \bar{v}^\delta - \frac{(1 - \delta)}{\delta(p - q)}.$$

Substituting into (7.5.5) and simplifying, we obtain

$$\bar{v}^\delta \leq 2 - \frac{(1 - p)}{(p - q)}.$$

Thus, independent of the discount factor, the payoff in every strongly symmetric PPE is bounded away from the payoff from EE. The imperfection in the monitoring (and restriction to strong symmetry) necessarily implies an efficiency loss.

Problem 7.6.21 asks you to prove that this upper bound is in fact tight for large δ (i.e, for sufficiently large δ, there is a strongly symmetric PPE whose payoff equals the upper bound).

Finally, the upper bound is consistent with (7.5.4) (use (7.5.3)).   ★

---

[12]Since we have not proved that the set of PPE payoffs is compact, we should define $\bar{v}^\delta$ as the supremum rather than maximum. The argument that follows can accommodate defining $\bar{v}^\delta$ as the supremum at the cost of introducing ε's at various points and making the logic opaque. So, I treat $\bar{v}^\delta$ as a maximum. Moreover, the set of PPE payoffs is compact (Mailath and Samuelson, 2006, Corollary 7.3.1), so this is without loss of generality.

[13]In a strongly symmetric equilibrium, payoffs are higher when the first action profile is EE.

Remark 7.5.1. The notion of PPE only imposes ex ante incentive constraints. If the stage game has a non-trivial dynamic structure, such as in Problem 7.6.20, then it is natural to impose additional incentive constraints.

♦

7.6 Problems

7.6.1. Suppose $G := \{(A_i, u_i)\}$ is an n-person normal form game and G^T is its T-fold repetition (with payoffs evaluated as the average). Let $A := \prod_i A_i$. The strategy profile, s, is *history independent* if for all i and all $1 \leq t \leq T - 1$, $s_i(h^t)$ is independent of $h^t \in A^t$ (i.e., $s_i(h^t) = s_i(\hat{h}^t)$ for all $h^t, \hat{h}^t \in A^t$). Let $N(1)$ be the set of Nash equilibria of G. Suppose s is history independent. Prove that s is a subgame perfect equilibrium *if and only if* $s(h^t) \in N(1)$ for all t, $0 \leq t \leq T - 1$ and all $h^t \in A^t$ ($s(h^0)$ is of course simply s^0). Provide examples to show that the assumption of history independence is needed in both directions.

7.6.2. Prove the infinitely repeated game with stage game given by matching pennies does not have a pure strategy Nash equilibrium for any δ.

7.6.3. Suppose $(\mathcal{W}, w^0, f, \tau)$ is a (pure strategy representing) finite automaton with $|\mathcal{W}| = K$. Label the states from 1 to K, so that $\mathcal{W} = \{1, 2, \ldots, K\}$, $f : \{1, 2, \ldots, K\} \to A$, and $\tau : \{1, 2, \ldots, K\} \times A \to \{1, 2, \ldots, K\}$. Consider the function $\Phi : \mathbb{R}^K \to \mathbb{R}^K$ given by $\Phi(v) = (\Phi_1(v), \Phi_2(v), \ldots, \Phi_K(v))$, where

$$\Phi_k(v) = (1 - \delta)u_i(f(k)) + \delta v_{\tau(k, f(k))}, \qquad k = 1, \ldots, K.$$

(a) Prove that Φ has a unique fixed point. [**Hint:** Show that Φ is a contraction.]

(b) Given an explicit equation for the fixed point of Φ.

(c) Interpret the fixed point.

7.6.4. This problem outlines a direct proof that $\widetilde{V}_i$ satisfies (7.1.2) (i.e., of the principle of optimality). For any $w \in \mathcal{W}$, denote by $s_j(w)$ player j's repeated game strategy described by the automaton $(\mathcal{W}, w, f_{-i}, \tau)$.

(a) Prove that player i has a best response to any strategy profile s_{-i}. That is, if we set

$$\tilde{V}_i(w) := \sup_{s_i \in S_i} (1 - \delta) \sum_{t=0}^{\infty} \delta^t u_i(a(s_i, s_{-i})) = \sup_{s_i \in S_i} U_i(s_i, s_{-i}),$$

prove that the sup can be replaced by a max. [**Hint:** The supremum is finite since A is finite. Fix w and let s_i^k be a strategy satisfying $U_i(s_i^k, s_{-i}(w))) > \tilde{V}_i(w) - 1/k$. Consider the sequence $(s_i^k)_{k=1}^{\infty}$. Since A_i is finite, there is a subsequence on which $(s_i^k(\varnothing))$ is constant. Extract further similar subsubsequences by period.]

(b) For all $w \in \mathcal{W}$, denote by $\tilde{s}_i(w)$ a best response to $s_{-i}(w)$ (whose existence is guaranteed by part (a)), so that

$$\tilde{V}_i(w) = (1 - \delta) \sum_{t=0}^{\infty} \delta^t u_i(a(\tilde{s}_i(w), s_{-i}(w))).$$

Prove that for any $w \in \mathcal{W}$,

$$\tilde{V}_i(w) \geq \max_{a_i \in A_i} \left\{ (1 - \delta) u_i(a_i, f_{-i}(w)) + \delta \tilde{V}_i(\tau(w, (a_i, f_{-i}(w)))) \right\}.$$

(c) Complete the verification that that $\tilde{V}_i$ satisfies (7.1.2).

7.6.5. A different proof (that can be generalized to more general infinite horizon games without a recursive structure) of the hard direction of Theorem 7.1.3 is the following:

Suppose $(s_1, \ldots, s_n)$ (with representing automaton $(\mathcal{W}, w^0, f, \tau)$) is not subgame perfect.

(a) Prove there is a history $\tilde{h}^{t'}$ and a player i with profitable deviation $\bar{s}_i$ that only disagrees with $s_i|_{\tilde{h}^{t'}}$ for the first T periods.

(b) Prove that either player i has a one shot deviation from $\bar{s}_i$, or player i has a different "$T - 1$"-period deviation. (The proof is then completed by induction).

7.6.6. Prove that the profile described in Example 7.1.5 is not a Nash equilibrium if $\delta \in [1/3, 1/2)$. [Hint: What is the payoff from always playing A?] Prove that it is Nash if $\delta \in [1/2, 1)$.

7.6.7. Suppose two players play the infinitely-repeated prisoner's dilemma displayed in Figure 7.6.1(a).

(a)

	E	S
E	1,1	$-\ell, 1+g$
S	$1+g, -\ell$	0,0

(b)

	E	S
E	2,2	$-1,5$
S	$5,-1$	0,0

Figure 7.6.1: Two prisoner's dilemmas. (a) A general prisoner's dilemma, where $\ell > 0$ and $g > 0$. (b) The prisoner's dilemma for Problem 7.6.8.

(a) For what values of the discount factor δ is grim trigger a subgame perfect equilibrium?

(b) Describe a simple automaton representation of the behavior in which player I alternates between E and S (beginning with E), player II always plays E, and any deviation results in permanent SS. For what parameter restrictions is this a subgame perfect equilibrium?

(c) For what parameter values of ℓ, g, and δ is tit-for-tat a subgame perfect equilibrium?

7.6.8. In this question, we reconsider the profile given in Figure 7.1.5, but for the prisoner's dilemma given in Figure 7.6.1(b).

(a) Prove that the profile given in Figure 7.1.5 is not a subgame perfect equilibrium for any δ.

(b) Prove that the profile given in Figure 7.6.2 is a subgame perfect equilibrium for large δ. What is the appropriate bound for δ?

7.6.9. Suppose the game in Figure 7.6.3 is infinitely repeated: Let δ denote the common discount factor for both players and consider the strategy profile that induces the outcome path $DL, UR, DL, UR, \cdots$, and that, after any unilateral deviation by the row player specifies the outcome path $DL, UR, DL, UR, \cdots$, and after any unilateral deviation by the column player, specifies the outcome path $UR, DL, UR, DL, \cdots$ (simultaneous deviations are ignored, i.e., are treated as if neither player had deviated).

(a) What is the simplest automaton that represents this strategy profile?

(b) Suppose $x = 5$. For what values of δ is this strategy profile subgame perfect?

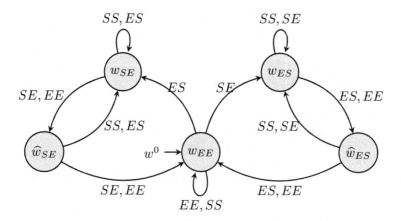

Figure 7.6.2: The automaton for Problem 7.6.8.

	L	R
U	2, 2	x, 0
D	0, 5	1, 1

Figure 7.6.3: The game for Problem 7.6.9.

	E	S
E	1, 2	−1, 3
S	2, −4	0, 0

Figure 7.6.4: The game for Problem 7.6.13.

(c) Suppose now $x = 4$. How does this change your answer to part (b)?

(d) Suppose $x = 5$ again. How would the analysis in part (b) be changed if the column player were short-lived (lived for only one period)?

7.6.10. Fix a stage game $G = \{(A_i, u_i)\}$ and discount factor δ. Let $\mathcal{E}^p(\delta) \subset \mathcal{F}^{p*}$ be the set of pure strategy subgame perfect equilibrium payoffs.

(a) Prove that every payoff $v \in \mathcal{E}^p(\delta)$ is decomposable on $\mathcal{E}^p(\delta)$.

(b) Suppose $\gamma : A \to \mathcal{E}^p(\delta)$ enforces the action profile a'. Describe a subgame perfect equilibrium in which a' is played in the first period.

(c) Prove that every payoff $v \in \mathcal{F}^{p*}$ decomposable on $\mathcal{E}^p(\delta)$ is in $\mathcal{E}^p(\delta)$.

7.6.11. Consider the prisoner's dilemma from Figure 7.1.1. Suppose the game is infinitely repeated with perfect monitoring. Recall that a *strongly symmetric* strategy profile (s_1, s_2) satisfies $s_1(h^t) = s_2(h^t)$ for all h^t. Equivalently, its automaton representation satisfies $f_1(w) = f_2(w)$ for all w. Let $\mathcal{W} = \{\delta v, v\}$, $v > 0$ to be determined, be the set of continuation promises. Describe a strongly symmetric strategy profile (equivalently, automaton) whose continuation promises come from $\mathcal{W}$ which is a subgame perfect equilibrium for some values of δ. Calculate the appropriate bounds on δ and the value of v (which may or may not depend on δ).

7.6.12. Describe the five state automaton that yields v^0 as a strongly symmetric equilibrium payoff with the indicated cycle in Figure 7.4.2.

7.6.13. Consider the (asymmetric) prisoner's dilemma in Figure 7.6.4. Suppose the game is infinitely repeated with perfect monitoring. Prove that for $\delta < \frac{1}{2}$, the maximum (average discounted) payoff to player 1 in any pure strategy subgame perfect equilibrium is 0, while for

$$
\begin{array}{c|c|c|}
 & L & R \\
\hline
T & 2,3 & 0,2 \\
\hline
B & 3,0 & 1,1 \\
\hline
\end{array}
$$

Figure 7.6.5: The game for Problem 7.6.14.

$\delta = \frac{1}{2}$, there are equilibria in which player 1 receives a payoff of 1. [**Hint:** First prove that, if $\delta \leq \frac{1}{2}$, in any pure strategy subgame perfect equilibrium, in any period, if player 2 chooses E then player 1 chooses E in that period.]

7.6.14. Consider the stage game in Figure 7.6.5, where player 1 is the row player and 2, the column player (as usual).

(a) Suppose the game is infinitely repeated, with perfect monitoring. Players 1 and 2 are both long-lived, and have the same discount factor, $\delta \in (0,1)$. Construct a three state automaton that for large δ is a subgame perfect equilibrium, and yields a payoff to player 1 that is close to $2\frac{1}{2}$. Prove that the automaton has the desired properties. [**Hint:** One state is only used off the path-of-play.]

(b) Now suppose that player 2 is short-lived (but maintain the assumption of perfect monitoring, so that the short-lived player in period t knows the entire history of actions up to t). Prove that player 1's payoff in any pure strategy subgame perfect equilibrium is no greater than 2 (the restriction to pure strategy is not needed—can you prove the result without that restriction?). For which values of δ is there a pure strategy subgame perfect equilibrium in which player 1 receives a payoff of precisely 2?

7.6.15. Suppose all players other than i are playing a public strategy in a repeated finite game of imperfect public monitoring. Prove that player i has a public strategy as a best reply. [**Hint:** Prove that player i faces a Markov decision problem with states given by the public histories.]

7.6.16. Prove that every pure strategy in a repeated game of imperfect public monitoring has a realization equivalent public pure strategy.

7.6.17. Fix a repeated finite game of imperfect public monitoring (as usual, assume Y is finite). Say that a player has a *profitable one-shot devi-*

$$\begin{array}{c|c|c}
 & h & \ell \\
\hline
H & 4,3 & 0,2 \\
\hline
L & x,0 & 3,1 \\
\end{array}$$

Figure 7.6.6: The game for Problem 7.6.19.

ation from the public strategy $(\mathcal{W}, w^0, f, \tau)$ if there is some history $h^t \in Y^t$ and some action $a_i \in A_i$ such that (where $w = \tau(w^0, h^t)$)

$$V_i(w) < (1 - \delta)u_i(a_i, f_{-i}(w)) + \delta \sum_y V_i(\tau(w, y))\rho(y \mid (a_i, f_{-i}(w))).$$

(a) Prove that a public strategy profile is a perfect public equilibrium if and only if there are no profitable one-shot deviations. (This is yet another instance of the one-shot deviation principle).

(b) Prove Lemma 7.5.1.

7.6.18. Consider the prisoner's dilemma game in Example 7.5.1.

(a) For what parameter values is the grim-trigger profile an equilibrium?

(b) An example of a forgiving grim-trigger profile is described by the automaton $(\widehat{\mathcal{W}}, \hat{w}^0, \hat{f}, \hat{\tau})$, where $\widehat{\mathcal{W}} = \{\hat{w}_{EE}, \hat{w}'_{EE}, \hat{w}_{SS}\}$, $\hat{w}^0 = \hat{w}_{EE}$, $\hat{f}(w_a) = a$, and

$$\hat{\tau}(w, y) = \begin{cases} \hat{w}_{EE}, & \text{if } w = \hat{w}_{EE} \text{ or } \hat{w}'_{EE}, \text{ and } y = \bar{y}, \\ \hat{w}'_{EE}, & \text{if } w = \hat{w}_{EE} \text{ and } y = \underline{y}, \\ \hat{w}_{SS}, & \text{otherwise.} \end{cases}$$

For what parameter values is this forgiving grim-trigger profile an equilibrium? Compare the payoffs of grim trigger and this forgiving grim trigger when both are equilibria.

7.6.19. Player 1 (the row player) is a firm who can exert either high effort (H) or low effort (L) in the production of its output. Player 2 (the column player) is a consumer who can buy either a high-priced product, h, or a low-priced product ℓ. The actions are chosen simultaneously, and payoffs are given in Figure 7.6.6. Player 1 is infinitely lived, discounts the future with discount factor δ, and plays the above game in every period with a different consumer (i.e., each consumer lives only one period). The game is one of *public monitoring*: while the actions of the consumers are public, the actions of the firm are not. Both the high-priced and low-priced products are *experience* goods of random quality, with the distribution over quality determined by the effort choice. The consumer learns the quality of the product after purchase (consumption). Denote by $\bar{y}$ the event that the product purchased is high quality, and by $\underline{y}$ the event that it is low quality (in other words, $y \in \{\underline{y}, \bar{y}\}$ is the quality signal). Assume the distribution over quality is independent of the price of the product:

$$\Pr(\bar{y} \mid a) = \begin{cases} p, & \text{if } a_1 = H, \\ q, & \text{if } a_1 = L, \end{cases}$$

with $0 < q < p < 1$.

(a) Describe the ex post payoffs for the consumer. Why can the ex post payoffs for the firm be taken to be the ex ante payoffs?

(b) Suppose $x = 5$. Describe a perfect public equilibrium in which the patient firm chooses H infinitely often with probability one, and verify that it is an equilibrium. [**Hint:** This can be done with one-period memory.]

(c) Suppose now $x \geq 8$. Is the one-period memory strategy profile still an equilibrium? If not, can you think of an equilibrium in which H is still chosen with positive probability?

7.6.20. A financial manager undertakes an infinite sequence of trades on behalf of a client. Each trade takes one period. In each period, the manager can invest in one of a large number of risky assets. By exerting effort ($a = E$) in a period (at a cost of $e > 0$), the manager can identify the most profitable risky asset for that period, which generates a high return of $R = H$ with probability p and a low return $R = L$ with probability $1 - p$. In the absence of effort ($a = S$), the manager cannot distinguish between the different risky assets. For simplicity, assume the manager then chooses the wrong asset, yielding the low return $R = L$ with probability 1; the cost of no effort is 0. In each period, the client chooses the level of the fee $x \in [0, \bar{x}]$ to be paid to the financial manager for that period. Note that there is an exogenous upper bound $\bar{x}$ on the fee that can be paid in a period. The client and financial manager are risk neutral, and so the client's payoff in a period is

$$u_c(x, R) = R - x,$$

while the manager's payoff in a period is

$$u_m(x,a) = \begin{cases} x - e, & \text{if } a = E, \\ x, & \text{if } a = S. \end{cases}$$

The client and manager have a common discount factor δ. The client observes the return on the asset prior to paying the fee, but does not observe the manager's effort choice.

(a) Suppose the client cannot sign a binding contract committing him to pay a fee (contingent or not on the return). Describe the unique sequentially rational equilibrium when the client uses the manager for a single transaction. Are there any other Nash equilibria?

(b) Continue to suppose there are no binding contracts, but now consider the case of an infinite sequence of trades. For a range of values for the parameters $(\delta, \bar{x}, e, p, H, \text{ and } L)$, there is a perfect public equilibrium in which the manager exerts effort on behalf of the client in *every* period. Describe it and the restrictions on parameters necessary and sufficient for it to be an equilibrium.

(c) Compare the fee paid in your answer to part (b) to the fee that would be paid by a client for a single transaction,

   i. when the client *can* sign a legally binding commitment to a fee schedule as a function of the return of that period, and

   ii. when the client can sign a legally binding commitment to a fee schedule as a function of effort.

(d) Redo part (b) assuming that the client's choice of fee level and the manager's choice of effort are simultaneous, so that the fee paid in period t cannot depend on the return in period t. Compare your answer with your answer to part (b).

7.6.21. In this question, we revisit the partnership game of Example 7.5.1. Suppose $3p - 2q > 1$. This question asks you to prove that for sufficiently large δ, any payoff in the interval $[0, \bar{v}]$, is the payoff of some strongly symmetric PPE equilibrium, where

$$\bar{v} = 2 - \frac{(1-p)}{(p-q)},$$

and that no payoff larger than $\bar{v}$ is the payoff of some strongly symmetric pure strategy PPE equilibrium. Strong symmetry implies it is enough to focus on player 1, and the player subscript will often be omitted.

(a) The action profile SS is trivially *enforced* by any constant continuation $\gamma \in [0, \bar{\gamma}]$ independent of y. Let $\mathcal{W}^{SS}$ be the set of values that can be obtained by SS and a constant continuation $\gamma \in [0, \bar{\gamma}]$, i.e.,

$$\mathcal{W}^{SS} = \{(1 - \delta)u_1(SS) + \delta\gamma : \gamma \in [0, \bar{\gamma}]\}.$$

Prove that $\mathcal{W}^{SS} = [0, \delta\bar{\gamma}]$. [This is almost immediate.]

(b) Recalling Definition 7.4.1, say that v is *decomposed* by EE on $[0, \bar{\gamma}]$ if there exists $\gamma^{\bar{y}}, \gamma^{\underline{y}} \in [0, \bar{\gamma}]$ such that

$$v =(1 - \delta)u_1(EE) + \delta\{p\gamma^{\bar{y}} + (1 - p)\gamma^{\underline{y}}\} \tag{7.6.1}$$
$$\geq(1 - \delta)u_1(SE) + \delta\{q\gamma^{\bar{y}} + (1 - q)\gamma^{\underline{y}}\}. \tag{7.6.2}$$

(That is, EE is *enforced* by the continuation promises $\gamma^{\bar{y}}, \gamma^{\underline{y}}$ *and* implies the value v.) Let $\mathcal{W}^{EE}$ be the set of values that can be decomposed by EE on $[0, \bar{\gamma}]$. It is clear that $\mathcal{W}^{EE} = [\gamma', \gamma'']$, for some γ' and γ''. Calculate γ' by using the smallest possible choices of $\gamma^{\bar{y}}$ and $\gamma^{\underline{y}}$ in the interval $[0, \bar{\gamma}]$ to enforce EE. (This will involve having the inequality (7.6.2) holding with equality.)

(c) Similarly, give an expression for γ'' (that will involve $\bar{\gamma}$) by using the largest possible choices of $\gamma^{\bar{y}}$ and $\gamma^{\underline{y}}$ in the interval $[0, \bar{\gamma}]$ to enforce EE. Argue that $\delta\bar{\gamma} < \gamma''$.

(d) As in Example 7.4.3, we would like all continuations in $[0, \bar{\gamma}]$ to be themselves decomposable using continuations in $[0, \bar{\gamma}]$, i.e., we would like

$$[0, \bar{\gamma}] \subset \mathcal{W}^{SS} \cup \mathcal{W}^{EE}.$$

Since $\delta\bar{\gamma} < \gamma''$, we then would like $\bar{\gamma} \leq \gamma''$. Moreover, since we would like $[0, \bar{\gamma}]$ to be the largest such interval, we have $\bar{\gamma} = \gamma''$. What is the relationship between γ'' and $\bar{v}$?

(e) For what values of δ do we have $[0, \bar{\gamma}] = \mathcal{W}^{SS} \cup \mathcal{W}^{EE}$?

(f) Let $(\mathcal{W}, w^0, f, \tau)$ be the automaton given by $\mathcal{W} = [0, \bar{v}]$, $w^0 \in [0, \bar{v}]$,

$$f(w) = \begin{cases} EE, & \text{if } w \in \mathcal{W}^{EE}, \\ SS, & \text{otherwise,} \end{cases}$$

and

$$\tau(w, y) = \begin{cases} \gamma^y(w), & \text{if } w \in \mathcal{W}^{EE}, \\ w/\delta, & \text{otherwise,} \end{cases}$$

where $\gamma^y(w)$ solves (7.6.1)–(7.6.2) for $w = v$ and $y = \bar{y}, \underline{y}$. For our purposes here, assume that $V(w) = w$, that is, the value to a player of being in the automaton with initial state w is precisely w. (From the argument of Lemma 7.4.2, this should be intuitive.) Given this assumption, prove that the automaton describes a PPE with value w^0.

Chapter 8

Topics in Dynamic Games

8.1 Dynamic Games and Markov Perfect Equilibria

As usual, we have n players. The action space for player i is A_i. The new ingredient is that there is a set of *game* (or Markov) states S, with typical state $s \in S$, which affects payoffs. In general, the action space may be state dependent (as in Example 8.1.1), but it simplifies notation to leave this dependence implicit.

Player i's flow payoff is given by

$$u_i : S \times A \to \mathbb{R},$$

with flow payoffs discounted at rate $\delta \in (0, 1)$.

States vary over time, with the state transition given by

$$q : S \times A \to S,$$

and an initial state $s_0 \in S$. (More generally, we can have random transitions, so that q maps from $S \times A$ into $\Delta(S)$, but deterministic transitions will suffice for an introduction.)

Example 8.1.1. Suppose players 1 and 2 fish from a common area. In each period t, there is a stock of fish of size $s^t \in \mathbb{R}_+$. This is the state.

In period t, player i attempts to extract $a_i^t \in [0, s^t]$ units of fish. In particular, if player i attempts to extract a_i^t, then player i actually extracts

$$\hat{a}_i^t = \begin{cases} a_i^t, & \text{if } a_1^t + a_2^t \le s^t, \\ \frac{a_i^t}{a_1^t + a_2^t} s^t, & \text{if } a_1^t + a_2^t > s^t. \end{cases}$$

Player i receives flow payoff

$$u_i(s, a) = \log \hat{a}_i, \qquad \forall (s, a).$$

The transition rule is

$$q(s, a) = 2 \max\{0, s - a_1 - a_2\},$$

that is, it is deterministic and doubles any leftover stock after extraction. The initial stock is fixed at some value s^0.   ★

The state is public and there is perfect monitoring of actions, so the history to period t is

$$h^t = (s^0, a^0, s^1, a^1, \ldots, s^{t-1}, a^{t-1}, s^t) \in (S \times A)^t \times S.$$

Let H^t denote the set of all *feasible* t-period histories (so that s^τ is consistent with $(s^{\tau-1}, a^{\tau-1})$ for all $1 \leq \tau \leq t$).[1] A pure strategy for i is a mapping

$$\sigma_i : \cup_t H^t \to A_i.$$

As for repeated games, every strategy profile σ can be represented as automaton $(\mathcal{W}, w^0, f, \tau)$, where the only change is that the transition function has an enlarged domain, so that

$$\tau : \mathcal{W} \times A \times S \to \mathcal{W},$$

where the triple $(w, a, s) \in \mathcal{W} \times A \times S$ represents, in period t, the automaton state in period t, the action profile in period t, and the incoming game state s in period $t+1$. Note that even for our case of deterministic transitions, the space S is needed, since the automaton state in period t may not identify the period t game state.[2]

For any history h^t, write the function that identifies the last state s^t by $s(h^t)$. Let $G(s)$ denote the dynamic game with initial state s. As usual, we have:

Definition 8.1.1. *The profile σ is a subgame perfect equilibrium if for all h^t, $\sigma|_{h^t} := (\sigma_1|_{h^t}, \ldots, \sigma_n|_{h^t})$ is a Nash equilibrium of the dynamic game $G(s(h^t))$.*

---

[1] For the case of deterministic transitions q, this simply means $s^\tau = q(s^{\tau-1}, a^{\tau-1})$, and so in *this* case, the states *other than* s_0 are redundant. If transitions are stochastic, then s^τ should lie in the support of $q(s^{\tau-1}, a^{\tau-1})$, and the states are not redundant.

[2] Since under deterministic transitions, $s^\tau = q(s^{\tau-1}, a^{\tau-1})$, we can drop the dependence on states by taking $\mathcal{W}$ as the set of all histories of action profiles. This trick does not work for simpler automata.

Different histories that lead to the same state are effectively "payoff equivalent." Loosely, a strategy is said to be Markov if at different histories that are effectively payoff equivalent, the strategy specifies identical behavior. See Maskin and Tirole (2001) for a discussion of why this may be a reasonable restriction. There are no general strong foundations for the restriction, though it is possible to provide one in some settings (Bhaskar, Mailath, and Morris, 2013).

Definition 8.1.2. *A strategy $\sigma_i : \cup_t H^t \to A_i$ is* Markov *if for all histories h^t and $\hat{h}^t$, if $s(h^t) = s(\hat{h}^t)$, then*

$$\sigma_i(h^t) = \sigma_i(\hat{h}^t).$$

If the above holds for histories h^t and $\hat{h}^\tau$ of possibly different length (so that $t \neq \tau$ is allowed), the strategy is stationary.

A profile is stationary Markov if it has an automaton representation $(\mathcal{W}, w^0, f, \tau)$ in which $\mathcal{W} = S$ and $\tau(w, a, s) = s$.

Restricting equilibrium behavior to Markov strategies:

Definition 8.1.3. *A* Markov perfect equilibrium *(MPE) is a strategy profile σ that is a subgame perfect equilibrium, and each σ_i is Markov.*

A repeated game has a trivial set of Markov states (since different histories lead to strategically equivalent subgames, see Remark 7.1.1), and the only Markov perfect equilibria involve specifying static Nash equilibria in each period.

> Note that while there is a superficial similarity between *Markov states s* and *automata states w* used in the theory of repeated games, they are very different. A Markov state is part of the description of the environment, while an automaton state is part of the description of a representation of a particular strategy profile (behavior). Markov states are fixed as we consider different behaviors (this is why the restriction to Markov strategies has bite), while automaton states are not.

Example 8.1.2 (Example 8.1.1 continued). Fix a symmetric stationary MPE. Let $V(s)$ denote the common equilibrium value from the state s (in an MPE, this must be independent of other aspects of the history).

The common equilibrium strategy is $a_1(s) = a_2(s)$.

It will turn out that in equilibrium, $a_1(s) + a_2(s) < s$, so we will ignore the rationing rule (see Problem 8.4.1(a)), and assume $\hat{a}_i = a_i$. One-shot deviation principle holds here, and so for each player i, $a_i(s)$ solves, for any $s \in S$, the Bellman equation:

$$a_i(s) \in \underset{\tilde{a}_i \in A_i,\ \tilde{a}_i \leq s}{\arg\max}\ (1 - \delta)\log(\tilde{a}_i) + \delta V(\max\{0, 2(s - \tilde{a}_i - a_j(s))\}).$$

Assuming V is differentiable, that the optimal choices satisfy

$$s > \tilde{a}_i + a_j(s) > 0,$$

and imposing $\tilde{a}_i = a_j(s) = a(s)$ after differentiating, the implied first order condition is

$$\frac{(1 - \delta)}{a_i(s)} = 2\delta V'(2(s - 2a_i(s))).$$

To find an equilibrium, suppose

$$a_i(s) = ks,$$

for some k. Then we have

$$s^{t+1} = 2(s^t - 2ks^t) = 2(1 - 2k)s^t.$$

Given an initial stock s, in period t, $a_i^t = k[2(1 - 2k)]^t s$, and so

$$V(s) = (1 - \delta)\sum_{t=0}^{\infty} \delta^t \log\{k[2(1 - 2k)]^t s\}$$

$$= (1 - \delta)\sum_{t=0}^{\infty} \delta^t \log\{k[2(1 - 2k)]^t\} + \log s.$$

This implies V is indeed differentiable, with $V'(s) = 1/s$. Solving the first order condition, $k = \frac{1-\delta}{2-\delta}$, and so

$$a_i(s) = \frac{1 - \delta}{2 - \delta} s.$$

The MPE is not Pareto efficient (see Problem 8.4.3). If the players are sufficiently patient, the Pareto efficient outcome can be supported as a subgame perfect equilibrium (again, see Problem 8.4.3).

Finally, there is also a symmetric MPE in which $a_i(s) = s$ for all s. ★

Example 8.1.3 (Asynchronous move games). Consider the repeated prisoner's dilemma, but where player 1 moves in odd periods only and player

	E_2	S_2
E_1	$2, 2$	$-x, 3$
S_2	$3, -x$	$0, 0$

Figure 8.1.1: The stage game for Example 8.1.3, where $x > 0$.

2 moves in even periods only. The game starts with E_1 exogenously and publicly specified for player 1. The stage game is given in Figure 8.1.1.

This fits into the above formulation of a dynamic game, with $S = \{E_1, S_1, E_2, S_2\}$, $s^0 = E_1$,

$$q(s, a) = \begin{cases} a_1, & \text{if } s \in \{E_2, S_2\}, \\ a_2, & \text{if } s \in \{E_1, S_1\}, \end{cases}$$

$$u_1(s, a) = \begin{cases} g_1(a_1, s), & \text{if } s \in \{E_2, S_2\}, \\ g_1(s, a_2), & \text{if } s \in \{E_1, S_1\}, \end{cases}$$

where g_i describes the stage game payoffs from the PD, and

$$u_2(s, a) = \begin{cases} g_2(s, a_2), & \text{if } s \in \{E_1, S_1\}, \\ g_2(a_1, s), & \text{if } s \in \{E_2, S_2\}. \end{cases}$$

In particular, when the current state is player 1's action (i.e., we are in an even period), 1's choice is irrelevant and can be ignored.[3]

Grim trigger is

$$\sigma_i^{GT}(h^t) = \begin{cases} E_i, & \text{if } h^t = E_1 \text{ or always } E, \\ S_i, & \text{otherwise.} \end{cases}$$

Need to check two classes of information sets: when players are supposed to play E_i, and when they are supposed to play S_i (applying one-shot optimality):

1. Optimality of E_i after all E's:

$$2 \geq 3(1 - \delta) + \delta \times 0$$

$$\iff \delta \geq \frac{1}{3}.$$

---

[3]In this formulation, strategies in the infinite-horizon game should not depend on the irrelevant choice.

2. The optimality of S_i after any S_1 or S_2 is trivially true for all δ. If the current state is S_j, then

$$0 \geq (-x)(1 - \delta) + \delta \times (-x)(1 - \delta) + \delta^2 \times 0.$$

If the current state is E_j, then

$$3(1 - \delta) + \delta \times 0 \geq (2)(1 - \delta) + \delta \times (-x)(1 - \delta) + \delta^2 \times 0.$$

This equilibrium is not an equilibrium in Markov strategies (and so is not an MPE).

Supporting effort using Markov pure strategies requires a "tit-for-tat" like behavior:

$$\hat{\sigma}_i(h^t) = \begin{cases} E_i, & \text{if } s^t = E_j, \\ S_i, & \text{if } s^t = S_j. \end{cases}$$

For $t \geq 1$, everything is symmetric. The value when the current state is E_j is

$$V_i(E_j) = 2,$$

while the payoff from a one-shot deviation is

$$3(1 - \delta) + \delta \times 0 = 3(1 - \delta),$$

and so the deviation is not profitable if (as before) $\delta \geq \frac{1}{3}$.

The value when the current state is S_j is

$$V_i(S_j) = 0,$$

while the payoff from a one-shot deviation is (since under the Markov strategy, a deviation to E_i triggers perpetual $E_1 E_2$; the earlier deviation is "forgiven")

$$-x(1 - \delta) + \delta \times 2 = (2 + x)\delta - x.$$

The deviation is not profitable if

$$(2 + x)\delta - x \leq 0$$

$$\iff \delta \leq \frac{x}{2 + x}.$$

Note that

$$\frac{x}{2 + x} \geq \frac{1}{3} \iff x \geq 1.$$

Thus, $\hat{\sigma}$ is an MPE (inducing the outcome path $(E_1 E_2)^\infty$) if $x \geq 1$ and

$$\frac{1}{3} \leq \delta \leq \frac{x}{2 + x}. \tag{8.1.1}$$

★

8.2 Disappearance of Monopoly Power and the Coase Conjecture

A seller and buyer are bargaining over a potential sale. In each period, the seller makes an offer that the buyer accepts or rejects. If the buyer accepts, the game is over, and if the the buyer rejects, play moves to the next period, and the seller makes a new offer. Game is described as a bargaining game of one-sided offers.

The seller's cost (value) is publicly known and equals zero.

The buyer's value for the good v is known only to the buyer, and is uniformly distributed on $[0, 1]$.

8.2.1 One and Two Period Example

In the one-period game, the seller is making a take-it-or-leave-it offer of p. The buyer accepts the offer of p if $v > p$ and rejects if $v < p$.

The seller chooses p to maximize

$$p \Pr\{\text{sale}\} = p(1 - p),$$

i.e., chooses $p = 1/2$, for a payoff of $1/4$. It turns out that the seller can do no better, no matter what game the seller designs (i.e., this is the optimal seller mechanism; this is easily shown using standard mechanism design techniques, see Section 10.3). A critical feature of the optimal seller mechanism is that the seller *commits* to not making any further sales (something that is guaranteed in the one-period game).

Suppose now there are two periods, with common discount factor $\delta \in (0, 1)$, and *no* seller commitment. If the seller chose $p^0 = 1/2$ in the first period, and buyers with $v > 1/2$ buy in period 0, then buyers with value $v \in [0, 1/2]$ are left. The lack of seller commitment means that, when there is no sale in the first period, the seller finds it optimal to try again and sell to the buyer at a price $p^1 = 1/4$ (since this is the last period, there is no further issue of commitment). But then buyer $v = 1/2$ strictly prefers to wait till period 1, and so by continuity so do some buyers with $v > 1/2$.

Thus, the statically optimal price is not an equilibrium price in the first period, because the seller cannot commit to not trying to make a sale in the second period if no sale occurred in the inital period.

The analysis of the two-period game is straightforward, but we first must be more careful (explicit) about the equilibrium concept. We model the two-period game as a Bayesian game. A strategy for the seller is the pair (p^0, σ_s), where $p^0 \in \mathbb{R}_+$ is the price in period 0 and $\sigma_s : \mathbb{R}_+ \to \mathbb{R}_+$ specifies the period 1 price after any *rejected* period 0 price (the game ends after acceptance). A strategy for the buyer of type $v \in [0,1]$ is the pair (σ_v^0, σ_v^1), where $\sigma_v^0 : \mathbb{R}_+ \to \{A, R\}$ specifies accept (A) or reject (R), in period 0 as a function of the period 0 price and $\sigma_v^1 : \mathbb{R}_+^2 \to \{A, R\}$ specifies accept or reject in period 1 as a function of the rejected period 0 price and the period 1 price.

The notion of Bayes-Nash equilibrium is clear: the seller and every buyer type are each playing a best reply to the behavior of the players. For the seller, this means choosing an optimal sequence of prices, given the accept-reject behavior of the buyers. For the buyer of type v, this means optimally accepting or rejecting the initial expected price, given the second period price the buyer expects, and if optimally rejecting the initial price, optimally accepting or rejecting the expected period 1 price.

What about sequential rationality (perfect Bayesian equilibrium)? We first require the buyer of type v accept any period 1 price strictly less than v and reject any period 1 price strictly larger than v. In period 1, the only possible out of equilibrium information set for the seller if he had not deviated in period 0, is that in the equilibrium, all buyers were supposed to accept p^0, and yet the price is rejected. In that event, we require the seller maximize his expected payoff given some belief over buyer valuations (this turns out not to be an issue, since there is always some type who will reject in the initial period).

Finally, we need to consider the possibility that the seller deviated in the initial period. Buyers should respond optimally, given their beliefs of the continuation play of the seller. The optimal continuation play of the seller depends on how the seller believes the buyers will respond (and on the seller beliefs over the buyer types)!

Suppose the seller offers p^0 in the initial period. While this setting is best thought of as a Bayesian game, this *two* period game is strategically equivalent to the game where Nature determines the buyer's type *after* the initial price has been chosen.[4] In that extensive form, the Nash equilibrium constraints in the subgame with root p^0 capture exactly the restrictions we want for perfect Bayesian equilibria. Note first that all buyers reject $p^0 > 1$, so we can restrict attention to $p^0 \in [0,1]$. Fix an equilibrium of this subgame and suppose buyers $v < \kappa$ don't buy in period 0. Then, $p^1 = \kappa/2$ in that equilibrium. If $0 < \kappa < 1$, then κ should be indifferent between

---

[4]The issue this specification (trick) deals with is the appropriate specification of seller beliefs after the seller deviates. In particular, it is natural to require that the seller still update beliefs using Bayes' rule after a deviation, but almost PBE does not require this (Problem 5.4.9 illustrates). This trick will not work for longer horizons and periods other than the initial period. In that case, as we will do in Section 8.2.2, the perfect Bayesian restrictions need to be imposed directly.

purchasing in period 0 and period 1, so that

$$\kappa - p^0 = \delta(\kappa - p^1)$$
$$= \delta\kappa/2$$
$$\implies \kappa = \frac{2p^0}{2 - \delta} =: \kappa(p^0).$$

Thus if $p^0 \in (0, 1)$, the unique continuation play is that buyers with value strictly greater than $\kappa(p^0)$ accept the offer, buyers with value strictly less than $\kappa(p^0)$ reject the offer, and receive an offer of $\kappa(p^0)/2$ in the last period. It doesn't matter what we specify for buyer with value $v = \kappa(p^0)$ since he is indifferent and the seller assigns that type zero probability. It is also easy to see that in equilibrium, an initial offer of $p^0 = 0$ must be accepted by almost all buyers. (So, for a continuum of types, the equilibrium behavior in the subgame turns out to be simple. It is more subtle with discrete types, see Problem 8.4.6.)

Since we have a unique specification of continuation play for each p^0, we can now solve for the optimal p^0. Since $\kappa(\cdot)$ is a one-to-one function of p^0 in the relevant range of $[0, 1]$, we solve for the seller's optimal κ. The seller's payoff (as a function of κ) is

$$\kappa(1 - \delta/2)(1 - \kappa) + \delta\kappa^2/4.$$

The first-order condition is

$$(1 - \delta/2) - 2(1 - \delta/2)\kappa + \delta\kappa/2 = 0$$
$$\implies \kappa = \frac{2 - \delta}{4 - 3\delta} \; (< 1)$$
$$\implies p^0 = \frac{(2 - \delta)^2}{8 - 6\delta} < \frac{1}{2}.$$

The resulting payoff is

$$\frac{(2 - \delta)^2}{4(4 - 3\delta)} < \frac{1}{4}.$$

Finally, if $\delta = 1$, then delay is not costly for either buyer or seller, and the seller can achieve his statically optimal profits. Note that there is still a lack of commitment in the initial period, so this requires no sales in the initial period. Any price satisfying $p^0 \geq \frac{1}{2}$ is consistent with equilibrium.

8.2.2 Infinite Horizon

Seller makes an offer p^t in each period $t = 0, 1, 2, \ldots$.

After each offer, buyer *Accepts* or *Rejects*.

If there is agreement in period t at price p^t, the payoff to the seller is

$$u_s = \delta^t p^t,$$

and the payoff to the buyer is

$$u_b = \delta^t \left(v - p^t \right).$$

We are interested in an equilibrium of the following form, suggested by the equilibrium in Section 8.2.1:

If p^t offered in period t, types $v \geq \lambda p^t$ accept and types $v < \lambda p^t$ reject, where $\lambda > 1$.

If at time t, seller's posterior beliefs are uniform on $[0, \kappa]$, seller offers $p^t(\kappa) = \gamma \kappa$, where $\gamma < 1$.

Under this description, there is *skimming*: higher valuation buyers buy before lower valuation buyers, and prices monotonically decline.

Since the above holds on and off the path of play, the resulting equilibrium is a perfect Bayesian equilibrium.

It is natural to treat κ as a state variable, and so the equilibrium is Markov.

Under this profile, $p^0 = \gamma$ and seller's posterior entering period 1 is $[0, \gamma \lambda]$, so in order for profile to be well defined, $\gamma \lambda < 1$. Thus, $p^1 = \gamma(\gamma \lambda) = \gamma^2 \lambda$ and seller's posterior entering period 2 is $[0, \gamma^2 \lambda^2]$. Prices are thus falling exponentially, with $p^t = \gamma^{t+1} \lambda^t$.

Let $U_s(\kappa)$ be the discounted expected value to the seller, when his posterior beliefs are uniform on $[0, \kappa]$. Then

$$U_s(\kappa) = \max_p \left\{ \frac{(\kappa - \lambda p)}{\kappa} \times p + \delta \frac{\lambda p}{\kappa} U_s(\lambda p) \right\},$$

or

$$W_s(\kappa) = \max_p (\kappa - \lambda p) p + \delta W_s(\lambda p), \qquad (8.2.1)$$

where $W_s(\kappa) = \kappa U_s(\kappa)$. If W_s is differentiable, then $p(\kappa)$ solves the first order condition,

$$\kappa - 2\lambda p(\kappa) + \delta \lambda W_s'(\lambda p(\kappa)) = 0.$$

The envelope theorem applied to (8.2.1) gives

$$W_s'(\kappa) = p(\kappa) = \gamma \kappa,$$

so that

$$W_s'(\lambda p(\kappa)) = p(\lambda p(\kappa)) = \gamma \lambda p(\kappa) = \lambda \gamma^2 \kappa.$$

Substituting,

$$\kappa - 2\lambda\gamma\kappa + \delta\lambda^2\gamma^2\kappa = 0,$$

or

$$1 - 2\lambda\gamma + \delta\lambda^2\gamma^2 = 0. \tag{8.2.2}$$

Turning to the buyer's optimality condition, a buyer with valuation $v = \lambda p$ must be indifferent between accepting and rejecting, so

$$\lambda p - p = \delta\left(\lambda p - \gamma\lambda p\right),$$

or

$$\lambda - 1 = \delta\lambda\left(1 - \gamma\right). \tag{8.2.3}$$

Solving (8.2.2) for $\lambda\gamma$ yields

$$\gamma\lambda = \frac{2 \pm \sqrt{4 - 4\delta}}{2\delta} = \frac{1 \pm \sqrt{1 - \delta}}{\delta}.$$

Since we know $\gamma\lambda < 1$, take the negative root,

$$\gamma\lambda = \frac{1 - \sqrt{1 - \delta}}{\delta}.$$

Substituting into (8.2.3),

$$\lambda = \delta\lambda + \sqrt{1 - \delta},$$

or

$$\lambda = \frac{1}{\sqrt{1 - \delta}},$$

so that

$$\gamma = \sqrt{1 - \delta} \times \frac{\left(1 - \sqrt{1 - \delta}\right)}{\delta} = \frac{\sqrt{1 - \delta} - (1 - \delta)}{\delta} =: \gamma(\delta).$$

The equilibrium is not unique. It is the only stationary equilibrium.[5]
The Coase conjecture (Coase, 1972) is:

As the commitment problem becomes more severe (because the monopoly seller can (not commit to not) make a new offer more quickly after a rejection), the seller is effectively facing more competition from his future self,

---

[5]This is called the *no-gap case*, because there are buyer types arbitrarily close to the seller's cost (of 0), allowing a simple recursive formulation. In the *gap case*, where buyer types are bounded away from the seller's cost, the game has a unique equilibrium, because at some point (when the seller's uncertainty is sufficiently small) the seller prefers to sell to all remaining buyer types at a price equal to the minimum buyer type (this arises, for example, in Problem 8.4.6 for some parameter values).

and so the price charged in the initial period converges to the competitive price, and trade becomes efficient.

We make the Coase conjecture precise as follows:

Let τ denote *real* time, Δ the length of a period, and r the rate of time discount, so that $\delta = e^{-r\Delta}$. Thus, as $\Delta \to 0$, we have $\delta \to 1$. In this setting, since the seller has zero cost, the competitive price is 0.

Theorem 8.2.1. *For all $\tau > 0$, $\varepsilon > 0$, and for all $v \in (0, 1]$, there exists $\bar{\Delta} > 0$, such that in the stationary equilibrium, for all $\Delta \in (0, \bar{\Delta})$,*

$$p^0 < \varepsilon$$

and buyer of type v buys before τ.

Proof. Since $\lim_{\delta \to 1} \gamma(\delta) = 0$ and $p^0 = \gamma(\delta)$, for Δ close to 0, we have p^0 close to 0.

If buyer with valuation v buys at or after τ, her utility is no more than $e^{-r\tau}v$, which is bounded away from v (as $\Delta \to 0$). Buying in period 0, she earns $v - \gamma(\delta)$, which converges to v as $\Delta \to 0$, and so for sufficiently small Δ, she prefers to buy immediately rather than wait till after r. ■

Note that this is not a uniform statement (since for all τ and all δ there exists v close to 0 such that v purchases after τ).

The classic reference on the Coase conjecture is Gul, Sonnenschein, and Wilson (1986).

8.3 Reputations

8.3.1 Two Periods

The stage game of the the chain store paradox from Example 2.2.1 is reproduced in Figure 8.3.1 (note that payoffs are written in a different order and so differently to facilitate the comparison with Figure 8.3.2). The game has two Nash equilibria: (In, Accommodate) and (Out, Fight). The latter violates backward induction.

In the chain store game, the game is played twice, against two different entrants (E_1 and E_2), with the second entrant E_2 observing the outcome of first interaction. The incumbent's payoff is the sum of payoffs in the two interactions.

The *chain store paradox*: The only backward induction (subgame perfect) outcome is that both entrants enter (play In), and the incumbent always accommodates. This is true for any finite chain store.

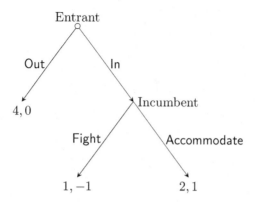

Figure 8.3.1: The stage game for the chain store. The first payoff is that of the incumbent, and the second is that of the entrant.

Now we introduce incomplete information of a very particular kind. In particular, we suppose the incumbent could be *tough*, ω_t. The tough incumbent receives a payoff of 2 from fighting and only 1 from accommodating. The other incumbent is *normal*, ω_n, with payoffs as described in Figure 8.3.1. Both entrants assign prior probability $\rho \in (0, \frac{1}{2})$ to the incumbent being ω_t.

Suppose first E_1 chooses O_1 (we use subscripts to indicate the period in which the choice is made). Then, in any sequential equilibrium, the analysis of the second period is just that of the static game of incomplete information with E_2's beliefs on the incumbent given by the prior,[6] and so E_2 optimally plays I_2, the normal type accommodates and the tough type fights.

We next analyze the behavior that follows E_1's choice of I_1, i.e., entry in the first market. Because in any sequential equilibrium, in the second period the normal incumbent accommodates and the tough incumbent fights, this behavior must be an equilibrium of the signaling game illustrated in Figure 8.3.2 (given the optimal play of the incumbents in the second market).

It is easy to verify that there are no pure strategy equilibria.

---

[6]This is immediate if the extensive form of the two-period chain store is specified as first E_1 chooses I_1 or O_1, with each choice leading to a move of nature which determines the type of the incumbent. If nature first determines the type of the incumbent, and then E_1 moves, there is no subgame following a choice of O_1 or I_1. Nonetheless, sequentiality implies that the choices of the incumbent of each type and of E_2 are as if they are playing an equilibrium of the subgame, for reasons analogous to those in Example 5.3.1. Problem 8.4.9 asks you to verify this claim.

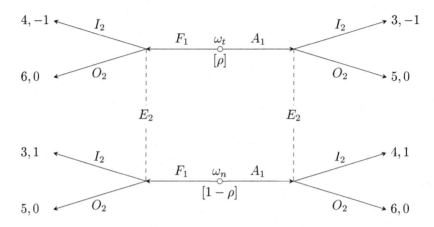

Figure 8.3.2: A signaling game representation of the subgame reached by E_1 entering. The first payoff is the payoff to the incumbent, and the second payoff is the payoff to E_2 *assuming* the incumbent plays F after a choice of I by E_2.

There is a unique mixed strategy equilibrium: Type ω_n plays $\alpha \circ F_1 + (1-\alpha) \circ A_1$, type ω_t plays F_1 for sure. Entrant E_2 enters for sure after A_1, and plays $\beta \circ I_2 + (1-\beta) \circ O_2$ after F_1.

Entrant E_2 is willing to randomize only if his posterior after F_1 that the incumbent is ω_t equals $\frac{1}{2}$. Since that posterior is given by

$$\Pr\{\omega_t \mid F_1\} = \frac{\Pr\{F_1 \mid \omega_t\}\Pr\{\omega_t\}}{\Pr\{F_1\}}$$
$$= \frac{\rho}{\rho + (1-\rho)\alpha},$$

solving

$$\frac{\rho}{\rho + (1-\rho)\alpha} = \frac{1}{2}$$

gives

$$\alpha = \frac{\rho}{1-\rho},$$

where $\alpha < 1$ since $\rho < \frac{1}{2}$.

Type ω_n is willing to randomize if

$$\underbrace{4}_{\text{Payoff from } A} = \underbrace{\beta 3 + 5(1-\beta)}_{\text{Payoff from } F},$$

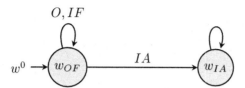

Figure 8.3.3: Automaton representation of Nash reversion in the complete information infinite horizon chain store.

i.e.,

$$\beta = \frac{1}{2}.$$

It remains to determine the behavior of entrant E_1. This entrant faces a probability of F_1 given by

$$\rho + (1 - \rho)\alpha = 2\rho.$$

Hence, if $\rho < \frac{1}{4}$, E_1 faces F_1 with sufficiently small probability that he enters. However, if $\rho \in (\frac{1}{4}, \frac{1}{2})$, E_1 faces F_1 with sufficiently high probability that he stays out. For $\rho = \frac{1}{4}$, E_1 is indifferent between O_1 and I_1, and so any specification of behavior is consistent with equilibrium.

8.3.2 Infinite Horizon

Suppose now there is an infinite horizon with the incumbent discounting at rate $\delta \in (0, 1)$ and a new potential entrant in each period.

Note first that in the complete information game, the outcome in which all entrants enter (play In) and the incumbent accommodates in every period is an equilibrium. Moreover, the profile in which all entrants stay out, any entry is met with F is a subgame perfect equilibrium, supported by the "threat" that play switches to the always-enter/always-accommodate equilibrium if the incumbent ever responds with A. The automaton representation is given in Figure 8.3.3.

Note that the relevant incentive constraint for the incumbent is conditional on I in state w_{OF} (since the incumbent does not make a decision when the entrant chooses O),[7] i.e.,

$$(1 - \delta) + \delta 4 \geq 2,$$

---

[7]This is similar to the collusion example discussed in Section 7.3.2. The stage game is not a simultaneous move game, and so the repeated game does *not* have perfect monitoring. In particular, the incumbent's choice between F and A is irrelevant (not observed) if the entrant plays O (as the putative equilibrium requires). Subgame perfection, however, requires that the incumbent's choice of F be optimal, given that the entrant had

i.e.,

$$\delta \geq \frac{1}{3}.$$

We now consider the *reputation game*, where the incumbent may be normal or tough.

The profile in which all entrants stay out, any entry is met with F is a subgame perfect equilibrium, supported by the "threat" that the entrants believe that the incumbent is normal and play switches to the always-enter/always-accommodate equilibrium if the incumbent ever responds with A.

Theorem 8.3.1. *Suppose the incumbent is either of type ω_n or type ω_t, and that type ω_t has prior probability less than $1/2$. Type ω_n must receive a payoff of at least $(1 - \delta) \times 1 + \delta \times 4 = 1 + 3\delta$ in any pure strategy Nash equilibrium in which ω_t always plays F.*

If type ω_t has prior probability greater than $1/2$, trivially there is never any entry and the normal has payoff 4.

Proof. In a pure strategy Nash equilibrium, either the incumbent always plays F, (in which case the entrants always stay out and the incumbent's payoff is 4), or there is a first period (say τ) in which the normal type accommodates, revealing to future entrants that he is the normal type (since the tough type plays F in every period). In such an equilibrium, entrants stay out before τ (since both types of incumbent are choosing F), and there is entry in period τ. After observing F in period τ, entrants conclude the firm is the ω_t type, and there is no further entry. An easy lower bound on the normal incumbent's equilibrium payoff is then obtained by observing that the normal incumbent's payoff must be at least the payoff from mimicking the ω_t type in period τ. The payoff from such behavior is at least as large as

---

played I. The principle of one-shot optimality applies here: The profile is subgame perfect if, conditional on I, it is optimal for the incumbent to choose F, given the specified continuation play (this is the same idea as that in Footnote 7 on page 192).

$$\underbrace{(1 - \delta) \sum_{\tau'=0}^{\tau-1} \delta^{\tau'} 4}_{\substack{\text{payoff in } \tau' < \tau \text{ from pooling} \\ \text{with } \omega_t \text{ type}}} \quad + \quad \underbrace{(1 - \delta)\delta^{\tau} \times 1}_{\substack{\text{payoff in } \tau \text{ from playing } F \text{ when} \\ A \text{ is myopically optimal}}}$$

$$+ \quad \underbrace{(1 - \delta) \sum_{\tau'=\tau+1}^{\infty} \delta^{\tau'} 4}_{\text{payoff in } \tau' > \tau \text{ from being treated as the } \omega_t \text{ type}}$$

$$\begin{aligned}
&= (1 - \delta^{\tau})4 + (1 - \delta)\delta^{\tau} + \delta^{\tau+1} 4 \\
&= 4 - 3\delta^{\tau}(1 - \delta) \\
&\geq 4 - 3(1 - \delta) = 1 + 3\delta.
\end{aligned}$$

∎

For $\delta > 1/3$, the outcome in which all entrants enter and the incumbent accommodates in every period is thus eliminated.

8.3.3 Infinite Horizon with Behavioral Types

In the reputation literature (see Mailath and Samuelson, 2006, Chapter 15, or Mailath and Samuelson, 2014, for an extensive introduction), it is standard to model the tough type as a *behavioral* type. In that case, the tough type is *constrained* to necessarily choose F. Then, the result is that in *any* equilibrium, $1 + 3\delta$ is the lower bound on the normal type's payoff. (The type ω_t from earlier in this section is an example of a *payoff* type.)

In fact, irrespective of the presence of other types, if the entrants assign positive probability to the incumbent being a tough behavioral type, for δ close to 1, the incumbent's payoff in *any* Nash equilibrium is close to 4 (this is an example of a *reputation effect*):

Suppose there is a set of types Ω for the incumbent. Some of these types are behavioral. One behavioral type, denoted $\omega_0 \in \Omega$, is the *Stackelberg*, or tough, type, who always plays F. The normal type is ω_n. Other types may include behavioral ω_k, who plays F in every period before k and A afterwards. Suppose the prior beliefs over Ω are given by μ.

Lemma 8.3.1. *Consider the incomplete information game with types Ω for the incumbent. Suppose the Stackelberg type $\omega_0 \in \Omega$ receives positive prior* *probability $\mu_0 > 0$. Fix a Nash equilibrium. Let h^t be a positive probability period-t history in which every entry results in F. The number of periods in h^t in which an entrant entered is no larger than*

$$k^* := -\frac{\log \mu_0}{\log 2}.$$

Proof. Denote by q_τ the probability that the incumbent plays F in period τ conditional on h^τ if entrant τ plays I. In equilibrium, if entrant τ does play I, then

$$q_\tau \le \frac{1}{2}.$$

(If $q_\tau > \frac{1}{2}$, it is not a best reply for the entrant to play I.) An upper bound on the number of periods in h^t in which an entrant entered is thus

$$k(t) := \#\{\tau : q_\tau \le \tfrac{1}{2}\},$$

the number of periods in h^t where $q_\tau \le \frac{1}{2}$. (This is an upper bound, and not the actual number, since the entrant is indifferent if $q_\tau = \frac{1}{2}$.)

Let $\mu_\tau := \Pr\{\omega_0|h^\tau\}$ be the posterior probability assigned to ω_0 after h^τ, where $\tau < t$ (so that h^τ is an initial segment of h^t). If entrant τ does not enter, $\mu_{\tau+1} = \mu_\tau$. If entrant τ does enter in h^t, then the incumbent fights and[8]

$$
\begin{aligned}
\mu_{\tau+1} = \Pr\{\omega_0|h^\tau, F\} &= \frac{\Pr\{\omega_0, F|h^\tau\}}{\Pr\{F|h^\tau\}} \\
&= \frac{\Pr\{F|\omega_0, h^\tau\}\Pr\{\omega_0|h^\tau\}}{\Pr\{F|h^\tau\}} \\
&= \frac{\mu_\tau}{q_\tau}.
\end{aligned}
$$

Defining

$$
\tilde{q}_\tau = \begin{cases} q_\tau, & \text{if there is entry in period } \tau, \\ 1, & \text{if there is no entry in period } \tau, \end{cases}
$$

we have, for all $\tau \le t$,

$$\mu_\tau = \tilde{q}_\tau \mu_{\tau+1},$$

Note that $\tilde{q}_\tau < 1 \implies \tilde{q}_\tau = q_\tau \le \frac{1}{2}$.
Then,

$$\mu_0 = \tilde{q}_0 \mu_1 = \tilde{q}_0 \tilde{q}_1 \mu_2$$

---

[8]Since the entrant's action is a function of h^τ only, it is uninformative about the incumbent and so can be ignored in the conditioning.

$$= \mu_t \prod_{\tau=0}^{t-1} \tilde{q}_\tau$$

$$= \mu_t \prod_{\{\tau : \tilde{q}_\tau \leq \frac{1}{2}\}} \tilde{q}_\tau$$

$$\leq \left(\frac{1}{2}\right)^{k(t)}.$$

Taking logs, $\log \mu_0 \leq k(t) \log \frac{1}{2}$, and so

$$k(t) \leq -\frac{\log \mu_0}{\log 2}.$$

∎

The key intuition here is that since the entrants assign prior positive probability (albeit small) to the Stackelberg type, they cannot be surprised too many times (in the sense of assigning low prior probability to F and then seeing F). Note that the upper bound is independent of t and δ, though it is unbounded in μ_0.

Theorem 8.3.2. *Consider the incomplete information game with types Ω for the incumbent. Suppose the Stackelberg type $\omega_0 \in \Omega$ receives positive prior probability $\mu_0 > 0$. In any Nash equilibrium, the normal type's expected payoff is at least $1 + 3\delta^{k^*}$. Thus, for all $\varepsilon > 0$, there exists $\bar{\delta}$ such that for all $\delta \in (\bar{\delta}, 1)$, the normal type's payoff in any Nash equilibrium is at least $4 - \varepsilon$.*

Proof. The normal type can guarantee histories in which every entry results in F by always playing F when an entrant enters. Such behavior yields payoffs that are no larger than the incumbent's Nash equilibrium payoffs in any equilibrium (if not, the incumbent has an incentive to deviate). Since there is positive probability that the incumbent is the Stackelberg type, the history resulting from always playing F after entry has positive probability. Applying Lemma 8.3.1 yields a lower bound on the normal types payoff of

$$\sum_{\tau=0}^{k^*-1} (1-\delta)\delta^\tau 1 + \sum_{\tau=k^*}^{\infty} (1-\delta)\delta^\tau 4 = 1 - \delta^{k^*} + 4\delta^{k^*} = 1 + 3\delta^{k^*}.$$

This can be made arbitrarily close to 4 by choosing δ close to 1. ∎

8.4 Problems

8.4.1. (a) Suppose (σ_1, σ_2) is an MPE of the fisheries game from Example 8.1.1 satisfying $\sigma_1(s) + \sigma(s) < s$ for all s. Prove that the profile remains an MPE of the dynamic game where payoffs are given by

$$
u_i(s, a) = \begin{cases} \log a_i, & \text{if } a_1 + a_2 \leq s, \\ \log\left\{\frac{a_i}{a_1+a_2}s\right\}, & \text{if } a_1 + a_2 > s. \end{cases}
$$

(b) Prove that

$$
a_i(s) = \frac{1-\delta}{2-\delta}s, \qquad i = 1, 2,
$$

does indeed describe an MPE of the fisheries game described in Example 8.1.1.

8.4.2. What is the symmetric MPE for the fisheries game of Example 8.1.2 when there are n players, and the transition function is given by

$$
q(s, a) = \alpha \max\left\{0, s - \sum_i a_i\right\},
$$

where $\alpha > 1$?

8.4.3. (a) In the MPE calculated in Example 8.1.2, for what values of the discount factor does the stock of fish grow without bound, and for which values does the stock decline to extinction?

(b) This MPE is inefficient, involving excess extraction. To see this, calculate the largest symmetric payoff profile that can by achieved when the firms choose identical Markov strategies, and prove that the efficient solution extracts less than does the MPE.

(c) Describe an efficient subgame perfect equilibrium for this game (it is necessarily non-Markov).

8.4.4. Consider the asynchronous move prisoner's dilemma from Section 8.1.

(a) Suppose $x \geq 1$. For some values of δ, there is a Markov perfect equilibrium in which players randomize at E between E and S, and play S for sure at S. Identify the bounds on δ and the probability of randomization for which the described behavior is an MPE.

(b) Suppose that the initial action of player 1 is not exogenously fixed. The game now has three states, the initial null state and E and S. At the initial state, both players choose an action, and then thereafter player 1 chooses an action in odd periods and player 2 in even periods. Suppose $x > 1$ and δ satisfies (8.1.1).

Coalition	1's payoff	2's payoff	3's payoff
$\{1,2\}$	9	3	0
$\{2,3\}$	0	9	3
$\{1,3\}$	3	0	9

Figure 8.4.1: Payoffs to players in each pairwise coalition for Problem 8.4.5. The excluded player receives a payoff of 0.

Prove that there is no pure strategy MPE in which the players choose E.

8.4.5. (A simplification of Livshits, 2002.) There are three players. In the initial period, a player i is selected randomly and uniformly to propose a coalition with one other player j, who can accept or reject. If j accepts, the game is over with payoffs given in Figure 8.4.1. If j rejects, play proceeds to the next period, with a new proposer randomly and uniformly selected. The game continues with a new proposer randomly and uniformly selected in each period until a proposal is accepted. Thus, the game is potentially of infinite horizon, and if no coalition is formed (i.e., there is perpetual rejection), all players receive a payoff of 0.

(a) Suppose $\delta < 3/4$. Describe a stationary pure strategy Markov perfect equilibrium. [Hint: In this equilibrium, every proposal is immediately accepted.]

(b) Suppose $\delta > 3/4$. Prove there is no Markov perfect equilibrium in stationary pure strategies. There is a stationary Markov perfect equilibrium in behavior strategies. What is it? [Hint: The randomization is on the part of the responder.]

(c) Suppose $3/4 < \delta < \sqrt{3/4}$. There are two nonstationary pure strategy Markov equilibria in which only the most valuable partnerships form. What are they? [Hint: If $\delta < \sqrt{3/4}$, then $\delta^2 < 3/4$.]

(d) Suppose $\delta \geq \frac{3}{7}$. Construct a nonstationary Markov perfect equilibrium in which in the first period, if 1 is selected, then 1 chooses 3 (who of course accepts).

8.4.6. Consider the model of Section 8.2, but assume the buyer's valuation v can only take on two values, 2 and 3. Moreover, the seller's beliefs assign probability α to the value 2. The seller's cost (value) is zero, and the buyer and seller have a common discount factor $\delta \in (0, 1]$.

(a) What are the unique perfect Bayesian equilibria (PBE) of the one period model (in this model, the seller makes a take-it-or-leave-it offer to the buyer)?

Consider now the two period model, that is, if the buyer rejects the offer in the initial period, then the seller make a final offer in period 1, after which the game ends. As in Section 8.2, determine the restrictions implied by perfect Bayesian equilibrium by assuming the buyer type is determined by nature after the seller has chosen the initial period price.

(b) What restrictions on period 1 pricing are implied by perfect Bayesian equilibrium concept?

(c) Prove that, in any PBE, both types of buyer must accept any first period price strictly smaller than 2.

(d) Prove that, in any PBE, if a (possibly non-equilibrium) first period price $p^0 > 2$ is rejected, then the seller's posterior in the beginning of the second period must assign probability at least α to the low value buyer.

(e) Suppose $\alpha = \frac{1}{2}$. Describe the unique pure strategy PBE.

(f) Suppose $\alpha = \frac{1}{4}$. Prove that there is no pure strategy PBE. [**Hint:** Suppose $\hat{p}^0$ is the first period price in a candidate pure strategy PBE. How should the seller respond to a rejection of a deviation to a price $p^0 \neq \hat{p}^0$?]

(g) Suppose $\alpha = \frac{1}{4}$. Suppose $\delta \neq 6/7$. Describe the unique PBE (from part 8.4.6(f), it is necessarily in mixed strategies).

(h) This game can be used to illustrate a shortcoming of almost perfect Bayesian equilibrium.[9] In the Bayesian game, nature first determines the type of the buyer, and then the seller chooses the initial price. The perfect Bayesian equilibrium concept is (should be) unaffected by the timing of nature's move. But the almost perfect Bayesian equilibria are *not* unaffected. In particular, the game with nature moving first has multiple almost perfect Bayesian equilibria. Prove this by verifying that

---

[9]Problem 5.4.9 is a stripped down version of this game.

$p^0 = 3 - \delta$ is part of an almost perfect Bayesian equilibrium for $\delta \in \left(\frac{6}{7}, \frac{9}{10}\right)$. (We already know from the above that for $\delta > \frac{6}{7}$, the game with nature moving second has a unique almost perfect Bayesian equilibrium with $p^0 \neq 3 - \delta$.)

8.4.7. As in the model of Section 8.2, there is an uninformed seller with cost zero facing a buyer with value uniformly distributed on $[0, 1]$. Suppose the seller has a rate of continuous time discounting of r_S (so the seller's discount factor is $\delta_S = e^{-r_S \Delta}$, where $\Delta > 0$ is the time between offers), while the buyer has a rate of continuous time discounting of r_B (so the buyer's discount factor is $\delta_B = e^{-r_B \Delta}$). Solve for an equilibrium of the infinite horizon game in which the uninformed sellers makes all the offers. What happens to the initial offer as $\Delta \to 0$?

8.4.8. Reconsider the two period reputation example (illustrated in Figure 8.3.2) with $\rho > \frac{1}{2}$. Describe all of the equilibria. Which equilibria survive the intuitive criterion?

8.4.9. Verify the claim in Footnote 6 on page 231. More specifically, suppose the two period reputation game in Section 8.3.1 is modeled as an extensive form in which nature first determines the type of the incumbent, and then E_1 chooses between O_1 and I_1.

(a) Prove that in any sequential equilibrium, after O_1, E_2 assigns probability ρ to the tough type.

(b) Prove that in any sequential equilibrium, after I_1, the choices of the incumbent of each type and E_2 are an equilibrium of the game displayed in Figure 8.3.2.

8.4.10. Describe the equilibria of the three period version of the reputation example.

8.4.11. Consider a stage game where player 1 is the row player and 2, the column player (as usual). Player 1 is one of two types ω_n and ω_0. Payoffs are given in Figure 8.4.2. The stage game is played twice, and player 2 is short-lived: a different player 2 plays in different periods, with the second period player 2 observing the action profile chosen in the first period. Describe all the equilibria of the game. Does the intuitive criterion eliminate any of them?

8.4.12. This is a continuation of Problem 7.6.14. Suppose now that the game with the long-lived player 1 and short-lived player 2's is a game of incomplete information. With prior probability $\rho \in (0, 1)$, player 1

	L	R
T	2, 3	0, 2
B	3, 0	1, 1

ω_n

	L	R
T	3, 3	1, 2
B	2, 0	0, 1

ω_0

Figure 8.4.2: The payoffs for Problem 8.4.11.

is a *behavioral type* who chooses T in every period, and with probability $1 - \rho$, he is a strategic or normal type as described above. Suppose $\rho > \frac{1}{2}$. Describe an equilibrium in which the normal type of player 1 has a payoff strictly greater than 2 for large δ.

8.4.13. (Based on Hu, 2014.) Reconsider the infinite horizon reputation game of Section 8.3.3. In addition to the endogenous signal of F and A, there is an exogenous signal $z \in \{z_0, z_1\}$, with

$$1 > \Pr\{z = z_0 \mid \omega_0\} := \alpha > \beta =: \Pr\{z = z_0 \mid \neg\omega_0\} > 0.$$

In period τ, entrant τ observes the history of entry decisions, the behavior of the incumbent in any period in which there was entry, and τ realizations of the exogenous signal.

Fix a Nash equilibrium.

(a) Prove that in any period in which there is entry, the probability of F, conditional on the incumbent not being ω_0, is no larger than $\frac{1}{2}$. Denote this probability by ϕ_τ.

(b) Prove that if the incumbent is not ω_0, and the entrants never enter, then with probability one, the entrants' posterior probability on ω_0 converges to zero.

(c) Let h^t be a positive probability period-t history in which every entry results in F, and there is entry in the last period. Provide an upper bound on the fraction of periods in h^t in which an entrant enters. [**Hint:** Express the odds ratio in period $\tau + 1$ after entry results in F in terms of the odds ratio in period τ, and use part 8.4.13(a).]

Chapter 9

Bargaining

9.1 Axiomatic Nash Bargaining

A bargaining problem is a pair (S, d), where $S \subset \mathbb{R}^2$ is compact and convex, $d \in S$, and there exists $s \in S$ such that $s_i > d_i$ for $i = 1, 2$. Let $\mathcal{B}$ denote the collection of bargaining problems. While d is often interpreted as a disagreement point, this is not the role it plays in the axiomatic treatment. It only plays a role in INV (where its role has the flavor of a normalization constraint) and in SYM. The appropriate interpretation is closely linked to noncooperative bargaining. It is *not* the value of an outside option!

Definition 9.1.1. *A bargaining solution is a function $f : \mathcal{B} \to \mathbb{R}^2$ such that $f(S, d) \in S$.*

9.1.1 The Axioms

1. **PAR (Pareto Efficiency):** If $s \in S$, $t \in S$, $t_i > s_i$, $i = 1, 2$, then
$$f(S, d) \neq s.$$

2. **IIA (Independence of Irrelevant Alternatives):** If $S \subset T$ and $f(T, d) \in S$, then
$$f(S, d) = f(T, d).$$

3. **INV (Invariance to Equivalent Utility Representations):** Given (S, d), and a pair of constants (α_i, β_i) with $\alpha_i > 0$ for each individual $i = 1, 2$, let (S', d') be the bargaining problem given by
$$S' = \{(\alpha_1 s_1 + \beta_1, \alpha_2 s_2 + \beta_2) : (s_1, s_2) \in S\}$$

and
$$d_i' = \alpha_i d_i + \beta_i, \quad i = 1, 2.$$

Then
$$f_i(S', d') = \alpha_i f_i(S, d) + \beta_i, \quad i = 1, 2.$$

4. **SYM (Symmetry):** If $d_1 = d_2$ and $(s_1, s_2) \in S \implies (s_2, s_1) \in S$, then
$$f_1(S, d) = f_2(S, d).$$

Remark 9.1.1. PAR and IIA are quite natural. SYM is also, but perhaps a little less so. The assumption that S is convex can be motivated by the assumption that the bargainers have access to lotteries, in which case INV is also natural.   ♦

9.1.2   Nash's Theorem

Theorem 9.1.1 (Nash, 1950a). *If $f : \mathcal{B} \to \mathbb{R}^2$ satisfies INV, SYM, IIA, and PAR, then*
$$f(S, d) = \underset{(d_1, d_2) \leq (s_1, s_2) \in S}{\arg\max} (s_1 - d_1)(s_2 - d_2) =: f^N(S, d).$$

The function f^N is called the *Nash bargaining solution.* If $S = \{s : s_1 + s_2 \leq 1\}$, then player I's Nash share is
$$s_1^* = \frac{1 + d_1 - d_2}{2}. \tag{9.1.1}$$

Proof. Leave as an exercise that f^N satisfies the four axioms.

Suppose that f satisfies the four axioms. Fix (S, d).

Step 1: Let $z = f^N(S, d)$. Then $z_i > d_i$, $i = 1, 2$. Apply the following affine transformations to move d to the origin $\mathbf{0} := (0, 0)$ and z to $(1/2, 1/2)$:
$$\alpha_i = \frac{1}{2(z_i - d_i)}; \quad \beta_i = \frac{-d_i}{2(z_i - d_i)}.$$

Denote the transformed problem $(S', \mathbf{0})$.

INV implies
$$f_i(S', \mathbf{0}) = \alpha_i f_i(S, d) + \beta_i$$

and
$$f_i^N(S', \mathbf{0}) = \alpha_i f_i^N(S, d) + \beta_i = \frac{1}{2}.$$

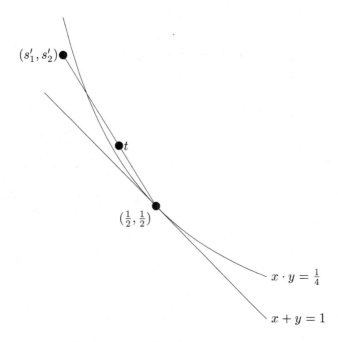

Figure 9.1.1: Illustration of step 2.

Note that $f_i(S, d) = f_i^N(S, d)$ if and only if $f_i(S', \mathbf{0}) = 1/2$.

Step 2: Claim - there does not exist $(s_1', s_2') \in S'$ such that $s_1' + s_2' > 1$.

If not, there exists $s' \in S'$ satisfying $s_1' + s_2' > 1$. Then convexity of S' implies $t = (1 - \varepsilon)(1/2, 1/2) + \varepsilon s' \in S'$, for all $\varepsilon \in (0, 1)$. Moreover, for ε small, $t_1 t_2 > 1/4$, contradicting $f^N(S', \mathbf{0}) = (1/2, 1/2)$ (see Figure 9.1.1).

Step 3: Define

$$T := \left\{ (s_1, s_2) \in \mathbb{R}^2 : s_1 + s_2 \leq 1, |s_i| \leq \max\left\{ |s_1'|, |s_2'| : s' \in S' \right\} \right\}$$

(see Figure 9.1.2). Then, by SYM and PAR, $f(T, \mathbf{0}) = (1/2, 1/2)$.

Step 4: Since $S' \subset T$, IIA implies $f(S', \mathbf{0}) = (1/2, 1/2)$. ∎

For the implications of weakening SYM to a form of individual rationality (and so obtaining the asymmetric Nash bargaining solution), see Problem 9.5.2.

9.2 Rubinstein (1982) Bargaining

Two agents bargain over a pie $[0, 1]$ using an alternating-offer protocol. Time is indexed by t, $t = 1, 2, \ldots$. A proposal is a division of the pie

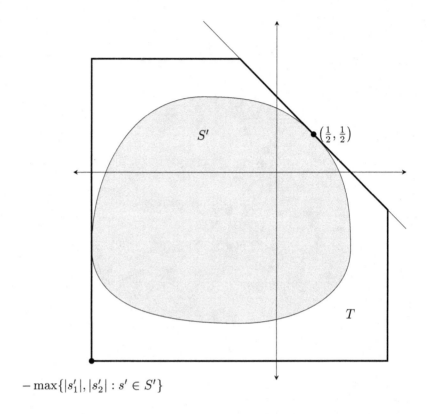

$- \max\{|s_1'|, |s_2'| : s' \in S'\}$

Figure 9.1.2: The bargaining set T.

$(x, 1-x)$, $x \geq 0$. The agents take turns to make proposals. Player I makes proposals on odd t and II on even t. If the proposal $(x, 1-x)$ is agreed to at time t, I's payoff is $\delta_1^{t-1}x$ and II's payoff is $\delta_2^{t-1}(1-x)$. Perpetual disagreement yields a payoff of $(0,0)$. Impatience implies $\delta_i < 1$.

Histories are $h^t \in [0,1]^{\tau-1}$.

Strategies for player I, $\tau_I^1 : \cup_{t\ \text{odd}}[0,1]^{t-1} \to [0,1]$, $\tau_I^2 : \cup_{t\ \text{even}}[0,1]^t \to \{A, R\}$, and for player II, $\tau_{II}^1 : \cup_{t\ \text{even}}[0,1]^{t-1} \to [0,1]$ and $\tau_{II}^2 : \cup_{t\ \text{odd}}[0,1]^t \to \{A, R\}$.

Note that histories have two interpretations, depending upon the context: For the proposal strategy as a function of history, the history is one in which all proposals have been rejected, while for the response strategy, the history is one in which all but the last proposal have been rejected and the current one is being considered.

9.2.1 The Stationary Equilibrium

All the subgames after different even length histories of rejected proposals are strategically identical. A similar comment applies to different odd
length histories of rejected proposals. Finally, all the subgames that follow
different even length histories of rejected proposals followed by the *same*
proposal on the table are strategically identical. Similarly, all the subgames
that follow different odd length histories of rejected proposals followed by
the *same* proposal on the table are strategically identical.

Consider first equilibria in history independent (or stationary) strategies. Recall that a strategy for player I is a pair of mappings, (τ_I^1, τ_I^2).
The strategy τ_I^1 is stationary if, for all $h^t \in [0,1]^{t-1}$ and $\hat{h}^{\hat{t}} \in [0,1]^{\hat{t}-1}$,
$\tau_I^1(h^t) = \tau_I^1(\hat{h}^{\hat{t}})$ (and similarly for the other strategies). Thus, if a strategy
profile is a stationary equilibrium with agreement, there is a pair (x^*, z^*),
such that I expects x^* in any subgame in which I moves first and expects
z^* in any subgame in which II moves first. In order for this to be an equilibrium, I's claim should make II indifferent between accepting and rejecting:
$1 - x^* = \delta_2(1 - z^*)$, and similarly I should be indifferent, so $z^* = \delta_1 x^*$.
[Proof: Consider the first indifference. Player I won't make a claim that II
strictly prefers to $1 - z^*$ next period,[1] so $1 - x^* \leq \delta_2(1 - z^*)$. If II strictly
prefers $(1 - z^*)$ next period, she rejects and gets $1 - z^*$ next period, leaving
I with z^*. But I can offer II a share $1 - z^*$ this period, avoiding the one
period delay.] Solving yields

$$x^* = (1 - \delta_2)/(1 - \delta_1\delta_2),$$

and

$$z^* = \delta_1(1 - \delta_2)/(1 - \delta_1\delta_2).$$

The stationary subgame perfect equilibrium (note that backward induction is not well defined for the infinite horizon game) is for I to always claim
x^* and accept any offer greater than or equal to z^*, and for II to always
offer z^* and always accept any claim less than or equal to x^*. Note the
implicit appeal to the one-shot deviation principle, which also holds here
(the proof is along similar lines, but simpler than, the proof of Theorem
7.1.3).

9.2.2 All Equilibria

While in principle, there could be nonstationary equilibria, it turns out
that there is only one subgame perfect equilibrium. The analysis will also

---

[1]If II does strictly I's claim of x^* to $1 - z^*$ next period, II would also prefer a
marginally more aggressive claim by I, implying I's claim could not have been optimal.

accommodate the possibility of delay in agreement (in other words, we will not assume that equilibrium offers necessarily result in agreement).

Denote by i/j the game in which i makes the initial proposal to j. Define

$$M_i = \sup \{i\text{'s discounted expected payoff}$$
$$\text{in any subgame perfect equilibrium of } i/j\}$$

and

$$m_i = \inf \{i\text{'s discounted expected payoff}$$
$$\text{in any subgame perfect equilibrium of } i/j\}.$$

We begin with two important observations:

1. In any subgame perfect equilibrium, i must accept any offer strictly larger than $\delta_i M_i$.

2. In any subgame perfect equilibrium, i must reject any offer strictly less than $\delta_i m_i$.

Claim 9.2.1. $m_j \geq 1 - \delta_i M_i$.

Proof. The claim is proved by proving that in every equilibrium, player j's payoff in j/i is at least $1 - \delta_i M_i$. Suppose not, that is, suppose there exists an equilibrium yielding a payoff $u_j < 1 - \delta_i M_i$ to j. But this is impossible, since j has a profitable deviation in such an equilibrium: offer $\delta_i M_i + \varepsilon$, ε small. Player i must accept, giving j a payoff of $1 - \delta_i M_i - \varepsilon > u_j$, for ε sufficiently small. ∎

Claim 9.2.2. $M_j \leq 1 - \delta_i m_i$.

Proof. If j makes an equilibrium offer that i accepts, then i must be getting at least $\delta_i m_i$. This implies that j gets no more than $1 - \delta_i m_i$.

The other possibility is that j makes an equilibrium offer that i rejects (in equilibrium). Then, in equilibrium, in the game i/j, i cannot offer j more than $\delta_j M_j$. In this case, j's payoff is no more than $\delta_j^2 M_j$.

So,

$$M_j \quad \leq \quad \max\{\underbrace{1 - \delta_i m_i}_{\text{if } i \text{ accepts}}, \underbrace{\delta_j^2 M_j}_{\text{if } i \text{ rejects}} \}$$
$$\implies \quad M_j \leq 1 - \delta_i m_i.$$

∎

Combining the above two claims,

$$M_j \leq 1 - \delta_i m_i \leq 1 - \delta_i(1 - \delta_j M_j)$$
$$\implies M_j \leq \frac{(1 - \delta_i)}{(1 - \delta_i \delta_j)}, \ M_i \leq \frac{(1 - \delta_j)}{(1 - \delta_i \delta_j)}.$$

This implies

$$m_i \geq 1 - \delta_j \frac{(1 - \delta_i)}{(1 - \delta_i \delta_j)} = \frac{(1 - \delta_j)}{(1 - \delta_i \delta_j)}$$

and so

$$m_i = M_i = \frac{(1 - \delta_j)}{(1 - \delta_i \delta_j)},$$

the stationary equilibrium payoffs derived in Section 9.2.1.

9.2.3 Impatience

In order to investigate the impact of reducing the bargaining friction intrinsic in impatience, we do the following:

Time is continuous, with each round of bargaining taking Δ units of time. If player i has discount rate r_i,

$$\delta_i = e^{-r_i \Delta}.$$

Player 1's share is then

$$x^*(\Delta) = \frac{1 - \delta_2}{1 - \delta_1 \delta_2} = \frac{1 - e^{-r_2 \Delta}}{1 - e^{-(r_1 + r_2)\Delta}}$$

and so

$$\begin{aligned}
\lim_{\Delta \to 0} x^*(\Delta) &= \lim_{\Delta \to 0} \frac{1 - e^{-r_2 \Delta}}{1 - e^{-(r_1 + r_2)\Delta}} \\
&= \lim_{\Delta \to 0} \frac{r_2 e^{-r_2 \Delta}}{(r_1 + r_2)e^{-(r_1 + r_2)\Delta}} \\
&= \frac{r_2}{r_1 + r_2},
\end{aligned}$$

where l'Hôpital's rule was used to get to the second line.

Note that the first mover advantage has disappeared (as it should). The bargaining outcome is determined by relative impatience, and so can be viewed as providing a basis for the asymmetric Nash bargaining solution (Problem 9.5.2).

An alternative basis for the asymmetric Nash bargaining solution is provided in Problem 9.5.7.

9.3   Outside Options

Player II has an outside option of value $(0, b)$.

9.3.1   Version I

Suppose player II can only select her outside option O when rejecting I's proposal, and receives b in that period. See Figure 9.3.1 for the extensive form.

Claim 9.3.1. $m_2 \geq 1 - \delta_1 M_1$.

Proof. Same argument as for Claim 9.2.1. ∎

Claim 9.3.2. $M_1 \leq 1 - b$, $M_1 \leq 1 - \delta_2 m_2$.

Proof. Since II can always opt out, $M_1 \leq 1 - b$. The second inequality, $M_1 \leq 1 - \delta_2 m_2$, follows by the same argument as for Claim 9.2.2. ∎

Claim 9.3.3. $m_1 \geq 1 - \max\{b, \delta_2 M_2\}$, $M_2 \leq 1 - \delta_1 m_1$.

Proof. If $b \leq \delta_2 M_2$, then the argument from Claim 9.2.1 shows that $m_1 \geq 1 - \delta_2 M_2$. If $b > \delta_2 M_2$, then II takes the outside option rather than rejecting and making a counterproposal. Thus, II's acceptance rule is accept any proposal of a share $> b$, and take the outside option for any proposal $< b$. Thus, I's payoffs is $1 - b$.

The remaining inequality, $M_2 \leq 1 - \delta_1 m_1$, is proved using the same argument as for Claim 9.2.2. ∎

Claim 9.3.4. $b \leq \delta_2(1-\delta_1)/(1 - \delta_1\delta_2) \implies m_i \leq (1-\delta_j)/(1 - \delta_1\delta_2) \leq M_i$.

Proof. Follows from the Rubinstein shares being equilibrium shares. ∎

Claim 9.3.5. $b \leq \delta_2(1-\delta_1)/(1 - \delta_1\delta_2) \implies m_i = (1-\delta_j)/(1 - \delta_1\delta_2) = M_i$

Proof. From Claim 9.3.2, $1 - M_1 \geq \delta_2 m_2$, and so from claim 9.3.1, $1 - M_1 \geq \delta_2(1 - \delta_1 M_1)$, and so $M_1 \leq (1 - \delta_2)/(1 - \delta_1\delta_2)$ and so we have equality.

From Claim 9.3.1, $m_2 \geq 1 - \delta_1(1 - \delta_2)/(1 - \delta_1\delta_2) = (1 - \delta_1)/(1 - \delta_1\delta_2)$, and so equality again.

From Claim 9.3.4, $\delta_2 M_2 \geq \delta_2(1 - \delta_1)/(1 - \delta_1\delta_2) \geq b$, and so by Claim 9.3.3, $m_1 \geq 1 - \delta_2 M_2 \geq 1 - \delta_2(1 - \delta_1 m_1)$. Thus, $m_1 \geq (1 - \delta_2)/(1 - \delta_1\delta_2)$, and so equality.

Finally, from Claim 9.3.3,

$$M_2 \leq 1 - \delta_1 m_1 = 1 - \delta_1(1 - \delta_2)/(1 - \delta_1\delta_2) = (1 - \delta_1)/(1 - \delta_1\delta_2).$$

∎

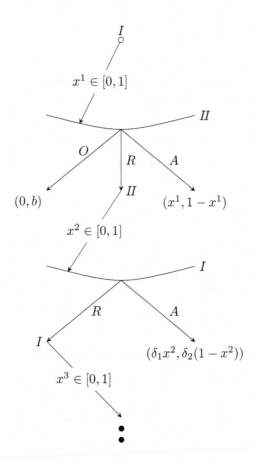

Figure 9.3.1: The first two periods when II can opt out only after rejecting I's proposal.

Thus, if $b \leq \delta_2(1 - \delta_1)/(1 - \delta_1\delta_2)$, equilibrium payoffs are uniquely determined. If $b < \delta_2(1 - \delta_1)/(1 - \delta_1\delta_2)$, then the subgame perfect equilibrium profile is also uniquely determined (player II never takes the outside option). If $b = \delta_2(1 - \delta_1)/(1 - \delta_1\delta_2)$, then there are multiple subgame perfect equilibrium profiles, which differ in whether player II takes the outside option or not after an unacceptable offer.

Claim 9.3.6. $b > \delta_2(1 - \delta_1)/(1 - \delta_1\delta_2) \implies m_1 \leq 1 - b \leq M_1$, $m_2 \leq 1 - \delta_1(1 - b) \leq M_2$.

Proof. Follows from the following being an equilibrium: I always proposes $1 - b$, and accepts any offer of at least $\delta_1(1 - b)$; II always proposes $\delta_1(1 - b)$ and accepts any claim of no more than $1 - b$, opting out if the claim is more than $1 - b$. ∎

Claim 9.3.7. $b > \delta_2(1 - \delta_1)/(1 - \delta_1\delta_2) \implies m_1 = 1 - b = M_1$, $m_2 = 1 - \delta_1(1 - b) = M_2$.

Proof. From Claim 9.3.2, $1 - M_1 \geq b$, i.e., $M_1 \leq 1 - b$, and so we have equality.

From Claim 9.3.1, $m_2 \geq 1 - \delta_1(1 - b)$, and so we have equality.

From Claim 9.3.6, $1 - b \geq m_1$ and so $(1 - \delta_2)/(1 - \delta_1\delta_2) > m_1$.

We now argue that $\delta_2 M_2 \leq b$. If $\delta_2 M_2 > b$, then $m_1 \geq 1 - \delta_2 M_2 \geq 1 - \delta_2(1 - \delta_1 m_1)$ and so $m_1 \geq (1 - \delta_2)/(1 - \delta_1\delta_2)$, a contradiction. Thus, $\delta_2 M_2 \leq b$.

From Claim 9.3.6 and 9.3.3, $1 - b \geq m_1 \geq 1 - \max\{b, \delta_2 M_2\} = 1 - b$ and so $m_1 = 1 - b$.

Finally, this implies $M_2 \leq 1 - \delta_1 m_1 = 1 - \delta_1(1 - b)$, and so equality. ∎

Thus, if $b > \delta_2(1 - \delta_1)/(1 - \delta_1\delta_2)$, equilibrium payoffs are uniquely determined. Moreover, the subgame perfect equilibrium profile is also uniquely determined (player II always takes the outside option after rejection):

$$b > \delta_2(1 - \delta_1)/(1 - \delta_1\delta_2) \iff b > \delta_2[1 - \delta_1(1 - b)].$$

9.3.2 Version II

If II can only select her outside option after I rejects (receiving b in that period, see Figure 9.3.2), then there are multiple equilibria. The equilibrium construction is a little delicate in this case. In fact, for many parameter constellations, there is no pure strategy stationary Markov perfect equilibrium.

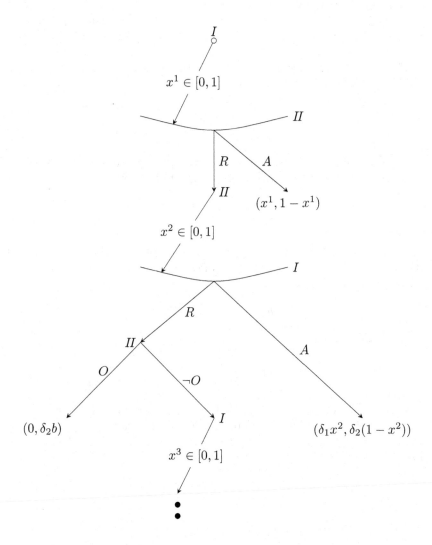

Figure 9.3.2: The first two periods when II can opt out only after I rejects his proposal.

There is, however, a stationary Markov perfect equilibrium in behavior strategies. To illustrate the role of the Markov assumption it is useful to discuss the construction. Note first that, in contrast to the game in Figure 9.3.1, there is only one state at which II can take the outside option. In a Markov strategy profile, I claims x^1 in every odd period, and II offers x^2 in every even period.

If in a putative pure strategy MPE, the outside option is not taken, then we must have, as in Rubinstein,

$$1 - x^1 = \delta_2(1 - x^2)$$
$$\text{and} \quad x^2 = \delta_1 x^1,$$

implying the Rubinstein shares for the two players. This can only be an equilibrium if it is indeed optimal for II not to take the outside option:

$$b \le \delta_2(1 - x^1) = \frac{\delta_2^2(1 - \delta_1)}{(1 - \delta_1\delta_2)}.$$

Suppose now II always takes the outside option. Then, $x^2 = 0$ and $x^1 = 1 - \delta_2$. This is an equilibrium only if it is indeed optimal for II to take the outside option:

$$b \ge \delta_2(1 - x^1) = \delta_2^2.$$

Suppose

$$\frac{\delta_2^2(1 - \delta_2)}{(1 - \delta_1\delta_2)} < b < \delta_2^2. \tag{9.3.1}$$

Then, there is no pure strategy MPE.

It is straightforward to calculate the stationary MPE in behavior strategies under (9.3.1). Let $\beta \in (0, 1)$ be the probability that the outside option is not taken. Then, if (x^1, x^2, β) describes an MPE, we must have (make sure you understand why we need each equality):

$$1 - x^1 = \delta_2(1 - x^2),$$
$$x^2 = \beta\delta_1 x^1,$$
$$\text{and} \quad b = \delta_2(1 - x^1).$$

We then have

$$x^1 = 1 - \frac{b}{\delta_2},$$
$$x^2 = 1 - \frac{b}{\delta_2^2},$$

$$\text{and} \qquad \beta = \frac{\delta_2^2 - b}{\delta_1 \delta_2 (\delta_2 - b)}.$$

It is easy to check that $\beta \in (0,1)$ under (9.3.1).

Note the critical role played by the timing of the decision on the outside option. In particular, even though I strictly prefers that II never take the outside option, I is not able to prevent it, since I's offer is only made after II decides not to take it. In Figure 9.3.1, on the other hand, I always has the option of offering $1 - x^1 > b$.

9.4 Exogenous Risk of Breakdown

We now consider a different modification of the alternating offer bargaining model of Rubinstein (1982). Suppose after any rejection there is a probability $1 - \theta$ of breakdown and in the even of breakdown, the outcome (d_1, d_2) is implemented. With probability θ, bargaining continues to the next round. To simplify notation, assume there is no discounting. Note that since always rejecting is a feasible strategy, $d_i \leq m_i \leq M_i$.

Claim 9.4.1. $m_j \geq 1 - \theta M_i - (1 - \theta) d_i$.

Proof. Note first that i must, in equilibrium, accept any offer strictly great than $\theta M_i + (1 - \theta) d_i$. Suppose $m_j < 1 - \theta M_i - (1 - \theta) d_i$. Then there would exists an equilibrium yielding a payoff $u_j < 1 - \theta M_i - (1 - \theta) d_i$ to j. But j can deviate in such a putative equilibrium, and offer $\theta M_i + (1 - \theta) d_i + \varepsilon$, ε small, which i accepts. This gives j a payoff of $1 - \theta M_i - (1 - \theta) d_i - \varepsilon > u_j$, for ε sufficiently small, and so the deviation is profitable. $\blacksquare$

Claim 9.4.2. $M_j \leq 1 - \theta m_i - (1 - \theta) d_i$.

Proof. In equilibrium, i rejects any offer strictly less than $\theta m_i + (1 - \theta) d_i$ and then i offers no more than $\theta M_j + (1 - \theta) d_j$. So,

$$\begin{aligned}
M_j &\leq \max\left\{1 - \theta m_i - (1 - \theta) d_i, \theta\left[\theta M_j + (1 - \theta) d_j\right] + (1 - \theta) d_j\right\} \\
&\implies M_j \leq 1 - \theta m_i - (1 - \theta) d_i,
\end{aligned}$$

since $M_j < \theta^2 M_j + (1 - \theta^2) d_j \iff M_j < d_j$. $\blacksquare$

The first claim implies

$$\begin{aligned}
M_j &\leq 1 - \theta(1 - \theta M_j - (1 - \theta) d_j) - (1 - \theta) d_i \\
&\implies M_j \leq \frac{(1 + \theta d_j - d_i)}{(1 + \theta)}, \; M_i \leq \frac{(1 + \theta d_i - d_j)}{(1 + \theta)}.
\end{aligned}$$

This implies

$$m_i \geq 1 - \theta \frac{(1 + \theta d_j - d_i)}{(1 + \theta)} - (1 - \theta) d_j = \frac{1 + \theta d_i - d_j}{(1 + \theta)} = M_i$$

and so

$$m_i = M_i = \frac{1 + \theta d_i - d_j}{(1 + \theta)}.$$

Now, we are interested in the payoffs as $\theta \to 1$, and

$$m_i \to \frac{1 + d_i - d_j}{2},$$

so that I's share is

$$x^* = \frac{1 + d_1 - d_2}{2},$$

which is (9.1.1).

For much more on bargaining, see Osborne and Rubinstein (1990).

9.5   Problems

9.5.1. A three person bargaining problem is a pair (S, d), where $S \subset \mathbb{R}^3$ is compact and convex, $d \in S$, and there exists $s \in S$ such that $s_i > d_i$ for $i = 1, 2, 3$. Let $\mathcal{B}$ denote the collection of bargaining problems, and as for the two person case, a bargaining solution is a function $f : \mathcal{B} \to \mathbb{R}^3$ such that $f(S, d) \in S$. The Nash axioms extend in the obvious manner. Prove that if a bargaining solution satisfies INV, SYM, IIA, and PAR, then

$$f(S, d) = \underset{\substack{(s_1, s_2, s_3) \in S, \\ d_i \leq s_i, i = 1, 2, 3}}{\arg \max} (s_1 - d_1)(s_2 - d_2)(s_3 - d_3).$$

9.5.2. Suppose f is a bargaining solution satisfying PAR, IIA, INV, and the following weakening of SYM to IR (individual rationality): $f_i(S, d) \geq d_i$ for $i = 1, 2$. Using the following steps, prove there exists $\alpha \in [0, 1]$ such that

$$f(S, d) = \underset{s \in S, s_i \geq d_i}{\arg \max} (s_1 - d_1)^\alpha (s_2 - d_2)^{1 - \alpha}. \qquad (9.5.1)$$

(This is the *asymmetric Nash bargaining solution*).

(a) Let $S^* := \{s : s_1 + s_2 \leq 1, s_i \geq 0\}$, and define

$$\alpha = f_1(S^*, \mathbf{0}).$$

Verify that PAR implies $1 - \alpha = f_2(S^*, \mathbf{0})$.

(b) Verify that
$$\alpha = \arg\max_{0 \le s_1 \le 1} s_1^\alpha (1 - s_1)^{1-\alpha}.$$

(c) Prove (9.5.1).

9.5.3. Two agents bargain over $[0, 1]$. Time is indexed by t, $t = 1, 2, \ldots, T$, T finite. A proposal is a division of the pie $(x, 1 - x)$, $x \ge 0$. The agents take turns to make proposals. Player I makes proposals on odd t and II on even t. If the proposal $(x, 1 - x)$ is agreed to at time t, I's payoff is $\delta_1^{t-1} x$ and II's payoff is $\delta_2^{t-1}(1 - x)$. Perpetual disagreement yields a payoff of $(0, 0)$. Impatience implies $\delta_i < 1$.

The game ends in period T if all previous proposals have been rejected, with each receiving a payoff of zero.

(a) Suppose T is odd, so that I is the last player to make a proposal. If $T = 1$, the player I makes a take-it-or-leave-it offer, and so in equilibrium demands the entire pie and II accepts. Prove that in the unique backward induction equilibrium, if there are k periods remaining, where k is odd and $k \ge 3$, I's proposal is given by

$$x_k = (1 - \delta_2) \sum_{r=0}^{\tau-1} (\delta_1 \delta_2)^r + (\delta_1 \delta_2)^\tau, \quad \tau = (k - 1)/2.$$

[**Hint:** First calculate x_1 (the offer in the last period), x_2, and x_3. Then write out the recursion, and finally verify that the provided expression satisfies the appropriate conditions.]

(b) What is the limit of x_T as $T \to \infty$?

(c) Suppose now that T is even, so that II is the last player to make a proposal. Prove that in the unique backward induction equilibrium, if there are k periods remaining, where k is even and $k \ge 2$, I's proposal is given by

$$y_k = (1 - \delta_2) \sum_{r=0}^{\tau-1} (\delta_1 \delta_2)^r, \quad \tau = k/2.$$

(d) What is the limit of y_T as $T \to \infty$?

9.5.4. (a) Give the details of the proof of Claim 9.3.4.

(b) Give the details of the proof of Claim 9.3.6.

9.5.5. We will use the finite horizon bargaining result from Problem 9.5.3 to give an alternative proof of uniqueness in the Rubinstein model.

   (a) Prove that in any subgame perfect equilibrium of the game in which I offers first, I's payoff is no more than x_k, for all k odd. [**Hint:** Prove by induction (the result is clearly true for $k = 1$).]

   (b) Prove that in any subgame perfect equilibrium of the game in which I offers first, I's payoff is no less than y_k, for all k even.

   (c) Complete the argument.

9.5.6. Let $G(n)$ denote Rubinstein's alternating offer bargaining game with discounting with the following change: The share going to player 1 must be an element of the finite set $A(n) = \{0, \frac{1}{n}, \ldots, \frac{(n-1)}{n}, 1\}$, where $n \in \{1, 2, 3, \ldots\}$. Suppose the two players have identical discount factors δ.

   (a) Show that for any n, there exists $\delta' \in (0, 1)$ such that for any $\delta \in (\delta', 1)$, the continuum case proof that there is a unique partition of the pie achievable in a subgame perfect equilibrium fails for $G(n)$. Show that any share to player 1 in $A(n)$ can be achieved in a subgame perfect equilibrium of $G(n)$.

   (b) Prove or provide a counterexample to the following claim: For any $\delta \in (0, 1)$ and any $\varepsilon > 0$, there exists n' such that for any $n > n'$ the share going to player 1 in any subgame perfect equilibrium is within ε of the Rubinstein solution.

   (c) Suppose

$$\frac{1}{1 + \delta} \neq \frac{k}{n} \qquad \forall k.$$

   Describe a symmetric stationary equilibrium that, for large n, approximates the Rubinstein solution.

9.5.7. Players 1 and 2 are bargaining over the pie $[0, 1]$ using the following modification of the alternating offer protocol. In each period, a biased coin determines who makes the offer, with player i selected with probability $p_i \in (0, 1)$, $p_1 + p_2 = 1$. If the offer is accepted, the offer is implemented and the game is over. If the offer is rejected, play proceeds to the next period, until agreement is reached. Players have the same discount factor $\delta \in (0, 1)$. Suppose each player is risk neutral (so that player i's utility from receiving x is given by $u_i(x) = x$). Describe a stationary subgame perfect equilibrium. Are there any other stationary subgame perfect equilibria? Argue that this protocol provides a noncooperative foundation for the asymmetric Nash

Bargaining solution as $\delta \to 1$. Describe its comparative statics (with respect to p_i, particularly near the boundaries) and provide some intuition.

9.5.8. There is a single seller who has a single object to sell (the seller's reservation utility is zero). There are two potential buyers, and they each value the object at 1. If the seller and buyer i agree to a trade at price p in period t, then the seller receives a payoff of $\delta^{t-1}p$, buyer i a payoff of $\delta^{t-1}(1 - p)$, and buyer $j \neq i$ a payoff of zero. Consider alternating offer bargaining, with the seller choosing a buyer to make an offer to (name a price to). If the buyer accepts, the game is over. If the buyer rejects, then play proceeds to the next period, when the buyer who received the offer in the preceding period makes a counter-offer. If the offer is accepted, it is implemented. If the offer is rejected, then the seller makes a new proposal to either the same buyer or to the other buyer. Thus, the seller is free to switch the identity of the buyer he is negotiating with after every rejected offer from the buyer.

(a) Suppose that the seller can only make proposals in odd-numbered periods. Prove that the seller's subgame perfect equilibrium payoff is unique, and describe it. Describe the subgame perfect equilibria. The payoffs to the buyers are not uniquely determined. Why not?

(b) Now consider the following alternative. Suppose that if the seller rejects an offer from the buyer, he can either wait one period to make a counteroffer to this buyer, or he can *immediately* make an offer to the other buyer. Prove that the seller's subgame perfect equilibrium payoff is unique, and describe it. Describe the subgame perfect equilibria. [Cf. Shaked and Sutton (1984).]

9.5.9. Extend alternating offer bargaining to three players as follows. Three players are bargaining over a pie of size 1. The game has perfect information and players take turns (in the order of their index). A proposed division (or offer) is a triple (x_1, x_2, x_3), with x_i being player i's share. All players must agree on the division. Play begins in period 1 with player 1 making a proposal, which player 2 either accepts or rejects. If 2 accepts, then 3 either accepts or rejects. If both accept, the offer is implemented. If either reject, the game proceeds to period 2, where player 2 makes a proposal, which player 3 either accepts or rejects. If 3 accepts, then 1 either accepts or rejects. If both accept, the offer is implemented in period 2. If either reject, then play proceeds to period 3, where 3 makes a proposal and 1 and 2 in turn accept or reject. Players rotate proposals and responses like this until agreement is reached. Player i has discount factor $\delta_i \in (0, 1)$.

(a) What is the stationary Markov equilibrium? Is it unique?

(b) [**Hard**] This game has many non-Markovian equilibria. Describe one.

9.5.10. (Based on Fernandez and Glazer, 1991.[2]) A firm and its employee union are negotiating a new wage contract. For simplicity assume the new contract will never be renegotiated. Normalize the value of output produced in each period to 1. The current wage is $\underline{w} \in [0, 1]$. The union and the firm alternate in making wage offers over discrete time: $t = 1, 2, \ldots$. In each odd period t, the union proposes a wage offer x. The firm then responds by either accepting (A) or rejecting (R). If the firm accepts the offer, negotiations are over and the newly agreed upon wage is implemented: $w_\tau = x$ and $\pi_\tau = 1 - x$ for all $\tau \geq t$ (where w_τ is the wage and π_τ is the profit in period τ). If the firm rejects the wage offer, the union decides whether to strike (s) or not to strike (n). If the union decides not to strike, workers work and receive the old wage $w_t = \underline{w}$, and the firm receives $\pi_t = 1 - \underline{w}$. If the union decides to strike, workers get $w_t = 0$ and the firm gets $\pi_t = 0$. After the union's strike decision, negotiations continue in the next period. In every even period, the firm offers the union a wage offer z. The union then responds by either accepting (A) or rejecting (R). Once again acceptance implies the newly agreed upon wage holds thereafter. If the union rejects, the union again must decide whether to strike (s) or not (n). After the union strike decision, negotiations continue in the next period.

Both the firm and workers discount future flow payoffs with the same discount factor $\delta \in (0, 1)$. The union's objective is to maximize workers' utility:

$$(1 - \delta) \sum_{t=1}^{\infty} \delta^{t-1} w_t.$$

The firm's objective is to maximize its discounted sum of profits:

$$(1 - \delta) \sum_{t=1}^{\infty} \delta^{t-1} \pi_t.$$

(a) If the union is *precommitted* to striking in every period in which there is a disagreement (i.e., the union does not have the option

---

[2]See Busch and Wen (1995) for a more general treatment.

of not striking), then there is a unique subgame perfect equilibrium of the game. Describe it (you are not asked to prove uniqueness).

(b) Explain why the option of not striking is not an outside option (in the sense of Section 9.3).

(c) Describe a stationary Markov equilibrium.

(d) Suppose $0 < \underline{w} \leq \frac{\delta^2}{1+\delta}$. There is a subgame perfect equilibrium which gives the same equilibrium outcome as that in part (a). Describe it and verify it is an equilibrium. (**Hint:** Use the equilibrium you found in part (c).)

(e) Describe an equilibrium with inefficient delay.

Chapter 10

Introduction to Mechanism Design[1]

10.1   A Simple Screening Example

How should a seller sell to a buyer with an unknown marginal willingness to pay? Specifically, suppose the buyer has preferences

$$\theta q - p,$$

where $\theta > 0$ is the buyer's marginal value, $q \geq 0$ is the quantity purchased, and p is the *total* price paid. The seller has costs given by the increasing convex $\mathcal{C}^2$ function

$$c(q),$$

satisfying $c'(0) = 0$, $c'' \geq 0$, and $\lim_{q \to \infty} c'(q) = \infty$.

If the seller knew the value of θ, then he would set $p = \theta q$ and choose q to maximize

$$p - c(q) = \theta q - c(q),$$

that is, choose q so that marginal cost $c'(q)$ equals θ (this is also the efficient quantity).

But what if the seller does not know the value of θ? Suppose the seller assigns probability $\alpha_H \in (0, 1)$ to $\theta = \theta_H$ and complementary probability $\alpha_L = 1 - \alpha_H$ to $\theta_L < \theta_H$. Such a situation is said to exhibit adverse selection: a relationship is subject to *adverse selection* if one of the participants

---

[1]I have shamelessly stolen the idea of introducing mechanism design in this way from Börgers (2015).

263

has private information, and uninformed participants are disadvantaged by this *hidden information* (another common term).

If the seller could price discriminate, how would he do so? If the seller price discriminates, he charges different prices for different quantities. Since there are two types, it is natural to think of the seller as offering two quantities, q_1 and q_2 and associated prices, p_1 and p_2, so that price is a function of quantity. But, in choosing these two quantities (and the associated prices), the seller is targeting the buyer type, so in what follows we think of the seller as offering a pair of *screening contracts* $\{(q_L, p_L), (q_H, p_H)\}$, with the intent that the buyer of type θ_t chooses the contract (q_t, p_t).

The seller chooses the pair of contracts to maximize

$$\alpha_L(p_L - c(q_L)) + \alpha_H(p_H - c(q_H))$$

subject to

$$\theta_L q_L - p_L \geq \theta_L q_H - p_H, \tag{IC$_L$}$$

$$\theta_H q_H - p_H \geq \theta_H q_L - p_L, \tag{IC$_H$}$$

$$\theta_L q_L - p_L \geq 0, \tag{IR$_L$}$$

$$\text{and} \quad \theta_H q_H - p_H \geq 0. \tag{IR$_H$}$$

Inequalities (IC$_L$) and (IC$_H$) are the *incentive compatibility* conditions that ensure the seller's targeting of buyer types is successful. Conditions (IR$_L$) and (IR$_H$) are the *individual rationality* or *participation constraints* that ensure that both buyer types at least weakly prefer to purchase. These last two constraints are illustrated in Figure 10.1.1.

It is immediate (both from the figure and algebraic manipulation of the IC and IR constraints) that (IR$_L$) and (IC$_H$) jointly imply (IR$_H$). Moreover, (IC$_L$) and (IC$_H$) imply $q_H \geq q_L$. Finally, if $q_L \neq q_H$, only one of (IC$_L$) and (IC$_H$) can bind.

The Langrangean for this maximization is

$$\mathcal{L} = \alpha_L(p_L - c(q_L)) + \alpha_H(p_H - c(q_H))$$
$$+ \lambda_L[\theta_L q_L - p_L - (\theta_L q_H - p_H)] + \lambda_H[\theta_H q_H - p_H - (\theta_H q_L - p_L)]$$
$$+ \mu_L[\theta_L q_L - p_L] + \mu_H[\theta_H q_H - p_H].$$

The first order conditions for an interior solution are

$$-\alpha_L c'(q_L) + \lambda_L \theta_L - \lambda_H \theta_H + \mu_L \theta_L = 0, \tag{10.1.1}$$

$$\alpha_L - \lambda_L + \lambda_H - \mu_L = 0, \tag{10.1.2}$$

$$-\alpha_H c'(q_H) - \lambda_L \theta_L + \lambda_H \theta_H + \mu_H \theta_H = 0, \text{ and} \tag{10.1.3}$$

$$\alpha_H + \lambda_L - \lambda_H - \mu_H = 0, \tag{10.1.4}$$

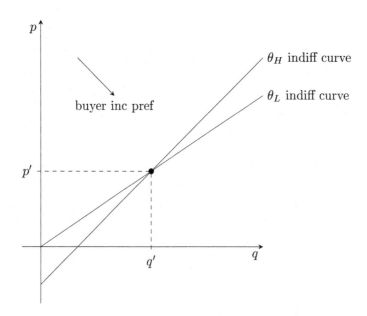

Figure 10.1.1: The buyer's indifference curves for different values of θ. Inequality (IR$_L$) binds for the contract (q', p'), since that buyer's indifference curve passes through the origin (and so the buyer is indifferent between the contract and not purchasing). Buyer θ_H strictly prefers the contract (q', p') to not purchasing, and so (IR$_H$) holds as a strict inequality.

together with the complementary slackness conditions on the multipliers.

From above, since (IR$_H$) does not bind, $\mu_H = 0$. Moreover, since only one of the two incentive constraints can bind, either λ_L or λ_H equal zero. But (10.1.4), $\mu_H = 0$, and $\lambda_L \geq 0$ imply $\lambda_H > 0$, so $\lambda_L = 0$. Then, (10.1.4) implies $\lambda_H = \alpha_H$, and so (10.1.3) implies q_H satisfies

$$c'(q_H) = \theta_H,$$

and so is the full information profit maximizing quantity. From $\lambda_H = \alpha_H$ and (10.1.2) we conclude $\mu_L = 1$. Finally, from (10.1.1), we have

$$-\alpha_L c'(q_L) - \alpha_H \theta_H + \theta_L = 0,$$

so that

$$c'(q_L) = \frac{\theta_L - \alpha_H \theta_H}{(1 - \alpha_H)}.$$

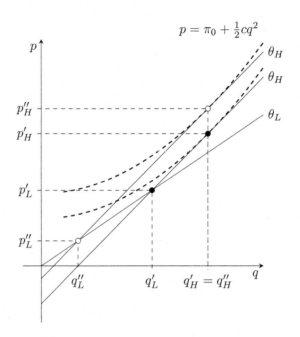

Figure 10.1.2: Optimal screening contracts for two different values of α, with $\alpha_H'' > \alpha_H'$, with quadratic costs. The two dashed curved lines are isoprofit loci for the seller.

This only implies an interior solution if $\theta_L > \alpha_H \theta_H$. If $\theta_L \le \alpha_H \theta_H$, $q_L = p_L = 0$, the seller is better off selling only to θ_H, and does not sell to θ_L (see Problem 10.5.1).

The prices p_L and p_H are then determined from (IR$_L$) and (IC$_H$) respectively (see Figure 10.1.2 for the case of quadratic costs, $c(q) = \frac{1}{2}cq^2$).

When the seller optimally sells to both types of buyer, buyer θ_H *strictly* prefers to purchase rather than not purchase, and this additional payoff (over zero) is called buyer θ_H's *information rent*.

Note that $q_L \searrow 0$ as α_H increases. As α_H increases, it is more valuable to the seller to sell to θ_H at a higher price. But in order to do so, it must make the low value contract less attractive to θ_H. Increasing the price on q_L does this but at the cost of losing sales to θ_L, since θ_L is already just indifferent between buying and not buying. But since θ_L values the good less than does θ_H, by simultaneously lowering q_L and p_L at just the right rate, the seller simultaneously maintains some sales to θ_L while making the contract less attractive to θ_H.

Rather than offering a pair of contracts, we can equivalently think of the seller as committing to the following *direct mechanism*: first the seller specifies which contract the buyer will receive as a function of the buyer's announcement of her value of θ, and then the buyer chooses which value of θ to report. If the seller's specification satisfies (IC$_L$), (IC$_H$), (IR$_L$), and (IR$_H$), then the buyer finds it optimal to both participate in the mechanism and truthfully report her valuation. This is a trivial example of the *revelation principle*, which we will return to several times (Theorems 11.2.1 and 12.1.1).

10.2 A Less Simple Screening Example

In this section, we suppose the seller has beliefs F over the buyer's valuation θ, with F having support $[\underline{\theta}, \overline{\theta}]$ and strictly positive density f. The possibilities for price discrimination are now significantly larger, since there are many more types. Rather than specifying price as a function of quantity, it is more convenient to apply the revelation principle (outlined at the end of the previous section) and model the seller as choosing *direct mechanism*, i.e., a menu of contracts $\{(q(\theta), p(\theta)) : \theta \in [\underline{\theta}, \overline{\theta}]\}$, with contract $(q(\theta), p(\theta))$ intended for buyer θ (but see Remark 10.2.1).

The seller chooses the pair of functions $(q, p) : [\underline{\theta}, \overline{\theta}] \to \mathbb{R}_+^2$ to maximize

$$\int [p(\theta) - c(q(\theta))] f(\theta) \, d\theta \tag{Π}$$

subject to

$$\theta q(\theta) - p(\theta) \geq \theta q(\hat{\theta}) - p(\hat{\theta}) \quad \forall \theta, \hat{\theta} \in [\underline{\theta}, \overline{\theta}] \tag{IC$_\theta$}$$

$$\text{and} \quad \theta q(\theta) - p(\theta) \geq 0 \quad \forall \theta \in [\underline{\theta}, \overline{\theta}]. \tag{IR$_\theta$}$$

Lemma 10.2.1. *Suppose a pair of functions (q, p) satisfies (IC$_\theta$), and define*

$$U(\theta) := \theta q(\theta) - p(\theta).$$

The function U is nondecreasing and convex, q is nondecreasing, and

$$U(\theta) = U(\underline{\theta}) + \int_{\underline{\theta}}^{\theta} q(\tilde{\theta}) \, d(\tilde{\theta}). \tag{10.2.1}$$

Proof. For each $\hat{\theta}$, the function defined by

$$g_{\hat{\theta}}(\theta) := \theta q(\hat{\theta}) - p(\hat{\theta})$$

is a nondecreasing affine function of θ.

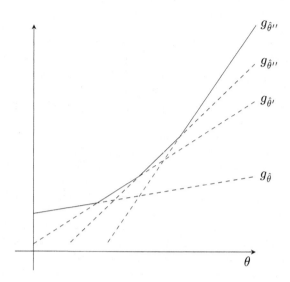

Figure 10.2.1: The upper envelope of affine functions is convex.

The conditions (IC_θ) imply (since $U(\theta) = g_\theta(\theta)$)

$$U(\theta) = \max_{\hat\theta \in [\underline\theta, \overline\theta]} g_{\hat\theta}(\theta) \qquad \forall \theta.$$

That is, U is the pointwise maximum (upper envelope) of the collection of nondecreasing affine functions $\{g_{\hat\theta} : \hat\theta \in [\underline\theta, \overline\theta]\}$. It is, therefore, nondecreasing and convex (see Figure 10.2.1).[2]

Every convex function on an open interval is absolutely continuous (Royden and Fitzpatrick, 2010, page 132, Corollary 17) and so differentiable almost everywhere. From the envelope characterization, the derivative of U at θ (when it exists) is given by $q(\theta)$ (Problem 10.5.5), and since it is the derivative of a convex function, q is nondecreasing when it is the derivative.[3] To prove that q is nondecreasing everywhere, add (IC_θ) to the same constraint reversing the roles of θ and θ' and rearrange to conclude

$$(\theta - \theta')(q(\theta) - q(\theta')) \geq 0.$$

---

[2] If each f_α is convex and $f(x) := \sup_\alpha f_\alpha(x)$, then $f(tx + (1-t)x') = \sup_\alpha f_\alpha(tx + (1-t)x') \leq \sup_\alpha\{tf_\alpha(x) + (1-t)f_\alpha(x')\} \leq t\sup_\alpha f_\alpha(x) + (1-t)\sup_\alpha f_\alpha(x') = tf(x) + (1-t)f(x')$.

[3] A more direct and intuitive (but less elegant) argument that implicitly assumes continuity of $q(\cdot)$ (as well as ignoring the issue of footnote 4 on page 269) is in Mas Colell, Whinston, and Green (1995, page 888). See also the proof of Lemma 12.2.1.

Finally, since U is absolutely continuous on $(\underline{\theta}, \overline{\theta})$, it is equal to the integral of its derivative.[4] That is, for all $\theta' < \theta'' \in (\underline{\theta}, \overline{\theta})$, we have

$$U(\theta'') = U(\theta') + \int_{\theta'}^{\theta''} q(\tilde{\theta}) \, d(\tilde{\theta}). \qquad (10.2.2)$$

Since U satisfies (IC_θ) and is nondecreasing, we have that for all $\theta^\dagger < \theta^\ddagger \in [\underline{\theta}, \overline{\theta}]$,

$$U(\theta^\ddagger) - (\theta^\ddagger - \theta^\dagger)q(\theta^\ddagger) \leq U(\theta^\dagger) \leq U(\theta^\ddagger), \qquad (10.2.3)$$

and so U is continuous on the closed interval $[\underline{\theta}, \overline{\theta}]$.[5] Consequently, (10.2.2) holds with $\theta' = \underline{\theta}$ and $\theta'' = \overline{\theta}$.

∎

It is immediate from (10.2.1) that U is continuous in θ, but as (10.2.3) reveals, continuity is a direct implication of (IC_θ).

Since the participation constraint is $U(\theta) \geq 0$, Lemma 10.2.1 implies that $U(\underline{\theta}) \geq 0$ is sufficient to imply participation for all θ. For the following result, it is enough to find a p for which $U(\underline{\theta}) = 0$ (something the firm's optimal (q, p) would presumably satisfy).

Lemma 10.2.2. *Suppose* $q : [\underline{\theta}, \overline{\theta}] \to \mathbb{R}_+$ *is a nondecreasing function. Then,* (q, p) *satisfies* (IC_θ) *and* (IR_θ), *where*

$$p(\theta) = \theta q(\theta) - \int_{\underline{\theta}}^{\theta} q(\tilde{\theta}) \, d\tilde{\theta} \qquad (10.2.4)$$

$$= \underline{\theta} q(\theta) + \int_{\underline{\theta}}^{\theta} [q(\theta) - q(\tilde{\theta})] \, d\tilde{\theta}.$$

Proof. Define

$$U^*(\theta) := \int_{\underline{\theta}}^{\theta} q(\tilde{\theta}) \, d\tilde{\theta} = \theta q(\theta) - p(\theta).$$

Then, U^* is a convex function with $U^*(\underline{\theta}) = 0$, and (IR_θ) is satisfied.

Moreover, since it is convex, for all $\hat{\theta}$,

$$U^*(\theta) \geq U^*(\hat{\theta}) + (\theta - \hat{\theta}) \frac{dU^*(\hat{\theta})}{d\theta}$$

---

[4]See Royden and Fitzpatrick (2010, Theorem 10, page 124). Absolute continuity is needed here, as there are functions for which the fundamental theorem of calculus fails. For an example of a strictly increasing function with zero derivative almost everywhere, see Billingsley (1995, Example 31.1).

[5]Continuity on the *open* interval follows from convexity, with continuity at $\underline{\theta}$ then being an implication of U being nondecreasing. There are nondecreasing convex functions that are discontinuous at $\overline{\theta}$, such as the function obtained from a continuous nondecreasing convex function by adding 1 to the function at $\overline{\theta}$.

$$= U^*(\hat{\theta}) + (\theta - \hat{\theta})q(\hat{\theta})$$
$$= \theta q(\hat{\theta}) - p(\hat{\theta}),$$

proving (IC$_\theta$).   ∎

Thus, to solve the seller's problem we need only maximize

$$\int_{\underline{\theta}}^{\overline{\theta}} \left\{ \theta q(\theta) - \int_{\underline{\theta}}^{\theta} q(\tilde{\theta}) \, d\tilde{\theta} - c(q(\theta)) \right\} f(\theta) \, d\theta$$

over all nondecreasing functions q.

We first apply integration by parts (i.e., $\int uv' = uv - \int u'v$) to the double integral:[6]

$$\int_{\underline{\theta}}^{\overline{\theta}} \int_{\underline{\theta}}^{\theta} q(\tilde{\theta}) \, d\tilde{\theta} \, f(\theta) \, d\theta = \left(F(\theta) \int_{\underline{\theta}}^{\theta} q(\tilde{\theta}) \, d\tilde{\theta} \right)\Bigg|_{\theta=\underline{\theta}}^{\overline{\theta}} - \int_{\underline{\theta}}^{\overline{\theta}} F(\theta)q(\theta) \, d\theta$$

$$= \int_{\underline{\theta}}^{\overline{\theta}} q(\tilde{\theta}) \, d\tilde{\theta} - \int_{\underline{\theta}}^{\overline{\theta}} F(\theta)q(\theta) \, d\theta$$

$$= \int_{\underline{\theta}}^{\overline{\theta}} (1 - F(\theta))q(\theta) \, d\theta.$$

Substituting into the seller's objective function, we have

$$\int_{\underline{\theta}}^{\overline{\theta}} \left\{ \left(\theta - \frac{(1 - F(\theta))}{f(\theta)} \right) q(\theta) - c(q(\theta)) \right\} f(\theta) \, d\theta.$$

Consider maximizing the integrand pointwise. Since the cost function c is convex, the first order condition is sufficient for an interior maximum. If

---

[6]An alternative derivation is via Fubini's theorem (where $\chi(\tilde{\theta}, \theta) = 1$ if $\tilde{\theta} \le \theta$ and 0 otherwise):

$$\int_{\underline{\theta}}^{\overline{\theta}} \int_{\underline{\theta}}^{\theta} q(\tilde{\theta}) \, d\tilde{\theta} \, f(\theta) \, d\theta = \int_{\underline{\theta}}^{\overline{\theta}} \int_{\underline{\theta}}^{\overline{\theta}} \chi(\tilde{\theta}, \theta)q(\tilde{\theta})f(\theta) \, d\tilde{\theta} d\theta$$

$$= \int_{\underline{\theta}}^{\overline{\theta}} \int_{\underline{\theta}}^{\overline{\theta}} \chi(\tilde{\theta}, \theta)q(\tilde{\theta})f(\theta) \, d\theta d\tilde{\theta}$$

$$= \int_{\underline{\theta}}^{\overline{\theta}} \int_{\tilde{\theta}}^{\overline{\theta}} q(\tilde{\theta})f(\theta) \, d\theta d\tilde{\theta}$$

$$= \int_{\underline{\theta}}^{\overline{\theta}} q(\tilde{\theta}) \int_{\tilde{\theta}}^{\overline{\theta}} f(\theta) \, d\theta d\tilde{\theta} = \int_{\underline{\theta}}^{\overline{\theta}} (1 - F(\tilde{\theta}))q(\tilde{\theta}) \, d\tilde{\theta}.$$

$q = q(\theta)$ maximizes the integrand, it satisfies the first order condition

$$\theta - \frac{1 - F(\theta)}{f(\theta)} = c'(q). \tag{10.2.5}$$

There remains the possibility that there may not be a strictly positive q solving (10.2.5), as would occur if the left side were negative.

Theorem 10.2.1. *Suppose the function*

$$\psi(\theta) := \theta - \frac{1 - F(\theta)}{f(\theta)} \tag{10.2.6}$$

is nondecreasing in θ. For all θ, denote by $q^*(\theta)$ the quantity given by

$$q^*(\theta) := \begin{cases} 0, & \psi(\theta) \le 0, \\ q(\theta), & \psi(\theta) > 0, \end{cases}$$

where $q(\theta)$ is the unique quantity q satisfying (10.2.5), and denote by $p^*(\theta)$ the right side of (10.2.4), i.e.,

$$p^*(\theta) := \theta q^*(\theta) - \int_{\underline{\theta}}^{\theta} q^*(\tilde{\theta}) \, d\tilde{\theta}.$$

The pair of functions (q^*, p^*) maximizes the seller's profits (Π) subject to (IC$_\theta$) and (IR$_\theta$). Moreover, any pair of functions $(\hat{q}^*, \hat{p}^*)$ that maximizes the seller's profits (Π) subject to (IC$_\theta$) and (IR$_\theta$) equals (q^*, p^*) almost surely.

Proof. If $\psi(\theta) > 0$, since the marginal cost is unbounded above, continuous with value 0 at 0, there exists a unique q solving (10.2.5).[7] Since ψ is nondecreasing in θ, the function q^* is also nondecreasing and so by Lemma 10.2.2, (q^*, p^*) is an admissible solution. Since $q^*(\theta)$ uniquely pointwise maximizes the integrand, it is the optimal quantity schedule. Since the optimal choices for the seller must leave the bottom buyer indifferent between accepting and rejecting the contract, $U(\underline{\theta}) = 0$ and the optimal price schedule is given by (10.2.4). ∎

The function ψ defined in (10.2.6) is called the *virtual type* or *virtual value*, and that assumption that ψ is increasing in θ is called the *regular case*. Note that there is no distortion at the top (since $F(\overline{\theta}) = 1$).

---

[7]If c' is bounded above, (10.2.5) may not have a solution. In particular, if $\psi(\theta)$ exceeds $c'(q)$ for all q, the integrand is unbounded at that θ, precluding the existence of a maximum over all $q \ge 0$. This is the only place the assumption that c' is unbounded above is used.

Moreover, every type θ earns an *information rent* of

$$\int_{\underline{\theta}}^{\theta} q^*(\tilde{\theta}) \, d\tilde{\theta}$$

(which is zero for $\theta = \underline{\theta}$).

Remark 10.2.1 (The Taxation Principle). In practice, it is more natural to think of the buyer as choosing a quantity q while facing a nonlinear pricing schedule, P (as in Section 10.1). Such a nonlinear schedule is said to *screen* buyers. The observation that one can always implement an optimal direct mechanism by an equivalent nonlinear pricing schedule is called the *taxation principle*. Problem 10.5.2 goes through the steps for the current setting. The taxation principle is an "inverse" of the revelation principle. ◆

10.3   The Take-It-or-Leave-It Mechanism

We continue with the previous section's example, but now consider the case of a seller selling *one* unit of a good to a buyer with unknown valuation. The preceding analysis applies once we interpret q as the probability of trade, which of course requires $q \in [0,1]$. Conditional on a report $\hat{\theta}$, there is a probability of trade $q(\hat{\theta})$ and a payment $p(\hat{\theta})$. The seller has an opportunity cost of provision of $c < \bar{\theta}$, so the expected cost is

$$c(q) = cq.$$

Suppose we are in the regular case, so that

$$\psi(\theta) = \theta - \frac{1 - F(\theta)}{f(\theta)}$$

is increasing in θ. Note that $E\psi(\theta) = \underline{\theta}$.

The seller chooses a nondecreasing probability of trade to maximize

$$\int_{\underline{\theta}}^{\bar{\theta}} \left\{ \theta - \frac{1 - F(\theta)}{f(\theta)} - c \right\} q(\theta) \, f(\theta) \, d\theta.$$

Denote by θ^* a value of θ that solves

$$\theta - \frac{1 - F(\theta)}{f(\theta)} = c;$$

note that $\theta^* > c$. The maximum is clearly achieved by choosing

$$q^*(\theta) = \begin{cases} 1, & \text{if } \theta \geq \theta^*, \\ 0, & \text{if } \theta < \theta^*. \end{cases}$$

From (10.2.4), it is immediate that

$$p^*(\theta) = \begin{cases} \theta^*, & \text{if } \theta \geq \theta^*, \\ 0, & \text{if } \theta < \theta^*. \end{cases}$$

In other words, the seller optimally chooses a price $p = \theta^*$, every buyer with a valuation above θ^* buys, and every buyer with a valuation below θ^* does not.

The conclusion that the take-it-or-leave-it mechanism is optimal does not require being in the regular case (though the proof does require more advanced techniques, see Section 2.2 of Börgers, 2015).

10.4   Implementation

We begin with a simple example of implementation.

Example 10.4.1. Two children (Bruce and Sheila) have to share a pie, which has a cherry on top. Both Bruce and Sheila would like more of the pie, as well as the cherry, with both valuing the cherry at $\theta \in (0,1)$. A division of the pie is denoted by $x \in [0,1]$, where x is the size of the slice with the cherry, with the other slice having size $1 - x$ (it is not possible to divide the cherry). An allocation is (x,i) where $i \in \{1,2\}$ is the recipient of the slice with the cherry. Payoffs are

$$v_j(x,i) = \begin{cases} \theta + x, & \text{if } j = i, \\ 1 - x, & \text{if } j \neq i, \end{cases}$$

for $i, j \in \{B, S\}$. If the parent knew that $\theta = 0$, then the parent can trivially achieve a fair (envy-free) division by cutting the pie in half. But suppose the parent does *not* know θ (but that the children do). How can the parent achieve an equitable (envy-free) division? The parent has the children play the following game: Bruce divides the pie into two slices, and Sheila then chooses which slice to take (with Bruce taking the remaining slice). It is straightforward to verify that in any subgame perfect equilibrium, the resulting division of the pie is envy-free: both Bruce and Sheila are indifferent over which slice they receive.   ★

Let $\mathcal{E}$ denote a set of environments and Z the outcome space.

Definition 10.4.1. *A social choice function* (scf) *is a function*

$$\xi : \mathcal{E} \to Z.$$

Example 10.4.2 (Social choice). Let Z be a finite set of social alternatives, $\{x, y, z, \ldots\}$. Let $\mathcal{R}$ denote the collection of preference orderings on Z; $R \in Z$ satisfies completeness and transitivity. We write xPy if xRy but not yRx (that is, x is strictly preferred to y). There are n members of society, so $\mathcal{E} = \mathcal{R}^n$.

A social choice function ξ satisfies *unanimity* if for all $e \in \mathcal{E}$ satisfying xP_iy for all $y \in Z$ and all i, we have $\xi(e) = x$.   ★

Example 10.4.3. A single indivisible good to be allocated to one of n people. Then $Z = \{0, 1, \ldots, n\} \times \mathbb{R}^n$, where $(j, t_1, \ldots, t_n) \in Z$ is the outcome where j receives the good (if $j = 0$, no one receives the good), and agent i pays the seller t_i (an amount that may be negative).

If player i values the good at v_i, then i's payoffs are

$$u_i(j, t_1, \ldots, t_n) = \begin{cases} v_i - t_i, & \text{if } i = j, \\ -t_i, & \text{if } i \neq j. \end{cases}$$

The set of environments is $\mathcal{E} = \prod_i V_i$, where V_i is the set of possible valuations for player i, so that $e = (v_1, \ldots, v_n)$.

Efficient scf: $\xi(e) = (j, t_1, \ldots, t_n)$ only if $v_j \geq \max_i v_i$.

Maximally revenue extracting and individually rational scf: $\xi(e) = (j, t_1, \ldots, t_n)$ only if $v_j \geq \max_i v_i$, $t_j = v_j$, and $t_i = 0$ for $i \neq j$.   ★

While the players know the details of the environment (though some of the information may be private), the "social planner" may not (Example 10.4.1 is a simple illustration). Nonetheless, the social planner wishes to "implement" a social choice function. Can the social planner arrange matters so that the players will reveal the details of the environment? This is the implementation problem.

More formally, a *game form* or *mechanism* Γ is described by $((A_i)_{i=1}^n, f)$, where A_i is player i's action set, and $f : A \to Z$ is the *outcome function* describing the induced outcome in Z for each action profile $a \in A := A_1 \times \cdots \times A_n$. If the mechanism has a nontrivial dynamic structure, then a_i is an extensive form strategy.

For each mechanism Γ and environment e, there will be a set of solutions or equilibria $\Sigma_\Gamma(e) \subset A$. In general, there are many choices possible for Σ_Γ, such as dominant strategy, Nash equilibrium, Bayes-Nash, subgame perfect, etc. The specification of Σ_Γ depends upon the nature of informational assumptions made about the players and the static/dynamic nature of Γ.

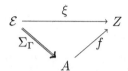

Figure 10.4.1: The mechanism Γ Σ-implements ξ if this diagram is a commutative diagram, in the sense of Definition 10.4.2.

Example 10.4.4. Suppose there are two players and three possibilities in Example 10.4.2. Voting by veto (Example 1.1.5) is an example of a game form, and using iterated deletion of dominated strategies gives, for each profile of preferences, an outcome. ★

We thus have the scf ξ which maps $\mathcal{E}$ to Z, the solution correspondence Σ_Γ, which maps $\mathcal{E}$ to A, and outcome function f, which maps A to Z (see Figure 10.4.1).

Definition 10.4.2. *The mechanism Γ weakly Σ-implements ξ if for all $e \in \mathcal{E}$, there exists a solution $\sigma_\Gamma(e) \in \Sigma_\Gamma(e)$ such that*

$$f(\sigma_\Gamma(e)) = \xi(e).$$

The mechanism Γ strongly (or fully) Σ-implements ξ if for all $e \in \mathcal{E}$, and for all solutions $\sigma_\Gamma(e) \in \Sigma_\Gamma(e)$,

$$f(\sigma_\Gamma(e)) = \xi(e).$$

If *implement* is used without an adjective, the writer typically means weakly implement.

Example 10.4.5. The second price auction strongly implements the efficient scf in the setting of Example 10.4.3 in weakly undominated strategies (Example 1.1.6), but it only implements (but *not* strongly so) in Nash equilibrium (Example 2.1.4). ★

Example 10.4.6 (Example 10.4.1 cont.). Fair division is strongly implemented by both subgame perfect equilibrium and (in the normal form) by the iterated deletion of weakly dominated strategies. ★

Remark 10.4.1. Implementation theory is part of mechanism design and fixes the social choice function and asks whether there is mechanism for which either (for weak implementation) there is an equilibrium or (for strong implementation) all equilibria result in outcomes consistent with the a priori fixed social choice function. Mechanism design does not necessarily fix the social choice function. A desired outcome or objective is first identified (examples include efficient trade, firm profits, social welfare, and gains from trade) and then solves for (designs) a mechanism with an equilibrium that either achieves or maximizes the objective (as appropriate) subject to the relevant resource and information constraints.   ◆

10.5   Problems

10.5.1. In this problem, we explicitly take into account the nonnegativity constraints in the simple screening example of Section 10.1. The Langrangean becomes (since the nonnegativity of prices is implied by the nonnegativity of quantities and IR, those constraints can be ignored)

$$\mathcal{L} = \alpha_L(p_L - c(q_L)) + \alpha_H(p_H - c(q_H))$$
$$+ \lambda_L[\theta_L q_L - p_L - (\theta_L q_H - p_H)] + \lambda_H[\theta_H q_H - p_H - (\theta_H q_L - p_L)]$$
$$+ \mu_L(\theta_L q_L - p_L) + \mu_H(\theta_H q_H - p_H) + \xi_L q_L + \xi_H q_H,$$

where ξ_L and ξ_H are the multipliers on the nonnegativity constraints $q_L \geq 0$ and $q_H \geq 0$, respectively.

(a) Describe the first order conditions and the complementary slackness conditions.

(b) Suppose $\theta_L < \alpha_H \theta_H$. Solve for the optimal pair of contracts (including the values of the multipliers).

(c) Suppose now that the marginal cost at zero is strictly positive, while maintaining all the other assumptions in Section 10.1. How does the analysis of the seller's optimal pair of contracts change?

10.5.2. Suppose (p, q) is an optimal direct mechanism for the less simple screening example of Section 10.2 and we are in the regular case.

(a) Prove that p is strictly increasing whenever q is (i.e., $\theta_1 < \theta_2$ and $q(\theta_1) < q(\theta_2)$ implies $p(\theta_1) < p(\theta_2)$).

(b) Prove that p is constant whenever q is (i.e., $\theta_1 < \theta_2$ and $q(\theta_1) = q(\theta_2)$ implies $p(\theta_1) = p(\theta_2)$).

(c) (The taxation principle.) Let $Q := q([\underline{\theta}, \overline{\theta}])$. Prove there exists a nonlinear pricing function $P : Q \to \mathbb{R}_+$ such that buyer θ optimally chooses $q = q(\theta)$ from Q.

10.5.3. Solve for the optional direct mechanism (p, q) for the less simple screening example of Section 10.2 when costs are quadratic, in particular, $c(q) = \frac{1}{2}q^2$, and θ has the following distributions.

(a) A uniform distribution on the interval $[0, 1]$.

(b) The type θ is distributed on $[1, 11]$ according to the following density:

$$f(\theta) = \begin{cases} 0.1, & \theta \in [1, 2], \\ 0.9(\theta - 1)^{-2}, & \theta \in (2, 10), \\ 0.1, & \theta \in [10, 11]. \end{cases}$$

10.5.4. Prove that the take-it-or-leave-it mechanism is optimal for the sale of a single item when the seller's beliefs are as described in Section 10.1.

10.5.5. Give the (few) details of the argument that $U'(\theta) = q(\theta)$ in the proof of Lemma 10.2.1.

10.5.6. Suppose a seller wishes to sell one unit of a good to a buyer of unknown valuation who is budget constrained. That is, assume the buyer in Section 10.3 cannot pay more than his budget of $b > 0$. Suppose we are in the regular case. If $b > \theta^*$, then this additional constraint is not binding. What is the firm's optimal mechanism if $b < \theta^*$?

10.5.7. Verify the claim in Example 10.4.1 that the division of the pie in any subgame perfect equilibrium is envy free (neither child strictly prefers one of the slices).

Chapter 11

Dominant Strategy Mechanism Design

11.1 Social Choice and Arrow's Impossibility Theorem

Let Z be a (finite) set of social alternatives, and n members of society.

A *preference ordering* R is a complete and transitive binary relation on Z, and let $\mathcal{R}$ be the set of all preference orderings on Z. Strict preference is defined by $zPy \iff (zRy \wedge \neg yRz)$.

A *profile* of preferences is $\mathbf{R} := (R_1, \ldots, R_n) \in \mathcal{R}^n$.

Definition 11.1.1. *A social welfare function is a function[1] f that assigns a preference ordering R to each profile $\mathbf{R} = (R_1, \ldots, R_n)$, i.e., it is a function $f : \mathcal{R}^n \to \mathcal{R}$. The preference ordering $f(\mathbf{R})$ is called the social ordering associated with $\mathbf{R}$.*

This definition requires that a social welfare function assign a transitive and complete social preference ordering to *every* profile. This is called *unrestricted domain*. (The fact that indifference is allowed is not important; it is important that all strict preferences are in the domain.)

Definition 11.1.2. *The social welfare function f satisfies non-dictatorship if there is no i such that for all $x, y \in Z$ and all $\mathbf{R} \in \mathcal{R}^n$,*

$$xP_iy \Rightarrow xPy,$$

---

[1] Some authors use the word functional when the range is not a subset of a Euclidean space.

where P is the strict social preference implied by $f(\mathbf{R})$.

Definition 11.1.3. *The social welfare function f satisfies* unanimity *if for all $x, y \in Z$ and $\mathbf{R} = (R_1, \ldots, R_n) \in \mathcal{R}^n$, if $x P_i y$ for all i, then $x P y$.*

Definition 11.1.4. *The social welfare function f satisfies* independence of irrelevant alternatives *(IIA) if $\mathbf{R} = (R_1, \ldots, R_n)$ and $\mathbf{R}' = (R'_1, \ldots, R'_n)$ agree on the choice between a pair x, y (i.e., $x R_i y \Leftrightarrow x R'_i y$ for all i),[2] then $f(\mathbf{R})$ and $f(\mathbf{R}')$ also agree on the choice between x and y (i.e., $x f(\mathbf{R}) y \Leftrightarrow x f(\mathbf{R}') y$).*

Example 11.1.1 (The Condorcet Paradox). Easy to verify that majority rule over two outcomes satisfies non-dictatorship, unanimity, and IIA.

Let $Z = \{x, y, z\}$ and $n = 3$ and consider the following (strict) preference orderings

$$P_1 \quad P_2 \quad P_3$$

$$x \quad y \quad z$$

$$y \quad z \quad x$$

$$z \quad x \quad y$$

Majority rule is not transitive (hence Condorcet, or voting, cycles). ★

Theorem 11.1.1 (Arrow's Impossibility Theorem). *If $|Z| \geq 3$, there exists no social welfare function with unrestricted domain satisfying non-dictatorship, unanimity, and IIA.*

The requirement that $|Z| \geq 3$ is clearly necessary, since majority rule is a nondictatorial social welfare function with unrestricted domain satisfying unanimity and IIA. In this case, IIA is of course trivial.

Proof. I give the elegant first proof from Geanakoplos (2005).

Suppose f is a social welfare function with unrestricted domain satisfying unanimity and IIA. We will show that it is dictatorial.

In the arguments that follow, note also the role of unrestricted domain, which guarantees that f is well defined for each defined preference profile.

Notation: For any preference profile $\mathbf{R}$, denote by R and P the weak and strict social preference implied by $f(\mathbf{R})$. Decorations are inherited, so for example, $P^{(j)}$ denotes the strict social preference implied by $f(\mathbf{R}^{(j)})$.

---

[2]But different agents may rank x and y differently.

Claim 11.1.1 (No compromise). *For any alternative $c \in Z$, suppose $\mathbf{R}$ is a profile in which, for all i, either xP_ic for all $x \neq c$ or cP_ix for all $x \neq c$ (i.e., each agent either ranks c at the bottom or at the top). Then, either xPc for all $x \neq c$ or cPx for all $x \neq c$.*

> *Proof.* Suppose not. Then there exist distinct $\alpha, \beta \in Z$, $\alpha, \beta \neq c$, such that
> $$\alpha R c R \beta.$$
> Let $\widehat{\mathbf{R}}$ denote the preference profile obtained from $\mathbf{R}$ by moving β to the top of every agent's ranking over alternatives other than c (if $cP_i\beta$, then $c\widehat{P_i}\beta$). In particular, β is ranked above α by all agents.
>
> Unanimity implies $\beta\widehat{P}\alpha$. But IIA implies $\alpha\widehat{R}c$ and $c\widehat{R}\beta$ (since neither the pairwise ranking of α and c, nor the pairwise ranking of β and c, has changed for any agent in going from $\mathbf{R}$ to $\widehat{\mathbf{R}}$), and so $\alpha\widehat{R}\beta$, a contradiction. □

Fix an alternative $c \in Z$, and suppose $\mathbf{R}$ is a profile in which xP_ic for all i and all $x \neq c$. Then unanimity implies xPc for all $x \neq c$.

I first argue that there must be an agent who is pivotal in a restricted sense. For each profile R_i, define a profile $R_i^{(j)}$ as follows:

$$xR_iy \implies xR_i^{(j)}y \quad \text{for all } x, y \neq c,$$

$$xP_i^{(j)}c \quad \text{if } i > j,$$

and

$$cP_i^{(j)}x \quad \text{if } i \leq j.$$

Finally, set $\mathbf{R}^{(j)} := (R_1^{(j)}, \ldots, R_n^{(j)})$; in words, agent by agent, the worst alternative c is moved to the top of the agent's preference ordering, without changing any other preference ordering.

Note that in $\mathbf{R}^{(n)}$, every agents ranks c strictly at the top, and so, again by unanimity, $cP^{(n)}x$ for all $x \neq c$.

Let j^* be the first index (lowest j) for which c is not ranked at the bottom by $f(\mathbf{R}^{(j)})$. By the above claim, c is then ranked at the top, and j^* is pivotal, at least at $\mathbf{R}$, in that

$$\forall x : xP^{(j^*-1)}c \text{ and } cP^{(j^*)}x. \tag{11.1.1}$$

I now argue that j^* is then even more pivotal:

Claim 11.1.2. *Suppose a, b are two distinct alternatives distinct from c. Then, for all $\widetilde{\mathbf{R}}$,*
$$a\widetilde{P}_{j^*}b \implies a\widetilde{P}b.$$

Proof. Suppose $a\widetilde{P}_{j*}b$. Let $\widetilde{R}_{j*}^{j^*}$ denote the preference ordering obtained from $\widetilde{R}_{j*}$ be moving c so that c is ranked between a and b (but the rankings of alternatives excluding c are unchanged). For $i \neq j^*$, let $\widetilde{R}_i^{j^*}$ denote the preference ordering obtained from $\widetilde{R}_i$ by moving c, for $i < j^*$, to the top of the agent's ordering, and, for $i > j^*$, to the bottom.

Applying IIA twice, we have

1. $a\widetilde{R}^{j^*}c \iff aR^{(j^*-1)}c$ (since for all i, $a\widetilde{R}_i^{j^*}c \iff aR_i^{(j^*-1)}c$) and

2. $c\widetilde{R}^{j^*}b \iff cR^{(j^*)}b$ (since, for all i, $c\widetilde{R}_i^{j^*}b \iff cR_i^{(j^*)}b$).

From (11.1.1), $aP^{(j^*-1)}c$ and $cP^{(j^*)}b$, so $a\widetilde{P}^{j^*}c\widetilde{P}^{j^*}b$. Transitivity then yields $a\widetilde{P}^{j^*}b$.

Finally, applying IIA again yields $a\widetilde{R}^{j^*}b \iff a\widetilde{R}b$ (since for all i, $a\widetilde{R}_i^{j^*}b \iff a\widetilde{R}_ib$), so $a\widetilde{P}b$. □

We thus have that j^* is dictatorial over comparisons that exclude c, and one decision on c, at **R**. It remains to show that j^* is dictatorial over all comparisons.

Fix an alternative $a \neq c$. Applying the reasoning for Claim 11.1.2 to a gives an agent $i^\dagger$ who is dictatorial over comparisons that exclude a. In particular, at $\mathbf{R}^{j^*-1}$ and $\mathbf{R}^{j^*}$, agent $i^\dagger$'s ranking of $b \notin \{a,c\}$ and c determines the social ranking of b and c. But we have already seen that this ranking is determined by i^*'s ranking, and so $i^\dagger = i^*$. ∎

Remark 11.1.1. The proof of Arrow's Theorem applies without change if the domain of the scf is instead the set of all strict preference orders on Z. Unrestricted domain is needed so that we can apply the scf to different strict preference orderings. ♦

11.2   Dominant Strategy Implementation and the Gibbard-Satterthwaite Theorem

We are now in a position to prove the Gibbard-Satterthwaite Theorem, introduced in Remark 1.1.3.

The setting is that of Example 10.4.2, except that we restrict attention to *strict* preference orderings. Let $\mathcal{P}$ denote the collection of all strict orders on Z, the finite set of outcomes. As usual, $P \in \mathcal{P}$ satisfies completeness, transitivity, and asymmetry (i.e., $xPy \Rightarrow \neg yPx$).

Setting $\mathcal{E} = \mathcal{P}^n$, a *social choice function* (scf) is now a function

$$\xi : \mathcal{P}^n \to Z.$$

Denote a preference profile by $\mathbf{P} := (P_1, \ldots, P_n) \in \mathcal{P}^n$.
There are many mechanisms a social planner could imagine using to attempt to implement a scf. A canonical mechanism is *the direct mechanism*, in which each player i reports their own preference ordering P_i, so that $A_i = \mathcal{P}$. It is then natural to focus on *truthful reporting* in the direct mechanism with outcome function $f = \xi$.

Theorem 11.2.1 (The revelation principle). *The scf ξ is dominant strategy implementable in some mechanism $\Gamma^* = ((A_i^*)_{i=1}^n, f^*)$ if and only if truthful reporting is a dominant strategy equilibrium of the direct mechanism $\Gamma^D = ((\mathcal{P}_i)_{i=1}^n, \xi)$.*

Proof. Clearly, if truthful reporting is a dominant strategy equilibrium of the direct mechanism, then ξ is clearly dominant strategy implementable in some mechanism.

We now argue the nontrivial direction.

Suppose ξ is dominant strategy implementable in the mechanism $\Gamma^* = ((A_i^*)_{i=1}^n, f^*)$. Then there exists a strategy profile $s^* := (s_1^*, \ldots, s_n^*)$, with $s_i^* : \mathcal{P} \to A_i^*$ such that for all $\mathbf{P}$, $(s_1^*(P_1), \ldots, s_n^*(P_n))$ is a dominant strategy equilibrium. That is,

$$\forall a_{-i} \ \nexists a_i' : f^*(a_i', a_{-i}) \ P_i \ f^*(s_i^*(P_i), a_{-i}). \tag{11.2.1}$$

Moreover, for all $\mathbf{P}$,

$$f^*(s^*(\mathbf{P})) = \xi(\mathbf{P}). \tag{11.2.2}$$

But (11.2.1) immediately implies

$$\forall \mathbf{P}_{-i} \ \nexists P_i' : f^*(s_i^*(P_i'), s_{-i}^*(\mathbf{P}_{-i})) \ P_i \ f^*(s_i^*(P_i), s_{-i}^*(\mathbf{P}_{-i})),$$

which, using (11.2.2), is equivalent to

$$\forall \mathbf{P}_{-i} \ \nexists P_i' : \xi(P_i', \mathbf{P}_{-i}) \ P_i \ \xi(P_i, \mathbf{P}_{-i}),$$

that is, truthful reporting is a dominant strategy equilibrium of the direct mechanism. ∎

Thus, when investigating whether an scf is dominant strategy implementable, it is enough to study whether truthful reporting is a dominant strategy equilibrium of the direct mechanism.

Note that the result does not say that every dominant strategy equilibrium of the direct mechanism implements ξ; that is false (see Problem

11.4.3). There may be other dominant strategy equilibria. Put differently, the revelation principle is *not* about strong implementation.

We next recall the notion of strategy proof from Remark 1.1.3.

Definition 11.2.1. *A scf ξ is strategy proof if announcing truthfully in the direct mechanism is a dominant strategy equilibrium. A scf is* manipulable *if it is not strategy proof, i.e., there exists an agent i, $\mathbf{P} \in \mathcal{P}^n$ and $P_i' \in \mathcal{P}$ such that*

$$\xi(P_i', \mathbf{P}_{-i}) \ P_i \ \xi(\mathbf{P}).$$

The revelation principle can be restated as: ξ is dominant strategy implementable in some mechanism if, and only if, it is strategy proof.

Definition 11.2.2. *A scf ξ is* dictatorial *if there is an i such that for all $\mathbf{P} \in \mathcal{P}^n$,*

$$[a P_i b \quad \forall b \neq a] \implies \xi(\mathbf{P}) = a.$$

Theorem 11.2.2 (Gibbard-Satterthwaite). *Every strategy-proof scf ξ satisfying $\mid \xi(\mathcal{P}^n) \mid \geq 3$ is dictatorial.*

Remark 11.2.1. Because of the dominant strategy revelation principle (Theorem 11.2.1), the Gibbard-Satterthwaite Theorem implies that the only nontrivial scf's that can be implemented in dominant strategies are dictatorial.   ♦

The remainder of this section proves Theorem 11.2.2 as an implication of Arrow's Impossibility Theorem (Theorem 11.1.1). There are direct proofs (see, for example, Benoit, 2000, and Reny, 2001). Suppose there exists a non-manipulable scf ξ satisfying $\mid \xi(\mathcal{P}^n) \mid \geq 3$.

Without loss of generality, assume $\xi(\mathcal{P}^n) = Z$. The first result is just a rewriting of definitions.

Lemma 11.2.1. *Suppose there exists a preference profile $\mathbf{P}$ and a preference P_i' such that $\xi(\mathbf{P}) = x$, $\xi(P_i', P_{-i}) = y$,and either $x P_i y$ and $x P_i' y$, or $y P_i x$ and $y P_i' x$. Then, ξ is manipulable.*

Lemma 11.2.2. *Suppose for profile $\mathbf{P}$, there exists a subset $B \subsetneq Z$ such that for all i, all $x \in B$, and all $y \in Z \setminus B$, $x P_i y$. Then, $\xi(\mathbf{P}) \in B$.*

Proof. Suppose not, that is, suppose there exist $\mathbf{P}$ and B satisfying the hypothesis, with $\xi(\mathbf{P}) \notin B$.

Choose $x \in B$. By assumption, $B \subset \xi(\mathcal{P}^n)$, and so there exists $\mathbf{P}'$ with $\xi(\mathbf{P}') = x$. Consider the following sequence of profiles,

$$\mathbf{P}^{(0)} := (P_1, P_2, ..., P_n) = \mathbf{P},$$
$$\mathbf{P}^{(1)} := (P_1', P_2, ..., P_n),$$
$$\vdots$$
$$\mathbf{P}^{(n)} := (P_1', P_2', ..., P_n') = \mathbf{P}'.$$

By assumption, $\xi(\mathbf{P}) \notin B$ and $\xi(\mathbf{P}') \in B$. Hence, there must be a first profile $\mathbf{P}^{(k)}$ for which $f(\mathbf{P}^{(k)}) \in B$. But then f is manipulable at $\mathbf{P}^{(k-1)}$ by individual k.   ∎

The idea is to construct a social welfare function F^* on Z, and then appeal to Theorem 11.1.1.

We need to specify either $xF^*(\mathbf{P})y$ on $yF^*(\mathbf{P})x$ for any $x, y \in Z$ and $\mathbf{P} \in \mathcal{P}^n$. We do this by "moving x and y to the top" of each person's preference order without changing their relative order, and using the social choice for this altered profile to rank x relative to y:

To "move x and y to the top" for a profile $\mathbf{P} = (P_1, ..., P_N)$, we proceed as follows. For each preference ordering P_i, define P_i' as follows:

1. $xP_i'y \iff xP_iy$,

2. for any $a, b \notin \{x, y\}$, $aP_i'b \iff aP_ib$, and

3. for any $a \in \{x, y\}$ and $b \notin \{x, y\}$, $aP_i'b$.

By Lemma 11.2.2, the social choice for the altered profile $\mathbf{P}'$ must be either x or y. This construction generates a complete ordering $F^*(\mathbf{P})$ on Z for every $\mathbf{P} \in \mathcal{P}^n$.

The mapping F^* is a social welfare function if $F^*(\mathbf{P})$ is transitive for all $\mathbf{P}$. Before proving transitivity, unanimity and IIA are verified.

Lemma 11.2.3. F^* *satisfies unanimity and IIA.*

Proof. IIA: Need to show that for all $\mathbf{P}$ and $\mathbf{P}' \in \mathcal{P}^n$ such that for a pair of alternatives $x, y \in Z$

$$[xP_iy \iff xP_i'y \;\forall i] \implies [xF^*(\mathbf{P})y \iff xF^*(\mathbf{P}')y].$$

We prove this by contradiction. Suppose the social choice when x and y are moved to the top in $\mathbf{P}$ is x and the social choice is y when x and y are moved to the top in $\mathbf{P}'$. We proceed by a chain argument similar to that in the proof of Lemma 11.2.2:

Let $\hat{\mathbf{P}}$ and $\hat{\mathbf{P}}'$ be the profiles obtained by moving x and y to the top in the profiles $\mathbf{P}$ and $\mathbf{P}'$, respectively. Consider the following sequence of profiles in which we change one $\hat{P}_i$ to $\hat{P}'_i$ at a time.

$$\hat{\mathbf{P}}^{(0)} := (\hat{P}_1, \hat{P}_2, ..., \hat{P}_n) = \hat{\mathbf{P}},$$

$$\hat{\mathbf{P}}^{(1)} := (\hat{P}'_1, \hat{P}_2, ..., \hat{P}_n),$$

$$\vdots$$

$$\hat{\mathbf{P}}^{(n)} := (\hat{P}'_1, \hat{P}'_2, ..., \hat{P}'_n) = \hat{\mathbf{P}}'.$$

By assumption, the social choice for $\hat{\mathbf{P}}^{(0)}$ is x and the social choice for $\hat{\mathbf{P}}^{(n)}$ is y. By Lemma 11.2.2, the social choice for each profile must be either x or y. Consider the first profile $\hat{\mathbf{P}}^{(k)}$ for which the social choice changes from x to y. Then, ξ must be manipulable at either $\hat{\mathbf{P}}^{(k-1)}$ or $\hat{\mathbf{P}}^{(k)}$ (since the ranking of x and y has not changed for any player).

It is straightforward to show that unanimity holds.   ■

Lemma 11.2.4. $F^*(\mathbf{P})$ *is transitive.*

Proof. Suppose not, i.e., suppose there exists $x, y, z \in Z$ with $xF^*(\mathbf{P})y$, $yF^*(\mathbf{P})z$ and $zF^*(\mathbf{P})x$. Consider the profile $\mathbf{P}'$ derived from $\mathbf{P}$ by "moving x, y, and z to the top." By Lemma 11.2.2, the social choice for this profile must be in $\{x, y, z\}$. Without loss of generality, assume it is x. Now consider another profile $\mathbf{P}''$ derived from $\mathbf{P}'$ by moving y to third place. Now, by IIA and the assumption that $zF^*(\mathbf{P})x$, we have $\xi(\mathbf{P}'') = z$.

Consider the sequence of profiles:

$$\mathbf{P}''^{(0)} := (P'_1, P'_2, ..., P'_n) = \mathbf{P}',$$

$$\mathbf{P}''^{(1)} := (P''_1, P'_2, ..., P'_n),$$

$$\vdots$$

$$\mathbf{P}''^{(n)} := (P''_1, P''_2, ..., P''_n) = \mathbf{P}''.$$

For every profile, the social choice is in $\{x, y, z\}$ by Lemma 11.2.2. There is a first profile $\mathbf{P}''^{(k)}$ for which the social choice is not x, i.e., the social choice is y or z. If the first change is to y, the scf is manipulable at $\mathbf{P}''^{(k)}$. If the first change is to z, then f is manipulable at either $\mathbf{P}''^{(k-1)}$ or $\mathbf{P}''^{(k)}$; in either case, this is a contradiction. Hence $F(\mathbf{P})$ is transitive.   ■

Hence by Arrow's Impossibility Theorem (Theorem 11.1.1, recall Remark 11.1.1), F^* is dictatorial. The argument is completed by showing that if i is a dictator for F^*, then i is also a dictator for the scf ξ, which is straightforward.

11.3 Efficiency in Quasilinear Environments

Suppose there is a finite set of social alternatives X. An outcome is $(x, t_1, \ldots, t_n) \in X \times \mathbb{R}^n =: Z$, where x is the social alternative and t_i is the transfer *to* player i. We do *not* require budget balance ($\sum t_i = 0$).

An environment is said to be *quasilinear* if all the players' Bernoulli utility functions are quasilinear, i.e., player i's payoff from the outcome $(x, t_1, \ldots, t_n)$ is[3]

$$v_i(x) + t_i,$$

for some $v_i : X \to \mathbb{R}$.

An environment is $\mathbf{v} = (v_1, \ldots, v_n)$, where $v_i : X \to \mathbb{R}$ is player i payoffs from each of the outcomes.[4] Since X is finite, we have $v_i \in \mathbb{R}^{|X|}$, and so the set of environments $\mathcal{E}$ is a subset of $\mathbb{R}^{n|X|}$. Assume $\mathcal{E} = \prod_i \mathcal{E}_i$; we allow $\mathcal{E}_i = \mathbb{R}^{|X|}$, but this is not necessary (except where explicitly noted).

Quasilinear environments allow for a transparent formulation of efficiency, which I now describe. In the absence of constraints on the size of transfers, it is meaningless to talk about the Pareto efficiency of outcomes. One sensible way to account for transfers with Pareto efficiency is to require total transfers be unchanged. That is, an outcome $z = (x, t_1, \ldots, t_n) \in Z$ is *Pareto efficient* if there is no outcome $z' = (x', t_1', \ldots, t_n') \in Z$ satisfying $\sum_i t_i = \sum_i t_i'$ and

$$v_i(x') + t_i' \geq v_i(x) + t_i$$

for all i, with strict inequality holding for at least one i.

It is straightforward to verify that the social choice function ξ is *efficient* if and only if

$$\xi_X(v_1, \ldots, v_n) \in \arg\max_{x \in X} \sum_i v_i(x), \qquad (11.3.1)$$

where $\xi(\mathbf{v}) = (\xi_X(\mathbf{v}), t_1(\mathbf{v}), \ldots, t_n(\mathbf{v}))$. Let $x^*(\mathbf{v})$ denote an efficient alternative for $\mathbf{v}$ (so that for an efficient ξ, $\xi_X = x^*$ for some x^*).

We are again interested in direct mechanisms where each player reports his or her valuation $v_i \in \mathcal{E}_i$. More specifically, we are interested in mechanisms $\Gamma^D = (\mathcal{E}, \xi)$, where the choice of social alternative is efficient.

In order to complete the specification of an efficient ξ, we need to specify the transfers.

---

[3]Since this payoff is the Bernoulli utility function, the agent is risk neutral with respect to transfers.

[4]For mnemonic purposes, it is convenient to denote an environment by $\mathbf{v}$ rather than e in this section.

Definition 11.3.1. *A* Groves mechanism *is given by* $\Gamma^G = (\mathcal{E}, \xi^G)$, *where* ξ^G *is efficient and*

$$t_i(\mathbf{v}) = \sum_{j \neq i} v_j(x^*(\mathbf{v})) + k_i(\mathbf{v}_{-i}),$$

for some collection of functions $(k_i)_i$, *where* $k_i : \mathcal{E}_{-i} \to \mathbb{R}$.

Theorem 11.3.1. *For all* $\mathbf{v} \in \mathcal{E}$, *truthtelling is a dominant strategy equilibrium of any Groves mechanism* Γ^G.

Proof. Player i's payoff from reporting $\hat{v}_i$ when his true valuation is v_i under $\Gamma^G = (\mathcal{E}, \xi^G)$ is

$$v_i(x^*(\hat{v}_i, \mathbf{v}_{-i})) + \sum_{j \neq i} v_j(x^*(\hat{v}_i, \mathbf{v}_{-i})) + k_i(\mathbf{v}_{-i}).$$

Since x^* is efficient, for all $\mathbf{v}_{-i}$, $x^*(v_i, \mathbf{v}_{-i})$) maximizes the above expression over possible $x^*(\hat{v}_i, \mathbf{v}_{-i})$. Hence, $\hat{v}_i = v_i$ is a best reply for all $\mathbf{v}_{-i}$, and so is a dominant strategy. ∎

Theorem 11.3.1 does not contradict the Gibbard-Satterthwaite Theorem, because the restriction to quasilinear preferences violates unrestricted domain.

The Groves mechanism pays each agent his or her externality: The transfer $t_i(\mathbf{v})$ only depends on i's announcement through the announcement's impact on the selection of the social alternative, and internalizes the resulting externality (i.e., impact on the payoffs of the other players).

An important special case of a Groves mechanism is obtained by setting

$$k_i(\mathbf{v}_{-i}) = -\sum_{j \neq i} v_j(x^*(\mathbf{v}_{-i})),$$

where $x^*(\mathbf{v}_{-i})$ is an efficient alternative for the $n - 1$ agents with payoff functions $\mathbf{v}_{-i}$. This is the *Clarke* or *pivot mechanism*. In this mechanism, the transfer is zero if the efficient alternative does not change, i.e., player i's transfer is only nonzero if player i is pivotal). Both the second price or Vickrey auction (Example 1.1.6) and the public good mechanism of Example 1.1.7 are instances of a pivot mechanism.

At first reading the setting and proof of Theorem 11.3.1 can seem a little abstract. An excellent exercise is to map the environment and discussion of Examples 1.1.6 and 1.1.7 to the discussion here.

The term *Vickrey-Clarke-Groves* or *VCG mechanism* is also used for either a Groves or pivot mechanism.

Remark 11.3.1 (Converse to Theorem 11.3.1). If $\mathcal{E} = \mathbb{R}^{n|X|}$, so that the environment is *rich*, every direct mechanism in which truthtelling is a dominant strategy equilibrium is a Groves mechanism (Mas-Colell, Whinston, and Green, 1995, Proposition 23.C.5).   ♦

A mechanism satisfies *ex post budget balance* if

$$\sum_i t_i(\mathbf{v}) = 0 \qquad \forall \mathbf{v} \in \mathcal{E}.$$

In general, Groves mechanisms do not satisfy budget balance (this is Green and Laffont (1979, Theorem 5.3)):

Theorem 11.3.2. *Suppose* $\mathcal{E} = \mathbb{R}^{n|X|}$. *Then every Groves mechanism violates ex post budget balance.*

11.4   Problems

11.4.1. Describe a social choice rule that is unanimous, nondictatorial, and strategy proof when there are two alternatives.

11.4.2. Consider the following social choice rule over the set $X = \{x, y, z\}$. There is an exogenously specified order $x \succ y \succ z$, and define $S(X') := \{a \in X' : a \succ a' \ \forall a' \in X' \setminus \{a\}\}$. Then,

$$f(\theta) = S(\{a \in X : a = t(\theta_i) \text{ for some } i\}).$$

Prove that f is unanimous and nondictatorial, but not strategy proof.

11.4.3. (a) Give an example of a scf ξ for which the direct mechanism with outcome function ξ has a dominant strategy equilibrium that does not implement ξ.

(b) Give an example of a nonconstant scf ξ for which the direct mechanism with outcome function ξ has both truthtelling as dominant strategy equilibrium implementing ξ and another dominant strategy equilibrium that does not implement ξ. [**Hint:** Suppose $n = 2$ and $Z = Z_1 \times Z_2$, with $\mathcal{E}$ being the collection of preferences for agents 1 and 2, where agent i's preferences over Z are determined by Z_i.]

11.4.4. The second-price sealed-bid auction is a pivot mechanism satisfying ex post budget balance. Reconcile this with Theorem 11.3.2.

11.4.5. There are $n \geq 3$ agents among whom two identical goods must be allocated. Agent i gets utility v_i from one unit of the good and is indifferent between the two goods; each agent is either allocated zero or one unit. Describe the Groves mechanism.

Chapter 12

Bayesian Mechanism Design

12.1 The Bayesian Revelation Principle

There are n players, $i = 1, \ldots, n$. Each player i has private information, denoted by $\theta_i \in \Theta_i$; player i's beliefs over the private information (types) of the other players is denoted by $p_i(\cdot|\theta_i) \in \Delta(\Theta_{-i})$, where $\Theta_{-i} := \prod_{j \neq i} \Theta_j$. As usual, $\Theta := \prod_i \Theta_i$.

The set of outcomes is denoted by Z, and $u_i(z, \theta)$ is player i's payoff from outcome z given type profile θ. Extend u_i to lotteries over outcomes in the standard manner: If $\lambda \in \Delta(Z)$,

$$u_i(\lambda, \theta) := Eu_i(z, \theta),$$

where the expectation is over z with respect to the distribution λ.

To state the following result in its most useful form, we allow for the possibility that the social choice function chooses a lottery over outcomes, i.e.,

$$\xi : \Theta \to \Delta(Z).$$

Recall that a special case of a mechanism is a *direct (revelation) mechanism*: $A_i = \Theta_i$ for each i. A direct mechanism is just a social choice function, and (at a slight abuse of notation) is also denoted ξ. In the following statement, the notation is from Section 10.4.

Theorem 12.1.1 (The Bayesian revelation principle). *Suppose s^* is a (possibly mixed) Bayes-Nash equilibrium of a game $(\{A_i\}_{i=1}^n, f)$. Truth telling is a pure strategy Bayes-Nash equilibrium of the direct mechanism ξ with $\xi(\theta) = f(s^*(\theta))$ for all θ.*

Proof. To simplify the notation, suppose the game and type spaces are finite. The (interim) payoff of type θ_i of player i from action a_i is

$$v_i(a_i, s^*_{-i}; \theta_i) = \sum_{\theta_{-i}} u_i(f(a_i, s^*_{-i}(\theta_{-i})), \theta) p_i(\theta_{-i}|\theta_i).$$

Since s^* is an equilibrium of the game given by $(\{A_i\}_{i=1}^n, f)$, for all i, $\theta_i \in \Theta_i$ and $a_i \in A_i$,

$$v_i(s^*_i(\theta_i), s^*_{-i}; \theta_i) \geq v_i(a_i, s^*_{-i}; \theta_i). \tag{12.1.1}$$

The payoff to type θ_i of player i from reporting type $\hat{\theta}_i$ in the direct mechanism is

$$v_i^D(\hat{\theta}_i; \theta_i) := \sum_{\theta_{-i}} u_i(\xi(\hat{\theta}_i, \theta_{-i}), \theta) p_i(\theta_{-i}|\theta_i)$$

$$= v_i(s^*_i(\hat{\theta}_i), s^*_{-i}; \theta_i).$$

Hence, (12.1.1) implies that for all i and all $\theta_i, \hat{\theta}_i \in \Theta_i$,

$$v_i^D(\theta_i; \theta_i) \geq v_i^D(\hat{\theta}_i; \theta_i),$$

i.e., truth-telling is an equilibrium of the direct mechanism.   ∎

Note that the revelation principle does not require a common prior. Most applications that I am aware of, however, do impose a common prior (as we will).

Theorem 12.1.1 has the following (the statement of which parallels Theorem 11.2.1) as an immediate corollary. The pedagogical advantage of Theorem 12.1.1 is that it emphasizes the role of the revelation principle in capturing all possible equilibrium behavior in all possible games.

Corollary 12.1.1. *The scf ξ is Bayes-Nash implementable in some game $((A_i)_{i=1}^n, f)$ if and only if truthful reporting is a Bayes-Nash equilibrium of the direct mechanism ξ.*

We have already seen an application of the Bayesian revelation principle in Section 10.2, where the revelation principle allowed us to identify the optimal selling mechanism. An important aspect of that application is that the optimal selling mechanism had a transparent implementation. The revelation principle only guarantees that the optimal mechanism has truthtelling as a Bayes-Nash equilibrium. There is no claim about uniqueness nor plausibility of that equilibrium. As we will see in Section 12.4

(on auctions), not all optimal mechanisms appear plausible, and so it is important to ask about the game form that implements the mechanism.

This issue of implementation is not relevant when the revelation principle is used to prove a *negative* result, such as Theorem 12.3.1: Proving that the truthtelling equilibrium of every incentive compatible direct mechanism fails to have a particular property obviously implies that any "plausible" equilibrium of any "nice" game also fails to have that property.

Remark 12.1.1 (Nash implementation). If all the information about the environment is common knowledge among the players (such as in Example 10.4.1), then Bayes-Nash equilibrium in Theorem 12.1.1 simplifies to Nash equilibrium. Problem 12.5.1 describes a social choice function that cannot be weakly implemented in dominant strategies, but can be easily weakly implemented in Nash equilibrium. ♦

12.2 Efficiency in Quasilinear Environments

We return to the setting of Section 11.3, where we impose additional structure. Player i's payoff from social alternative $x \in X$ is $v_i(x; \theta)$, where $\theta := (\theta_1, \ldots, \theta_n)$, θ_i is player i's type, independently distributed according to F_i with density f_i and support $[\underline{\theta}_i, \overline{\theta}_i] =: \Theta_i$ (so there is a common prior). Define $\Theta := \prod_i \Theta_i$. Note that quasilinearity implies risk neutrality, and allows a clean separation between the role of social alternatives and the transfers.

An allocation (outcome) is $(x, t) \in X \times \mathbb{R}^n$, where x is social alternative and t is the vector of transfers to the players. Note that we still do not impose ex post budget balance. The payoff to player i at θ is

$$v_i(x; \theta_i) + t_i.$$

As usual, extend v_i to lotteries over alternative be defining $v_i(\lambda, \theta) := Ev_i(x, \theta)$ for $\lambda \in \Delta(X)$.

A *direct mechanism* is a pair $\xi : \Theta \to \Delta(X \times \mathbb{R}^n)$. Denote the marginal distribution induced by $\xi(\theta)$ on X by $\xi_X(\theta)$.

Player i's *interim payoff* from reporting $\hat{\theta}_i$ is

$$U_i(\hat{\theta}_i; \theta_i) := E_{-i}[v_i(\xi_X(\hat{\theta}_i, \theta_{-i}); \theta_i) + t_i(\hat{\theta}_i, \theta_{-i})]$$
$$=: V_i(\hat{\theta}_i; \theta_i) + T_i(\hat{\theta}_i).$$

Truthtelling is a Nash equilibrium of the direct mechanism if it is *incentive compatible*:

$$U_i^*(\theta_i) := U_i(\theta_i; \theta_i) \geq U_i(\hat{\theta}_i; \theta_i) \qquad \forall \theta_i \in \Theta_i.$$

From quasilinearity, we have

$$U_i^*(\hat{\theta}_i) - U_i(\hat{\theta}_i; \theta_i) = V_i(\hat{\theta}_i; \hat{\theta}) - V_i(\hat{\theta}_i; \theta_i). \tag{12.2.1}$$

Lemma 12.2.1. *Suppose ξ is an incentive compatible mechanism and the implied interim valuation $V_i(\hat{\theta}_i; \theta_i)$ for each player is continuously differentiable in $(\hat{\theta}_i, \theta_i)$ at all points $\hat{\theta}_i = \theta_i$.[1] Then for all $\theta_i' \neq \theta_i''$,*

$$U_i^*(\theta_i'') = U_i^*(\theta_i') + \int_{\theta_i'}^{\theta_i''} \frac{\partial V_i(\tilde{\theta}_i; \tilde{\theta}_i)}{\partial \theta_i} d\tilde{\theta}_i. \tag{12.2.2}$$

The partial derivative in the integrand is with respect to player i's true type, and evaluated where player i reports truthfully.

Proof. Since ξ is incentive compatible,

$$U_i^*(\theta_i + \delta) \geq U_i(\theta_i; \theta_i + \delta) \quad \text{and} \quad U_i^*(\theta_i) \geq U_i(\theta_i + \delta; \theta_i),$$

and so

$$U_i^*(\theta_i + \delta) - U_i(\theta_i + \delta; \theta_i) \geq U_i^*(\theta_i + \delta) - U_i^*(\theta_i) \geq U_i(\theta_i; \theta_i + \delta) - U_i^*(\theta_i).$$

Applying (12.2.1), and dividing by δ gives

$$\frac{V_i(\theta_i + \delta; \theta_i + \delta) - V_i(\theta_i + \delta; \theta_i)}{\delta}$$

$$\geq \frac{U_i^*(\theta_i + \delta) - U_i^*(\theta_i)}{\delta}$$

$$\geq \frac{V_i(\theta_i; \theta_i + \delta) - V_i(\theta_i; \theta_i)}{\delta}.$$

Since V_i is continuously differentiable, the left and right sides both converge as $\delta \to 0$, and have the same limit, $\partial V_i(\theta_i; \theta_i)/\partial \theta_i$.[2] Hence, the middle term converges to the same limit.

Hence, U_i^* is differentiable everywhere with derivative $\partial V_i(\theta_i; \theta_i)/\partial \theta_i$. Moreover, since $\partial V_i(\theta_i; \theta_i)/\partial \theta_i$ is continuous, U_i^* is continuously differentiable, and so Lipshitz, and so (12.2.2) holds. ∎

---

[1] In general, assumptions on endogenous objects should be avoided. However, this assumption can be dropped in specific contexts, such as Section 10.2 and the next.

[2] The argument for the left side may need more detail:

$$\frac{V_i(\theta_i + \delta; \theta_i + \delta) - V_i(\theta_i + \delta; \theta_i)}{\delta}$$

$$= \frac{V_i(\theta_i + \delta; \theta_i + \delta) - V_i(\theta_i; \theta_i)}{\delta} + \frac{V_i(\theta_i, \theta_i) - V_i(\theta_i + \delta; \theta_i)}{\delta},$$

and the first term converges to the total derivative $dV_i(\theta_i; \theta_i)/d\theta_i = \partial V_i(\theta_i; \theta_i)/\partial \hat{\theta}_i + \partial V_i(\theta_i; \theta_i)/\partial \theta_i$, while the second term converges to $-\partial V_i(\theta_i; \theta_i)/\partial \hat{\theta}_i$.

Lemma 12.2.1 implies that the interim *payoffs* to a player i in any incentive compatible mechanism are completely determined by the determination of the social alternative ξ_X, up to a constant. The transfers can only affect the level of U_i^*, independent of θ_i. This does *not* imply that the transfers are independent of type. Recall that incentive compatibility is jointly determined by both the social alternative chosen and the transfers.

Consider now efficient direct mechanisms, i.e., mechanisms satisfying (11.3.1).

Theorem 12.2.1 (Krishna and Perry, 1998; Williams, 1999). *Suppose ξ is an efficient incentive compatible mechanism, and that the implied interim valuation $V_i(\hat{\theta}_i; \theta_i)$ for each player is continuously differentiable in $(\hat{\theta}_i, \theta_i)$ at all points $\hat{\theta}_i = \theta_i$. There exist constants k_i (independent of θ_{-i}) such that each player's interim payoff in the associated Groves mechanism (as defined in Definition 11.3.1) equals that player's interim payoff in ξ.*

Proof. Since $V(\hat{\theta}_i; \theta_i) = E_{-i}[v_i(\xi_X(\hat{\theta}_i, \theta_{-i}); \theta_i)]$, by Lemma 12.2.1, $U_i^*(\theta_i'') - U_i^*(\theta_i')$ is completely determined by ξ_X, which is determined by (11.3.1).

By Theorem 11.3.1, every Groves mechanism is incentive compatible. Moreover, every Groves mechanism with ξ_X has the same implied interim valuation V_i as ξ, and so Lemma 12.2.1 also applies to the interim payoffs in any Groves mechanism, and the resulting payoff is the same, being determined by ξ_X.

It remains to choose the constants: Fix θ_i' for each i and choose k_i so that i's interim payoff in the resulting Groves mechanism equals $U_i^*(\theta_i')$. ∎

The only role of the assumption that V_i is continuously differentiable is to ensure that (12.2.2) holds, so that $U_i^*(\theta_i'') - U_i^*(\theta_i')$ is completely determined by ξ_X. The following version of Theorem 12.2.1 makes this explicit.

Theorem 12.2.2. *Suppose every efficient incentive compatible mechanism satisfies (12.2.2). For any efficient incentive compatible mechanism ξ, there exist constants k_i (independent of θ_{-i}) such that each player's interim payoff in the associated Groves mechanism (as defined in Definition 11.3.1) equals that player's interim payoff in ξ.*

Remark 12.2.1. It turns out that in many quasilinear settings, *all* mechanisms (not just the efficient ones) that are Bayesian implementable also have a dominant strategy implementation (at the level of interim utilities). See Gershkov, Goeree, Kushnir, Moldovanu, and Shi (2013). ♦

12.3   Incomplete Information Bargaining

The economic setting is a generalization of Example 3.3.6. Buyer valuations
$v_b \in [\underline{v}_b, \bar{v}_b]$, distributed according to cdf F with density f; seller valuations
$v_s \in [\underline{v}_s, \bar{v}_s]$ distributed according to cdf G with density g. Buyer and seller
valuations are independent.

The quasilinearity (risk neutrality) of the buyer and seller imply that we
can restrict attention to to allocations (p, τ), where p is the probability of
trade and τ is the expected transfer from the buyer to the seller. (If there
is only an ex post transfer when there is trade, the transfer conditional on
a sale is τ/p).

The payoff to the buyer is $pv_b - \tau$, while the payoff to the seller is $\tau - pv_s$.
A direct mechanism is a pair

$$(p, \tau) : [\underline{v}_b, \bar{v}_b] \times [\underline{v}_s, \bar{v}_s] \to [0, 1] \times \mathbb{R}.$$

The expected gain from trade under a mechanism (p, τ) is given by

$$\iint (v_b - v_s) p(v_b, v_s)\ F(dv_b) G(dv_s). \tag{12.3.1}$$

12.3.1   The Impossibility of Ex Post Efficient Trade

Trading is *ex post efficient* if

$$p(v_b, v_s) = \begin{cases} 1, & \text{if } v_b > v_s, \\ 0, & \text{if } v_b < v_s. \end{cases}$$

See Figure 12.3.1.

This is an example of the model from the previous section, with the set
of social alternatives being $X = \{x_b, x_s\}$, with x_i meaning player i gets the
object. Ex post budget balance is imposed by assumption.

In the following theorem, all the bargaining games under consideration
have as outcomes (p, τ), with the outcomes evaluated as described above,
so delay is not costly (see Problem 12.5.7 for the extension to bargaining
with costly delay).

Theorem 12.3.1 (Myerson and Satterthwaite, 1983). *Suppose some trade
is ex post efficient (i.e., $\underline{v}_s < \bar{v}_b$), but that not all trade is ex post efficient
(i.e., $\underline{v}_b < \bar{v}_s$). In every equilibrium of any bargaining game with voluntary
participation, trade is necessarily ex post inefficient.*

By the revelation principle, any equilibrium of a bargaining game induces the same outcome as an incentive compatible direct mechanism. Voluntary participation implies the mechanism satisfies individual rationality.

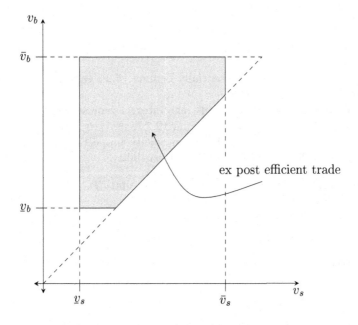

Figure 12.3.1: The region of ex post efficient trade.

Define

$$U_b(\hat{v}_b, v_b) := \int_{\underline{v}_s}^{\bar{v}_s} \{p(\hat{v}_b, v_s)v_b - \tau(\hat{v}_b, v_s)\} \, G(dv_s)$$

$$= \int_{\underline{v}_s}^{\bar{v}_s} p(\hat{v}_b, v_s) \, G(dv_s) \times v_b - \int_{\underline{v}_s}^{\bar{v}_s} \tau(\hat{v}_b, v_s) \, G(dv_s)$$

$$=: p_b(\hat{v}_b)v_b - \tau_b(\hat{v}_b) \tag{12.3.2}$$

and

$$U_s(\hat{v}_s, v_s) := \int_{\underline{v}_b}^{\bar{v}_b} \{\tau(v_b, \hat{v}_s) - p(v_b, \hat{v}_s)v_s\} \, F(dv_b)$$

$$=: \tau_s(\hat{v}_s) - p_s(\hat{v}_s)v_s. \tag{12.3.3}$$

Truth-telling is a Nash equilibrium of the direct mechanism (p, τ) if the mechanism is *incentive compatible*:

$$U_b(\hat{v}_b, v_b) \leq U_b^*(v_b) := U_b(v_b, v_b)$$

and

$$U_s(\hat{v}_s, v_s) \leq U_s^*(v_s) := U_s(v_s, v_s).$$

The mechanism satisfies *individual rationality* if

$$U_b^*(v_b) \geq 0 \quad \forall v_b \quad \text{and} \quad U_s^*(v_s) \geq 0 \quad \forall v_s.$$

The following lemma is essentially Lemma 10.2.1 (apply its proof separately to the buyer and seller).

If the densities are continuous, the integral representation of U_b^* and U_s^* can also be obtained from Lemma 12.2.1, as $V_b(\hat{v}_b, v_b) = G(\hat{v}_b)v_b$ and $V_s(\hat{v}_s, v_s) = (1 - F(\hat{v}_s))v_s$. The monotonicity properties of p_b and p_s are immediate implications of incentive compatibility.

Lemma 12.3.1. *Suppose (p, τ) is incentive compatible. Then,*

1. *p_b is nondecreasing,*

2. *p_s is nonincreasing,*

3. *$U_b^*(v_b) = U_b^*(\underline{v}_b) + \int_{\underline{v}_b}^{v_b} p_b(\tilde{v}_b)d\tilde{v}_b$, and*

4. *$U_s^*(v_s) = U_s^*(\bar{v}_s) + \int_{v_s}^{\bar{v}_s} p_s(\tilde{v}_s)d\tilde{v}_s$.*

Note also that this implies that if (p, τ) is incentive compatible, the individual rationality of (p, τ) is an implication of $U_b^*(\underline{v}_b) \geq 0$ and $U_s^*(\bar{v}_s) \geq 0$ (that is, we only need to impose the individual rationality for one type for buyer and seller).

Following Williams (1999),[3] we prove the result essentially applying Theorem 12.2.1. Myerson and Satterthwaite's (1983) original proof is explained after the proof of Lemma 12.3.2.

So suppose the result is false. Then there is an efficient direct mechanism satisfying ex post budget balance and interim individual rationality. Ex post budget balance implies

$$\int \tau_b(v_b)F(dv_b) = \int \tau_s(v_s)G(dv_s)$$

(note that this is simply ex ante budget balance). Since

$$\tau_b(v_b) = p_b(v_b)v_b - U_b^*(v_b)$$

and

$$\tau_s(v_s) = p_s(v_s)v_s + U_s^*(v_s),$$

we have

---

[3] See also Krishna (2002, §5.3).

$$\int p_b(v_b)v_b - U_b^*(v_b)F(dv_b) = \int p_s(v_s)v_s + U_s^*(v_s)G(dv_s) \iff$$

$$\iint (v_b - v_s)p(v_b, v_s)F(dv_b)G(dv_s) = \iint U_b^*(v_b) + U_s^*(v_s)F(dv_b)G(dv_s).$$

But p is efficient, and so budget balance requires

$$\mathcal{S} := \iint_{\{v_b > v_s\}} (v_b - v_s)F(dv_b)G(dv_s)$$

$$= \iint U_b^*(v_b) + U_s^*(v_s)F(dv_b)G(dv_s), \quad (12.3.4)$$

where $\mathcal{S}$ is the ex ante expected gains from trade, and of course $\mathcal{S} > 0$.

By Theorem 12.2.2 (which applies because of Lemma 12.3.1),[4] there is a Groves mechanism with constants k_b and k_s with the same interim payoffs. In the Groves mechanism the transfers t are

$$t_b(v_b, v_s) = \begin{cases} -v_s + k_b, & \text{if } v_b \geq v_s, \\ k_b, & \text{if } v_b < v_s, \end{cases}$$

and

$$t_s(v_b, v_s) = \begin{cases} v_b + k_s, & \text{if } v_b \geq v_s, \\ k_s, & \text{if } v_b < v_s, \end{cases}$$

Hence,

$$U_b^*(v_b) = \int_{\{v_b > v_s\}} [v_b - v_s]\, G(dv_s) + k_b$$

and so

$$\int U_b^*(v_b)\, F(dv_b) = \iint_{\{v_b > v_s\}} [v_b - v_s]\, G(dv_s)F(dv_b) + k_b = \mathcal{S} + k_b.$$

We similarly have

$$\int U_s^*(v_s)G(dv_s) = \mathcal{S} + k_s,$$

and so from (12.3.4),

$$\mathcal{S} = 2\mathcal{S} + k_b + k_s \iff k_b + k_s = -\mathcal{S} < 0.$$

Individual rationality requires $U_b^*(\underline{v}_b) \geq 0$ and $U_s^*(\bar{v}_s) \geq 0$. We have

$$U_b^*(\underline{v}_b) = \int_{\{v_s < \underline{v}_b\}} [\underline{v}_b - v_s]\, G(dv_s) + k_b$$

---

[4]If the densities g and f are continuous, then Theorem 12.2.1 can be applied directly.

and
$$U_s^*(\bar{v}_s) = \int_{\{v_b > \bar{v}_s\}} [v_b - \bar{v}_s]\, F(dv_b) + k_s.$$

If $\underline{v}_b = \underline{v}_s$ and $\bar{v}_b = \bar{v}_s$, then $U_b^*(\underline{v}_b) = k_b \geq 0$ and $U_s^*(\bar{v}_s) = k_s \geq 0$, and we have a contradiction (since $k_b + k_s < 0$). For the more general case where $\underline{v}_b = \underline{v}_s$ and $\bar{v}_b = \bar{v}_s$ may not hold, see Appendix 14.3. The idea is the same, but the calculation is a little tedious.

12.3.2   Maximizing Ex Ante Gains From Trade

Since ex post efficient trade is impossible, it is natural to ask about mechanisms which maximize the ex ante gains from trade. We first obtain a necessary condition that any incentive compatible mechanism must satisfy (under a mild technical assumption).

Lemma 12.3.2. *Suppose $f(v_b) > 0$ for all $v_b \in [\underline{v}_b, \bar{v}_b]$ and $g(v_s) > 0$ for all $v_s \in [\underline{v}_s, \bar{v}_s]$. If (p, τ) is incentive compatible, then*

$$\iint \left\{ \left[v_b - \frac{1 - F(v_b)}{f(v_b)} \right] - \left[v_s + \frac{G(v_s)}{g(v_s)} \right] \right\} p(v_b, v_s)\, F(dv_b) G(dv_s)$$
$$= U_b^*(\underline{v}_b) + U_s^*(\bar{v}_s). \quad (12.3.5)$$

The integrand of the left side of (12.3.5) is called the *virtual surplus*, while
$$v_b - \frac{1 - F(v_b)}{f(v_b)}$$

is the buyer's *virtual value* and

$$v_s + \frac{G(v_s)}{g(v_s)}$$

is the seller's *virtual value*.

Proof. Since
$$\tau_b(v_b) = p_b(v_b)v_b - U_b^*(v_b),$$

Lemma 12.3.1 implies

$$\int_{\underline{v}_b}^{\bar{v}_b} \tau_b(v_b)\, F(dv_b) = \int_{\underline{v}_b}^{\bar{v}_b} p_b(v_b)v_b f(v_b) dv_b$$
$$- U_b^*(\underline{v}_b) - \int_{\underline{v}_b}^{\bar{v}_b} \int_{\underline{v}_b}^{v_b} p_b(\tilde{v}_b) d\tilde{v}_b f(v_b) dv_b.$$

Integrating the last term by parts gives[5]

$$\int_{\underline{v}_b}^{\bar{v}_b} \int_{\underline{v}_b}^{v_b} p_b(\tilde{v}_b) d\tilde{v}_b f(v_b) dv_b$$

$$= \left. \int_{\underline{v}_b}^{v_b} p_b(\tilde{v}_b) d\tilde{v}_b \ F(v_b) \right|_{v_b=\underline{v}_b}^{\bar{v}_b} - \int_{\underline{v}_b}^{\bar{v}_b} p_b(v_b) F(v_b) dv_b$$

$$= \int_{\underline{v}_b}^{\bar{v}_b} p_b(v_b)(1 - F(v_b)) dv_b.$$

Thus,

$$\int_{\underline{v}_b}^{\bar{v}_b} \tau_b(v_b) \ F(dv_b) = \int_{\underline{v}_b}^{\bar{v}_b} p_b(v_b) v_b f(v_b) dv_b$$

$$- U_b^*(\underline{v}_b) - \int_{\underline{v}_b}^{\bar{v}_b} p_b(v_b)(1 - F(v_b)) dv_b$$

$$= \int_{\underline{v}_b}^{\bar{v}_b} \left[v_b - \frac{(1 - F(v_b))}{f(v_b)} \right] p_b(v_b) \ F(dv_b) - U_b^*(\underline{v}_b). \quad (12.3.6)$$

A similar calculation on the seller side yields

$$\int_{\underline{v}_s}^{\bar{v}_s} \tau_s(v_s) \ G(dv_s) = \int_{\underline{v}_s}^{\bar{v}_s} \left[v_s + \frac{G(v_s)}{g(v_s)} \right] p_s(v_s) \ G(dv_s) + U_s^*(\bar{v}_s). \quad (12.3.7)$$

Since

$$\int_{\underline{v}_b}^{\bar{v}_b} \int_{\underline{v}_s}^{\bar{v}_s} \tau(v_b, v_s) \ F(dv_b) G(dv_s) = \int_{\underline{v}_b}^{\bar{v}_b} \tau_b(v_b) \ F(dv_b)$$

$$= \int_{\underline{v}_s}^{\bar{v}_s} \tau_s(v_s) \ G(dv_s),$$

the right sides of (12.3.6) and (12.3.7) are equal, and rearranging gives (12.3.5). ∎

Myerson and Satterthwaite's (1983) original proof of Theorem 12.3.1 involved showing that the left side of (12.3.5) is strictly negative for the ex post efficient trading rule (violating individual rationality).

We now prove a converse to Lemmas 12.3.1 and 12.3.2.

Lemma 12.3.3. *Suppose $f(v_b) > 0$ for all $v_b \in [\underline{v}_b, \bar{v}_b]$ and $g(v_s) > 0$ for all $v_s \in [\underline{v}_s, \bar{v}_s]$. Suppose $p : [\underline{v}_b, \bar{v}_b] \times [\underline{v}_s, \bar{v}_s] \to [0, 1]$ is a probability of trade function satisfying*

---

[5]Equivalently, we could apply Fubini, as in footnote 6 on page 270.

1. p_b is nondecreasing,

2. p_s is nonincreasing, and

3. the expected virtual surplus (given by the left side of (12.3.5)) is non-negative.

There exists a transfer function $\tau : [\underline{v}_b, \bar{v}_b] \times [\underline{v}_s, \bar{v}_s] \to [0, 1]$ such that (p, τ) is incentive compatible and individually rational.

Proof. Inspired by Lemma 10.2.2 (in particular, by(10.2.4)), it is natural to look for a transfer function τ that satisfies

$$\tau_b(v_b) = p_b(v_b)v_b - \int_{\underline{v}_b}^{v_b} p_b(v)dv$$

$$\text{and} \quad \tau_s(v_s) = p_s(v_s)v_s + \int_{v_s}^{\bar{v}_s} p_s(v)dv.$$

But in general this might be asking too much, since it implies that both buyer $\underline{v}_b$ and seller $\bar{v}_s$ have zero payoff (and with ex post budget balance, that seems like too big a coincidence).

This suggests introducing a free constant k and considering a transfer function of the form[6]

$$\tau(v_b, v_s) := k + p_b(v_b)v_b - \int_{\underline{v}_b}^{v_b} p_b(v)dv + p_s(v_s)v_s + \int_{v_s}^{\bar{v}_s} p_s(v)dv.$$

The level of k has no impact upon incentives, and simply serves to reallocate value between the buyer and the seller. The proof of Lemma 10.2.2 then implies that, for any k, the resulting direct mechanism (p, t) satisfies incentive compatibility for both buyer and seller, and so Lemma 12.3.2 holds.

It remains to verify individual rationality. It suffices to show that k can be chosen so that, under the mechanism, $U_b^*(\underline{v}_b) \geq 0$ and $U_s^*(\bar{v}_s) \geq 0$.

By hypothesis, the expected virtual surplus is nonnegative, and so by Lemma 12.3.2,

$$U_b^*(\underline{v}_b) + U_s^*(\bar{v}_s) \geq 0.$$

Since k simply reallocates value between the buyer and seller, there clearly exist values of k which imply the resulting mechanism satisfies individual rationality (for example, choose k so that $U_s^*(\bar{v}_s) = 0$, or to evenly split the expected virtual surplus between the bottom buyer and top seller).   ∎

---

[6]This functional form is convenient,since it does not require detailed knowledge of p. But it may require transfers even when trade does not occur. The only requirement is that the transfer function have the correct interim transfers, so other transfer functions are possible. For an illustration, see Example 12.3.1.

The problem of maximizing the ex ante surplus reduces to choosing a mechanism to maximize (12.3.1) subject to the conditions in Lemma 12.3.3. We proceed by considering the relaxed problem where we ignore the monotonicity condition (much as we did in Section 10.2).

The Lagrangean for the relaxed problem is

$$\mathcal{L} := \iint (v_b - v_s) p(v_b, v_s) \, F(dv_b) G(dv_s)$$

$$+ \lambda \iint \left\{ \left[v_b - \frac{1 - F(v_b)}{f(v_b)} \right] - \left[v_s + \frac{G(v_s)}{g(v_s)} \right] \right\}$$

$$\times p(v_b, v_s) \, F(dv_b) G(dv_s)$$

$$= (1 + \lambda) \iint \left\{ \left[v_b - \alpha \frac{1 - F(v_b)}{f(v_b)} \right] - \left[v_s + \alpha \frac{G(v_s)}{g(v_s)} \right] \right\}$$

$$\times p(v_b, v_s) \, F(dv_b) G(dv_s),$$

where $\alpha := \lambda / (1 + \lambda)$. Define the α-virtual values as

$$\mathcal{V}_b(v_b, \alpha) := v_b - \alpha \frac{1 - F(v_b)}{f(v_b)} \quad \text{and} \quad \mathcal{V}_s(v_s, \alpha) := v_s + \alpha \frac{G(v_s)}{g(v_s)}.$$

As the relaxed program is is a linear program (the objective and constraint functions are both linear in p), if (p^*, λ^*) is a solution, then

$$p^*(v_b, v_s) = \begin{cases} 0, & \text{if } \mathcal{V}_b(v_b, \alpha^*) < \mathcal{V}_s(v_s, \alpha^*), \\ 1, & \text{if } \mathcal{V}_b(v_b, \alpha^*) > \mathcal{V}_s(v_s, \alpha^*), \end{cases}$$

where $\alpha^* = \lambda^* / (1 + \lambda^*)$.

As in Section 10.2, a sufficient condition for p^* to satisfy the monotonicity conditions in Lemma 12.3.3 is that the distribution functions are regular:

$$v_b - \frac{1 - F(v_b)}{f(v_b)} \quad \text{and} \quad v_s + \frac{G(v_s)}{g(v_s)}$$

are increasing. In the regular case, any solution to the relaxed problem is a solution to the original problem. As in Sections 10.2 and 10.3, more advanced techniques handle the nonregular case.

From Theorem 12.3.1, we know ex post efficiency is not implementable, and so $\lambda^* > 0$. The precise value of λ^* is determined by the requirement that the expected virtual surplus under p^* is zero.

Example 12.3.1. Suppose v_b, v_s are both uniform draws from $[0, 1]$. The α-virtual values are

$$\mathcal{V}_b(v_b, \alpha) = (1 + \alpha)v_b - \alpha \quad \text{and} \quad \mathcal{V}_s(v_s, \alpha) := (1 + \alpha)v_s,$$

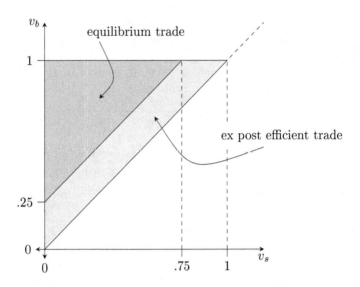

Figure 12.3.2: The trading regions for Example 12.3.1. The darker region
is the trading region in the linear equilibrium of the double
auction, which maximizes the ex ante gains from trade.

and

$$\mathcal{V}_b(v_b, \alpha) > \mathcal{V}_s(v_s, \alpha) \iff v_b - v_s > \frac{\alpha}{1 + \alpha}.$$

Evaluating the expected virtual surplus of a trading rule p which trades if
and only if $\mathcal{V}_b(v_b, \alpha) > \mathcal{V}_s(v_s, \alpha)$ yields

$$\int_0^1 \int_0^1 \{[2v_b - 1] - 2v_s\} \, p(v_b, v_s) \, dv_b dv_s$$

$$= \int_0^{1/(1+\alpha)} \int_{v_s + \alpha/(1+\alpha)}^1 \{[2v_b - 1] - 2v_s\} \, dv_b dv_s$$

$$= \frac{3\alpha - 1}{6(1 + \alpha)^3}.$$

Thus, the expected virtual surplus is strictly positive for $\alpha > \frac{1}{3}$, strictly
negative for $\alpha < \frac{1}{3}$, and and equal to 0 if $\alpha = \frac{1}{3}$, and so $\alpha^* = \frac{1}{3}$. Trade
under the ex ante surplus maximizing mechanism occurs if $v_b > v_s + \frac{1}{4}$ and
does not occur if $v_b < v_s + \frac{1}{4}$. See Figure 12.3.2.

It is worth noting that the linear equilibrium of the double auction
(Example 3.3.6) maximizes ex ante welfare, so the double auction implements the ex ante welfare maximizing trading rule. But because there are

other equilibria which do not (Problem 3.6.12), the double auction does not strongly implement the ex ante welfare maximizing trading rule.

It is instructive to compare the transfer function implied by the proof of Lemma 12.3.3 and the transfer function implied by the linear equilibrium of the double auction. Since

$$p_b(v_b) = \max\left\{0, v_b - \tfrac{1}{4}\right\} \text{ and } p_s(v_s) = \max\left\{0, \tfrac{3}{4} - v_s\right\},$$

and $U_b^*(\underline{v}_b) = 0$ and $U_s^*(\bar{v}_s) = 0$, Lemma 12.3.1 implies

$$\tau_b(v_b) = \max\left\{0, \tfrac{1}{2}v_b^2 - \tfrac{1}{32}\right\} \text{ and } \tau_s(v_s) = \max\left\{0, \tfrac{9}{32} - \tfrac{1}{2}v_s^2\right\}. \quad (12.3.8)$$

The transfer function implied by the proof of Lemma 12.3.3 is

$$\tau(v_b, v_s) = -\tfrac{9}{64} + \tau_b(v_b) + \tau_s(v_s)$$

$$= \begin{cases} \tfrac{7}{64} + \tfrac{1}{2}(v_b^2 - v_s^2), & v_b \geq \tfrac{1}{4}, v_s \leq \tfrac{3}{4}, \\ -\tfrac{11}{64} + \tfrac{1}{2}v_b^2, & v_b \geq \tfrac{1}{4}, v_s > \tfrac{3}{4}, \\ \tfrac{9}{64} - \tfrac{1}{2}v_s^2, & v_b < \tfrac{1}{4}, v_s \leq \tfrac{3}{4}, \\ -\tfrac{9}{64}, & v_b < \tfrac{1}{4}, v_s > \tfrac{3}{4}. \end{cases}$$

Observe that under this transfer function, transfers occur even when there is no trade. In contrast, the transfer function implied by the linear equilibrium of the double auction is

$$\tau(v_b, v_s) = \begin{cases} \tfrac{1}{6} + \tfrac{1}{3}(v_s + v_b), & v_b \geq v_s + \tfrac{1}{4}, \\ 0, & \text{otherwise}, \end{cases}$$

and transfers only occur when there is trade. Of course, both transfer function imply the same interim transfers (12.3.8). ★

12.4 Independent Private Values Auctions

We return to the setting of Example 10.4.3. Agent i's valuation, $v_i(\theta) = \theta_i$, with θ_i distributed according to F_i with support $[\underline{\theta}_i, \bar{\theta}_i]$ and strictly positive density f_i. Agents' types are independent.

Outcomes as a function of type profiles are

$$(\rho, t) : \prod_i [\underline{\theta}_i, \bar{\theta}_i] \to \Delta(\{0, 1, \ldots, n\}) \times \mathbb{R}^n,$$

where $\rho(\theta)$ is the probability distribution over who obtains the object (with 0 meaning the seller retains the good), and $t_i(\theta)$ is the transfer from bidder

i to the seller. As in Section 12.3, quasilinearity of payoffs implies restricting attention to deterministic transfers as a function of the type profile is without loss of generality.

Suppose $\rho_i(\theta)$ is the probability that i receives the object, so that $\sum_{i=1,\ldots,n} \rho_i(\theta) \leq 1$. Player i's expected ex post utility under the direct mechanism (ρ, t), from reporting $\hat{\theta}_i$ is

$$\rho_i(\hat{\theta}_i, \theta_{-i})\theta_i - t_i(\hat{\theta}_i, \theta_{-i}),$$

and i's interim utility from reporting $\hat{\theta}_i$ is[7]

$$\int_{\theta_{-i}} \rho_i(\hat{\theta}_i, \theta_{-i})\theta_i - t_i(\hat{\theta}_i, \theta_{-i}) \prod_{j \neq i} dF_j(\theta_j) =: p_i(\hat{\theta}_i)\theta_i - T_i(\hat{\theta}_i). \quad (12.4.1)$$

Theorem 12.4.1 (Revenue Equivalence). *Any two equilibria of any two auctions yielding identical probabilities of winning (i.e., ρ), and identical interim utilities for type $\underline{\theta}_i$ of each bidder i gives equal expected revenue to the seller.*

In particular, all efficient auctions are revenue equivalent.

Proof. Note that (12.4.1) has the same form as (12.3.2) and (12.3.3).

By the revelation principle, it suffices to consider incentive compatible and individually rational direct direct mechanisms. Fix such a direct mechanism (ρ, t). By applying the same logic as for Lemma 10.2.1, p_i is nondecreasing and

$$U_i^*(\theta_i) := p_i(\theta_i)\theta_i - T_i(\theta_i) = U_i^*(\underline{\theta}_i) + \int_{\underline{\theta}_i}^{\theta_i} p_i(\tilde{\theta}_i)d\tilde{\theta}_i. \quad (12.4.2)$$

The expected revenue of the seller is

$$\int \cdots \int \sum_i t_i(\theta) \prod_j dF_j(\theta_j) = \sum_i \int_{\theta_i} T_i(\theta_i) \, dF_i(\theta_i). \quad (12.4.3)$$

But, from (12.4.2),

$$T_i(\theta_i) = p_i(\theta_i)\theta_i - U_i^*(\underline{\theta}_i) - \int_{\underline{\theta}_i}^{\theta_i} p_i(\tilde{\theta}_i) \, d\tilde{\theta}_i, \quad (12.4.4)$$

and so expected revenue of the seller is completely determined by p_i and $U_i^*(\underline{\theta}_i)$ for each i.   ∎

---

[7]Note that in (12.4.1), we are *defining* the notation

$$p_i(\theta_i) := \int_{\theta_{-i}} \rho_i(\theta_i, \theta_{-i}) \prod_{j \neq i} dF_j(\theta_j) \text{ and } T_i(\theta_i) := \int_{\theta_{-i}} t_i(\theta_i, \theta_{-i}) \prod_{j \neq i} dF_j(\theta_j).$$

Remark 12.4.1. Revenue equivalence requires risk neutrality (quasilinear utilities) of the bidders. The first price auction yields more revenue than the second price auction when bidders are risk averse. For the general proof, see Krishna (2002, Section 4.1); it is straightforward to compare the expected revenues for the uniform case in Problem 12.5.4. ◆

We now consider *optimal auctions*, i.e., auctions that maximize revenue.[8] Since the seller's expected revenue has the form given in (12.4.3), applying integration by parts (or Fubini) to (12.4.4) in a familiar way yields the following expression for expected revenue:

$$\sum_i \int_{\underline{\theta}_i}^{\bar{\theta}_i} \left\{ \theta_i - \frac{1 - F_i(\theta_i)}{f_i(\theta_i)} \right\} p_i(\theta_i) f_i(\theta_i) d\theta_i. \tag{12.4.5}$$

It is worth noting that this expression can be rewritten directly in terms of ρ as

$$\sum_i \int_{\underline{\theta}_1}^{\bar{\theta}_1} \cdots \int_{\underline{\theta}_n}^{\bar{\theta}_n} \left\{ \theta_i - \frac{1 - F_i(\theta_i)}{f_i(\theta_i)} \right\} \rho_i(\theta) \prod_j f_j(\theta_j) d\theta_j. \tag{12.4.6}$$

Moreover, the logic in the proof of Lemma 10.2.2 and the immediately following discussion shows that if p_i is non-decreasing for all i, then the associated ρ is part of an incentive compatible mechanism. Hence, the revenue maximizing mechanism satisfying incentive compatibility and individual rationality is the solution to the following problem:

Choose $\rho : \prod_i [\underline{\theta}_i, \bar{\theta}_i] \to \Delta(\{0, 1, \dots, n\})$ to maximize (12.4.6), subject to p_i being non-decreasing for all i. To obtain individual rationality, set $T_i(\underline{\theta}_i) := p_i(\underline{\theta}_i)\underline{\theta}_i$, which implies $U_i^*(\underline{\theta}_i) = 0$.

Suppose we are in the *regular case*, so that i's virtual valuation

$$\psi_i(\theta_i) := \theta_i - \frac{1 - F_i(\theta_i)}{f_i(\theta_i)}$$

is increasing in θ_i for all i.[9]

Theorem 12.4.2. *Suppose $n \geq 2$ and all bidders' virtual valuations ψ_i are increasing. Any allocation rule ρ^* satisfying*

$$\rho_i^*(\theta) > 0 \implies \psi_i(\theta_i) = \max_j \psi_j(\theta_j) > 0,$$

---

[8]Of course, a seller may have other objectives, such as efficiency or encouraging diversity of ownership.

[9]As before, more advanced techniques (some introduced by Myerson (1981) in the context of optimal auctions) handle the nonregular case.

and the implied transfers with $T_i(\underline{\theta}_i) := p_i^*(\underline{\theta}_i)\underline{\theta}_i$ is an incentive-compatible individually-rational mechanism maximizing revenue.

Proof. As usual, we proceed by first ignoring the monotonicity constraints on p_i, and pointwise maximizing the integrand in (12.4.6). It is here that expressing expected revenue in the form (12.4.6) is important, since the allocation probabilities are of the form $\rho_i(\theta)$ and not $p_i(\theta_i)$, as in (12.4.5). The constraint that $\rho(\theta)$ is a probability distribution on $\{0, 1, \ldots, n\}$ with pointwise maximization forces $\rho_i(\theta) = 0$ for any i for which $\psi_i(\theta_i) < \max_j \psi_j(\theta_j)$. There is no comparable restriction on $\{p_i(\theta_i) : i = 1, \ldots, n\}$.

It is straightforward to verify that the implied p_i^* are non-decreasing when virtual values are increasing, and so ρ^* maximizes revenue. ∎

Observe that under ρ^*, if $\psi_i(\theta_i) < 0$, then $\rho_i^*(\theta_i, \theta_{-i}) = 0$ for all θ_{-i}.

In general, optimal auctions need not always sell the good (this occurs when $\psi_i(\theta_i) < 0$ for all i). This should not be a surprise, given the analysis of the take-it-or-leave-it mechanism in Section 10.3.

Perhaps more interestingly, even when the good is sold, the bidder who values the good most highly need not receive the good. Rather, it is the bidder with the highest *virtual* valuation (see Example 12.4.1). So, in general, revenue maximization induces two different types of inefficiency: trade does not occur when it should, and the wrong bidder receives the good.

For the general asymmetric case, the optimal auction cannot be implemented by any standard auction.[10]

For simplicity, we assume $n = 2$, and consider the symmetric case, $F_1 = F_2 =: F$ and $f_1 = f_2 =: f$. Moreover, we suppose the common $\psi_i =: \psi$ are strictly increasing and continuous, and let $\theta = r^*$ solve $\psi(\theta) = 0$. Symmetry also implies that $\psi(\theta_1) > \psi(\theta_2) \iff \theta_1 > \theta_2$, so that a bidder with the higher virtual valuation also has the higher value. The highest value bidder always receives the good when it is sold, and revenue maximization only induces inefficiency through trade not occurring when it should.

The implied probability of acquiring the good is

$$p_i(\theta_i) = \begin{cases} 0, & \text{if } \theta_i \leq r^*, \\ F(\theta_i), & \text{if } \theta_i > r^*, \end{cases}$$

---

[10]The issue is not symmetry per se. An auction is *symmetric* if all bidders have the same action spaces and a permutation of an action profile results in a similar permutation of the outcome profile. Deb and Pai (2017) show it is typically possible to implement asymmetric allocations (such as the optimal auction allocation) using equilibria of symmetric auctions. However, the symmetric auction used is not a standard auction and the complicated details of the transfer rule depend on the specific value distributions.

which is non-decreasing as required.
The associated interim transfers are given by

$$T_i(\theta_i) = p_i(\theta_i)\theta_i - \int_{\underline{\theta}}^{\theta_i} p_i(\tilde{\theta}_i) \, d\tilde{\theta}_i$$

$$= \begin{cases} 0, & \text{if } \theta_i \leq r^*, \\ F(\theta_i)\theta_i - \int_{r^*}^{\theta_i} F(\tilde{\theta}_i) \, d\tilde{\theta}_i, & \text{if } \theta_i > r^*. \end{cases}$$

For $\theta_i > r^*$, applying integration by parts,

$$T_i(\theta_i) = F(\theta_i)\theta_i - F(\tilde{\theta}_i)\tilde{\theta}_i \Big|_{r^*}^{\theta_i} + \int_{r^*}^{\theta_i} \tilde{\theta}_i f(\tilde{\theta}_i) \, d\tilde{\theta}_i$$

$$= F(r^*)r^* + \int_{r^*}^{\theta_i} \tilde{\theta}_i f(\tilde{\theta}_i) \, d\tilde{\theta}_i$$

$$= \int_{\underline{\theta}}^{\theta_i} \max\{r^*, \tilde{\theta}_i\} \, f(\tilde{\theta}_i) \, d\tilde{\theta}_i.$$

The transfers have the nice representation as the expected value of

$$t_i(\theta_i, \theta_j) = \begin{cases} \max\{r^*, \theta_j\}, & \text{if } \theta_i > \max\{r^*, \theta_j\}, \\ 0, & \text{if } \theta_i \leq \max\{r^*, \theta_j\}. \end{cases}$$

These transfers are the payments in a second price auction with a reserve price of r^*.

The allocation rule and expected payment of the lowest type are the same as in a symmetric equilibrium of the first-price sealed-bid auction with a reservation price of r^*. Applying revenue equivalence, a symmetric equilibrium is given by, for $\theta_i \leq r^*$, setting $\sigma_i(\theta_i) = 0$ and for $\theta_i > r^*$, setting

$$\sigma_i(\theta_i) = \frac{T_i(\theta_i)}{p_i(\theta_i)}$$

$$= \frac{1}{F(\theta_i)} \int_{\underline{\theta}}^{\theta_i} \max\{r^*, \tilde{\theta}_i\} \, f(\tilde{\theta}_i) \, d\tilde{\theta}_i$$

$$= E\left[\max\{r^*, \tilde{\theta}_j\} \mid \tilde{\theta}_j \leq \theta_i\right]. \tag{12.4.7}$$

Our final example illustrates the impact of asymmetries.

Example 12.4.1. Suppose $n = 2$, v_1 is uniformly distributed on $[1, 2]$ and v_2 is independently and uniformly distributed on $[1, 3]$. It is a straightforward application of the material we have seen to verify that the following sealed-bid auction is optimal: bidder 1 wins if $b_1 \geq b_2$ and pays

$\max\{1, b_2 - \frac{1}{2}\}$ in the event of winning, and bidder 2 wins if $b_2 > b_1$ and pays b_1 in the event of winning (see Problem 12.5.11). ★

12.5 Problems

12.5.1. Suppose $Z = \{a, b\}$, $n = 3$, and $\Theta_i = \{\theta_i^a, \theta_i^b\}$. Player i's utility is

$$u_i(z, \theta) = \begin{cases} 1, & \theta = \theta_i^z, \\ 0, & \theta \neq \theta_i^z. \end{cases}$$

Describe *all* the pure strategy Nash equilibrium outcomes of the complete information revelation game ξ, where the outcome is determined by the majority, i.e., $\xi(\theta_1, \theta_2, \theta_3) = z$ where $|\{i : \theta_i = \theta_i^z\}| \geq 2$.

12.5.2. What is the direct mechanism corresponding to the symmetric equilibrium derived in Example 3.3.1 for the sealed bid first price auction? Verify that truthtelling is a Nash equilibrium of the direct mechanism. Compare the direct mechanism with the VCG mechanism (i.e., the second price auction) and relate this to Theorem 12.2.1.

12.5.3. There is a single good to be allocated to one of two agents. Agents have quasilinear preferences, agent i values the good at v_i, with v_i being uniformly and independently distributed on $[0, 1]$. Consider the direct mechanism which awards the good to the agent with the highest report, and specifies that the agent receiving the good pays double the other agent's report.

(a) Prove that truthful reporting is not a Nash equilibrium.

(b) Describe an efficient Nash equilibrium, and the associated direct mechanism from the revelation principle.

12.5.4. (Based on Jehle and Reny, 2011, Problem 9.4) There are N buyers participating in an auction. Each buyer's value is an i.i.d. draw from an interval according to the distribution F, with a density. Buyers are *risk-averse*: If a buyer with value v wins the object at a price of of $b < v$, then her utility is $(v - b)^{\frac{1}{\alpha}}$ where $\alpha > 1$ is fixed and common to all bidders. In this problem, we will use mechanism design tools to solve for the symmetric Bayes-Nash equilibrium of the first price auction. Denote the equilibrium bid of a buyer of value v by $b_\alpha(v)$. Assume that $b_\alpha(v)$ is strictly increasing in v.

(a) Argue that a first price auction with risk-averse bidders whose values are independently distributed according to F is equivalent to a first-price auction with *risk-neutral* bidders whose values are independently distributed according to the distribution F^α.

(b) Use (12.4.2) and the appropriate second-price auction to provide a formula for $b_\alpha(v)$.

(c) It remains to verify that the formula just obtained is indeed an equilibrium of first price auction with risk averse buyers (which follows from part (a)).

12.5.5. (Dissolving a partnership, Cramton, Gibbons, and Klemperer, 1987) A symmetric version of bargaining: Two players equally own an asset and are bargaining to dissolve the joint ownership. Suppose v_i is the private value that player i assigns to the good, and suppose v_1 and v_2 are independently and uniformly drawn from a common interval $[\underline{v}, \bar{v}]$. Efficiency requires that player i receive the asset if $v_i > v_j$. The opportunity cost to player j if player i is assigned the asset is $\frac{1}{2}v_j$.

(a) Prove that efficient dissolution is possible by a mechanism that satisfies ex ante budget balance and individual rationality, and describe the Groves mechanism that achieves it. Prove that the Groves mechanism cannot satisfy ex post budget balance.

(b) It is possible to efficiently dissolve the partnership while still satisfying ex post budget balance. We consider two different games in which the two players simultaneously submit bids, with the higher bidder winning the object (a tie is resolved by a coin flip). The games differ in the payment rules:

 i. The winner pays the loser the winning bid.

 ii. Each player (winning or loosing) pays the other player their bid.

Both games clearly satisfy ex post budget balance. Suppose the values are uniformly drawn from the interval $[\underline{v}, \underline{v} + 1]$. For each game: Solve for the symmetric equilibrium in increasing differentiable strategies and prove it satisfies voluntary participation (note that a bid of $b_i = 0$ in the second game will typically violate individual rationality). [Efficiency is obvious, since the equilibria are in symmetric increasing strategies. The second payment rule can be extended to partnerships with more than 2 members, see Cramton, Gibbons, and Klemperer (1987).]

(c) Provide a sufficient condition for part (a) to extend to asymmetric general distributions.

12.5.6. (Public good provision, Mailath and Postlewaite, 1990) A community of size n must decide on whether to provide a public good. The per capita cost of the public good is $c > 0$. Every agent's valuation for the good is independently and uniformly drawn from the interval $[\underline{v}, \bar{v}]$, where $0 \leq \underline{v} < c < \bar{v}$.

(a) Prove that ex post efficient provision is impossible if ex ante budget balance and voluntary participation is required.

(b) Solve for the mechanism which maximizes the probability of public provision while respecting ex ante budget balance and voluntary participation.

(c) (**Hard**) Prove that the probability of provision goes to zero as the community becomes arbitrarily large when both ex ante budget balance and voluntary participation are required.

12.5.7. Consider again the alternating-offer identical-discount-factors bargaining model of Section 9.2, with the interpretation that player 1 is a seller with valuation v_s and player 2 is a buyer with valuation v_b. If the valuations are common knowledge, and $v_s > v_b$, then no trade is possible, while if $v_s < v_b$, we get the Rubinstein (1982) split of the surplus $v_b - v_s$. Suppose now there is two-sided incomplete information as described in Section 12.3.

(a) Prove that in *every* Nash equilibrium, there is delay to agreement, and that the ex ante expected cost of delay is bounded away from zero as $\delta \to 1$. [**Hint:** This is an implication of Theorem 12.3.1.]

(b) Does your answer to part (a) change if the game is altered so that only the seller can make an offer to the buyer?

(c) Suppose now it is commonly known that the seller's valuation is 0, and the buyer's valuation is uniformly drawn from the interval $[0, 1]$. Does your answer to part (b) change?

12.5.8. For the setting of Section 12.3, denote by p_x the probability of trade function given by

$$p_x(v_b, v_s) = \begin{cases} 1, & v_b \geq 1 - x, v_s \leq x, \\ 0, & \text{otherwise.} \end{cases}$$

Suppose v_b and v_s are uniformly and independently distributed on $[0, 1]$.

(a) For what values of x is there a transfer function τ_x for which (p_x, τ_x) is incentive compatible and individually rational?

(b) Which value of x maximizes the ex ante gains from trade? Compare the ex ante gains from trade under that trading rule p_x with the ex ante gains from trade from the mechanism calculated in Example 12.3.1.

(c) The interim utility of the buyer with value 0 is necessarily 0 in the ex ante welfare maximizing mechanism. Describe a direct mechanism that gives the 0-value buyer an interim utility of $1/27$.

(d) The interim utility of the buyer with value 0 is necessarily 0 in every equilibrium of the form analyzed in Problem 3.6.12. Reconcile this with your answer to part (c).

12.5.9. (Regulating a monopolist, Baron and Myerson, 1982) A public utility commission (the regulator) is charged with regulating a natural monopoly. The cost function of the natural monopoly is given by

$$C(q, \theta) = \begin{cases} 0, & \text{if } q = 0, \\ K + \theta q, & \text{if } q > 0, \end{cases}$$

where q is the quantity produced, $K > 0$ is the publicly known fixed cost, and $\theta \in (0, 1)$ is marginal cost. The inverse demand curve for the good is

$$p(q) := \max\{1 - 2q, \ 0\}.$$

Suppose there are no income effects for this good, so that consumer surplus is given by

$$V(q) = \int_0^q p(\tilde{q}) \ d\tilde{q} - p(q)q.$$

The regulator determines the firm's regulated price and subsidy, as well as whether the firm is allowed to operate at all. The firm cannot be forced to operate. As in the standard monopoly problem, without loss of generality, the regulator chooses q (with the *regulated* price given by $p(q)$).

The firm wishes to maximize expected profits,

$$\Pi(q) = p(q)q - C(q, \theta) + s,$$

where s is the subsidy. The regulator maximizes the weighted sum of consumer surplus and firm profits net of the subsidy,

$$V(q) + \alpha\Pi(q) - s, \qquad \alpha \in [0, 1].$$

(a) Suppose the marginal cost $\theta > 0$ is publicly known. Solve the regulator's problem. Interpret the resulting q, p and s.

(b) Suppose now the marginal cost θ is known only to the monopoly, with the regulator's beliefs over θ being described by the distribution function F, with support $[\underline{\theta}, \bar{\theta}]$ and strictly positive density f. Suppose F satisfies:

$$\theta + \frac{F(\theta)}{f(\theta)} \quad \text{is increasing in } \theta$$

and that

$$K \leq \frac{(1-\bar{\theta})^2}{4}.$$

What is the regulator's optimal regulated price and subsidy (which includes the decision on whether to let the firm operate)? Interpret the resulting mechanism.

12.5.10. Verify that the strategy given in (12.4.7) is a symmetric equilibrium strategy of the symmetric first-price sealed-bid auction with reserve price r^* (recall the hint from Problem 3.6.3).

12.5.11. Verify the claim in Example 12.4.1 via the following steps:

(a) Calculate the allocation rule in the optimal auction.

(b) Prove that it is a dominant strategy for bidder 1 to bid $v_1 + \frac{1}{2}$ and for bidder 2 to bid v_2 in the auction described in Example 12.4.1.

(c) Complete the verification.

Chapter 13

Principal Agency

13.1 Introduction

A principal hires an agent for some task.

If hired, the agent chooses an effort that results in some output (revenue) π for the principal.

Suppose the agent can choose effort level $e \in E \subset \mathbb{R}_+$.

The resulting profit is $\pi \in [\underline{\pi}, \bar{\pi}] = \Pi \subset \mathbb{R}$.

$\pi \sim F(\cdot \mid e)$, F has density $f(\cdot \mid e)$ with $f(\pi \mid e) > 0$, for all π and e.

Assume $F(\cdot \mid e)$ is ordered by *first order stochastic dominance*. That is,

$$e' < e'' \implies F(\pi \mid e') \geq F(\pi \mid e'') \qquad \forall \pi \in \Pi$$

(see Figure 13.1.1).

The principal pays a wage w to agent.

Agent has utility

$$v(w) - c(e),$$

where v is smooth strictly concave increasing function of w (so $v' > 0$ and $v'' < 0$) and c is a nonnegative smooth strictly convex increasing function of e (so $c \geq 0$ and $c', c'' > 0$). Agent is *risk averse*.

Principal is *risk neutral*, with payoffs $\pi - w$.

13.2 Observable Effort

Principal chooses $\tilde{w} : E \times \Pi \to \mathbb{R}$. Write $\mathcal{W}$ for the set of such wage schemes.

Agent chooses $\tilde{e} : \mathcal{W} \to E \cup \{\text{Reject}\}$.

Nature chooses $\pi \in \Pi$, according to $F(\cdot \mid e)$.

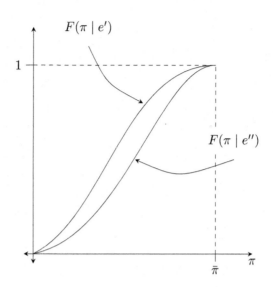

Figure 13.1.1: The distribution function $F(\cdot \mid e'')$ first order stochastically dominates $F(\cdot \mid e')$ if $e'' > e'$.

Principal's utility is $\pi - \tilde{w}(\tilde{e}(\tilde{w}), \pi)$.

Agent's utility is $v(\tilde{w}(\tilde{e}(\tilde{w}), \pi)) - c(\tilde{e}(\tilde{w}))$.

Principal can induce any e by choosing $\tilde{w}(e', \pi) = 0$ for $e' \neq e$ and all π, and choosing $\tilde{w}(e, \pi)$ appropriately. (This is a *forcing contract*.) Accordingly, if e is the effort level the principal wishes to induce, denote the associated wage profile by $\tilde{w}_e$.

The principal's problem is to choose $(e, \tilde{w}_e)$ to maximize

$$\int_{\underline{\pi}}^{\bar{\pi}} (\pi - \tilde{w}_e(\pi)) f(\pi \mid e) d\pi$$

subject to

$$\int_{\underline{\pi}}^{\bar{\pi}} v(\tilde{w}_e(\pi)) f(\pi \mid e) d\pi - c(e) \geq \bar{u}.$$

The inequality constraint is the *individual rationality* (IR) or *participation constraint*.

We can treat this as a two-step optimization: We first ask, for any e, what is least cost wage profile that induces that effort level e. Then, we optimize over the effort levels.

Suppose then the principal wished to induce an effort level e. The principal's objective, given e, is to choose the wage schedule $\hat{w}$ yielding an expected utility of at least $\bar{u}$ at least cost. That is, the principal chooses $\hat{w} : \Pi \to \mathbb{R}_+$ to minimize

$$\int_{\underline{\pi}}^{\bar{\pi}} \hat{w}(\pi) f(\pi \mid e) d\pi$$

subject to

$$\int_{\underline{\pi}}^{\bar{\pi}} v(\hat{w}(\pi)) f(\pi \mid e) d\pi - c(e) \geq \bar{u}.$$

The minimand is linear in $\hat{w}$, the constraint is strictly concave in $\hat{w}$, so the solution is characterized by the first order condition,

$$-f(\pi \mid e) + \gamma v'(\hat{w}(\pi)) f(\pi \mid e) = 0 \qquad \forall \pi \in \Pi,$$

where $\gamma \geq 0$ is the Lagrange multiplier on the IR constraint. Thus,

$$v'(\hat{w}(\pi)) = \frac{1}{\gamma}, \qquad \forall \pi \in \Pi,$$

i.e., $\hat{w}(\pi) = w^*$ for all $\pi \in \Pi$. The risk averse agent is *fully insured*. Since the agent's constraint must hold with equality at an optima, we have

$$w^* = v^{-1}(\bar{u} + c(e)).$$

Thus,

$$S(e) = \int_{\underline{\pi}}^{\bar{\pi}} \pi f(\pi \mid e) d\pi - v^{-1}(\bar{u} + c(e))$$

is the profit from inducing the effort level e.

The optimal effort e^* maximizes $S(e)$.

13.3 Unobservable Effort (Moral Hazard)

Suppose effort is not observable.

The principal chooses $\tilde{w} : \Pi \to \mathbb{R}$. Write $\mathcal{W}$ for the set of such wage schemes. The principal cannot condition the wage on the effort choice of the agent. The principal-agent relationship is said to be subject to *moral hazard*. The term originated in the insurance literature, and refers to situations where a participant may have an incentive to act against the interests of others (such as insurance lowering the incentives for the insured to exercise care). Another common term used to describe such a situation is *hidden action*.

The agent chooses $\tilde{e} : \mathcal{W} \to E \cup \{\text{Reject}\}$.

Nature chooses $\pi \in \Pi$ (according to $F(\cdot \mid e)$).
The principal's ex post utility is $\pi - \tilde{w}(\pi)$.
The agent's ex post utility is $v(\tilde{w}(\pi)) - c(\tilde{e}(\tilde{w}))$.
Consider first the problem of finding the cost minimizing wage schedule
to induce e. That is, choose $\hat{w} : \Pi \to \mathbb{R}_+$ to minimize

$$\int_{\underline{\pi}}^{\bar{\pi}} \hat{w}(\pi) f(\pi \mid e) d\pi \tag{13.3.1}$$

subject to *incentive compatibility* (IC),

$$e \in \arg\max_{e' \in E} \int_{\underline{\pi}}^{\bar{\pi}} v(\hat{w}(\pi)) f(\pi \mid e') d\pi - c(e'), \tag{13.3.2}$$

and *individual rationality* (IR),

$$\int_{\underline{\pi}}^{\bar{\pi}} v(\hat{w}(\pi)) f(\pi \mid e) d\pi - c(e) \geq \bar{u}. \tag{13.3.3}$$

Suppose $E = \{e_L, e_H\}$, with $e_L < e_H$, $c(e_L) = 0$, and $c(e_H) = c_H > 0$.
Suppose $e = e_L$. This is straightforward, since $\hat{w}(\pi) = v^{-1}(\bar{u}) =: w^\dagger$
minimizes the minimand subject to the IR constraint, and IC constraint is
trivially satisfied.
Suppose $e = e_H$. We must find $\hat{w}$ minimizing

$$\int_{\underline{\pi}}^{\bar{\pi}} \hat{w}(\pi) f(\pi \mid e_H) d\pi$$

subject to *incentive compatibility*,

$$\int_{\underline{\pi}}^{\bar{\pi}} v(\hat{w}(\pi)) f(\pi \mid e_H) d\pi - c_H \geq \int_{\underline{\pi}}^{\bar{\pi}} v(\hat{w}(\pi)) f(\pi \mid e_L) d\pi, \tag{IC}$$

and *individual rationality*,

$$\int_{\underline{\pi}}^{\bar{\pi}} v(\hat{w}(\pi)) f(\pi \mid e_H) d\pi - c_H \geq \bar{u}. \tag{IR}$$

Since both the left and right sides of (IC) are concave in $\hat{w}(\pi)$, the first-order
conditions are not obviously sufficient. But a change of variables from $\hat{w}(\pi)$
to $\hat{u}(\pi) := u(\hat{w}(\pi))$ ($\hat{u}$ is the promised agent utility from the wage function
$\hat{w}$) shows that this is OK (see Problem 13.6.1).
Writing μ for multiplier of the (IC) constraint (γ is, as before, the
multiplier on (IR)), the solution is characterized by the first order condition:

$$-f(\pi \mid e_H) + \gamma v'(\hat{w}(\pi))f(\pi \mid e_H) +$$
$$\mu v'(\hat{w}(\pi))[f(\pi \mid e_H) - f(\pi \mid e_L)] = 0.$$

That is,

$$\frac{1}{v'(\hat{w}(\pi))} = \gamma + \mu\left[1 - \frac{f(\pi \mid e_L)}{f(\pi \mid e_H)}\right].$$

If $\mu = 0$, (IC) fails (because the implied wage scheme is constant, i.e., independent of π). If $\gamma = 0$, the right side must fall to 0 which is impossible. Let $\bar{w}$ solve $v'(\bar{w}) = 1/\gamma$. Now

$$\hat{w}(\pi) \begin{pmatrix} < \\ = \\ > \end{pmatrix} \bar{w} \iff \frac{f(\pi \mid e_L)}{f(\pi \mid e_H)} \begin{pmatrix} > \\ = \\ < \end{pmatrix} 1$$

Since $[v'(w)]^{-1}$ is increasing in w, it is only under *monotone likelihood ratio property* (MLRP),

$$\frac{f(\pi \mid e_L)}{f(\pi \mid e_H)} \quad \text{decreasing in } \pi,$$

that we are guaranteed that $\hat{w}(\pi)$ as an increasing function of π. This is *strictly* stronger than first order stochastic dominance.

Example 13.3.1. $\Pi = [0, 1]$, $f(\pi \mid e_L) = 2(1 - \pi)$, $f(\pi \mid e_H) = 1$, $v(x) = \ln(x)$. Note that MLRP is satisfied.
 The first-order condition yields

$$\hat{w}(\pi) = \gamma + \mu\left[1 - \frac{f(\pi \mid e_L)}{f(\pi \mid e_H)}\right]$$
$$= \gamma + \mu[1 - 2(1 - \pi)]$$
$$= \gamma - \mu + 2\mu\pi,$$

and so wage is linear in output. ★

The cost of inducing high effort under *observable effort* is the wage $w^* = v^{-1}(\bar{u} + c(e_H))$. Jensen's inequality implies

$$\int_{\underline{\pi}}^{\bar{\pi}} v(\hat{w}(\pi))f(\pi \mid e_H)d\pi < v\left(\int_{\underline{\pi}}^{\bar{\pi}} \hat{w}(\pi)f(\pi \mid e_H)d\pi\right),$$

and so (IR) implies that the cost of inducing high effort satisfies

$$\int_{\underline{\pi}}^{\bar{\pi}} \hat{w}(\pi) f(\pi \mid e_H) > v^{-1}(\bar{u} + c_H) = w^*.$$

That is, the expected cost of inducing high effort when effort is unobservable is higher than when it is observable. The need to impose risk on the risk averse agent (in order to induce high effort) requires a higher average wage to maintain the agent's expected utility.

Finally, inducing high effort is worthwhile if

$$\int_{\underline{\pi}}^{\bar{\pi}} (\pi - \hat{w}(\pi)) f(\pi \mid e_H) d\pi \geq 0$$

and (recalling the determination of the constant wage $w^{\dagger}$ to induce e_L)

$$\int_{\underline{\pi}}^{\bar{\pi}} (\pi - \hat{w}(\pi)) f(\pi \mid e_H) d\pi \geq \int_{\underline{\pi}}^{\bar{\pi}} \pi f(\pi \mid e_L) d\pi - v^{-1}(\bar{u}).$$

Remark 13.3.1 (Continuum of effort levels). When e can take on any value in a continuum (such as $[\underline{e}, \bar{e}]$ or $\mathbb{R}_+$), (13.3.2) describes an an uncountable infinity of IC constraints. It is attractive to consider the relaxed problem of minimizing 13.3.1 subject to (13.3.3) and the first order (or local) condition

$$\int_{\underline{\pi}}^{\bar{\pi}} v(\hat{w}(\pi)) \frac{\partial f(\pi \mid e)}{\partial e} \, d\pi = c'(e).$$

This is called the *first order approach*, and in general, solutions to the relaxed problem do not solve the original problem. See Laffont and Martimort (2002, Section 5.1.3) for a discussion.   ◆

Example 13.3.2. Suppose $E = \{0, e_l, e_H\}$, and

$$c(e) = \begin{cases} 0, & e = 0, \\ c_L, & e = e_L, \\ c_H, & e = e_H, \end{cases}$$

with $c_H > 2c_L$. Output is binary, 0 and $y > 0$, with the probability distribution

$$\Pr(y|e) = \begin{cases} 1/4, & e = 0, \\ 1/2, & e = e_L, \\ 3/4, & e = e_H. \end{cases}$$

Finally, $\bar{u} = 0$.

As in Problem 13.6.1, I first solve for the optimal utility promises to induce the various efforts. Denote the contract by $(\underline{v}, \bar{v})$, with $\underline{v}$ the promised utility after no output and $\bar{v}$ the promised utility after y. Moreover, because there are only two outputs (and so two unknowns), two constraints suffice to determine the values of the unknowns. There is no need to analyze the first order conditions (the objective of the principal is not irrelevant, since otherwise the constraints need not bind).

Inducing $e = e_H$: The cost minimizing contract inducing $e = e_H$ satisfies the two IC constraints

$$\tfrac{3}{4}\bar{v} + \tfrac{1}{4}\underline{v} - c_H \geq \tfrac{1}{2}\bar{v} + \tfrac{1}{2}\underline{v} - c_L,$$

$$\tfrac{3}{4}\bar{v} + \tfrac{1}{4}\underline{v} - c_H \geq \tfrac{1}{4}\bar{v} + \tfrac{3}{4}\underline{v},$$

and IR,

$$\tfrac{3}{4}\bar{v} + \tfrac{1}{4}\underline{v} - c_H \geq 0.$$

Suppose the first IC constraint and IR are binding. Then

$$\tfrac{1}{4}(\bar{v} - \underline{v}) = c_H - c_L$$
$$\tfrac{3}{4}\bar{v} + \tfrac{1}{4}\underline{v} = c_H.$$

Solving gives

$$\underline{v} = 3c_L - 2c_H \quad \text{and} \quad \bar{v} = 2c_H - c_L.$$

Since $c_H > 2c_L$, $\underline{v} < 0$.

Under this contract, the expected utility from $e = 0$ is

$$\tfrac{3}{4}(3c_L - 2c_H) + \tfrac{1}{4}(2c_H - c_L) = 2c_L - c_H < 0,$$

and so the agent does not wish to choose $e = 0$.

Inducing $e = e_L$: The cost minimizing contract inducing $e = e_L$ satisfies the two IC constraints

$$\tfrac{1}{2}\bar{v} + \tfrac{1}{2}\underline{v} - c_L \geq \tfrac{3}{4}\bar{v} + \tfrac{1}{4}\underline{v} - c_H,$$

$$\tfrac{1}{2}\bar{v} + \tfrac{1}{2}\underline{v} - c_L \geq \tfrac{1}{4}\bar{v} + \tfrac{3}{4}\underline{v},$$

and IR,

$$\tfrac{1}{2}\bar{v} + \tfrac{1}{2}\underline{v} - c_L \geq 0.$$

Suppose the second IC constraint and IR are binding. Then

$$\tfrac{1}{4}(\bar{v} - \underline{v}) = c_L$$
$$\tfrac{1}{2}\bar{v} + \tfrac{1}{2}\underline{v} = c_L.$$

Solving gives

$$\underline{v} = -c_L \quad \text{and} \quad \bar{v} = 3c_L.$$

Under this contract, the expected utility from $e = e_H$ is

$$\tfrac{3}{4}(3c_L) + \tfrac{1}{4}(-c_L) - c_H = 2c_L - c_H < 0,$$

and again the agent does not wish to choose $e = 0$.

The optimal contract: In order to determine which effort the principal wishes to induce, we need to specify the agent's utility function.

The agent's utility function is given by

$$v(w) = \begin{cases} w, & w \geq 0, \\ 2w, & w < 0. \end{cases}$$

The utility function is illustrated in Figure 13.3.1 (the piecewise linear formulation allows us to explicitly solve for the wage payments, while maintaining concavity of the agent's utility function).

The expected wage payment under the e_H-inducing contract is

$$\tfrac{3}{4}(2c_H - c_L) + \tfrac{1}{4}\tfrac{1}{2}(3c_L - 2c_H) = \tfrac{1}{8}(10c_H - 3c_L).$$

The expected wage payment under the e_L-inducing contract is

$$\tfrac{1}{2}(3c_L) + \tfrac{1}{2}\tfrac{1}{2}(-c_L) = \tfrac{5}{4}c_L.$$

Zero effort is trivially induced by the constant contract $\underline{v} = \bar{v} = 0$. The principal wishes to induce high effort if

$$\tfrac{3}{4}y - \tfrac{1}{8}(10c_H - 3c_L) \geq \max\{\tfrac{1}{4}y, \tfrac{1}{2}y - \tfrac{5}{4}c_L\},$$

which is satisfied if

$$y > \frac{1}{4} \max\left\{ 10c_H - 3c_L, \ 20c_H - 26c_L \right\}. \tag{13.3.4}$$

13.4   Unobserved Cost of Effort (Adverse Selection)

Suppose effort is observable, but there is asymmetric information about the cost of effort: In particular, $c(e, \theta)$, where $\theta \in \Theta$ is known by the agent, but

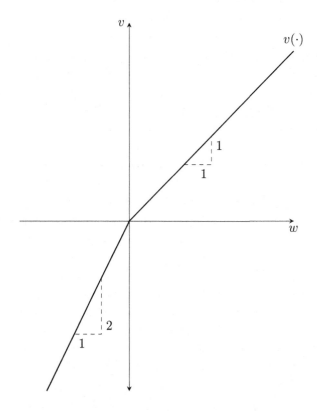

Figure 13.3.1: The agent's utility function for Example 13.3.2.

not the principal. Otherwise, the model is as in Section 13.2. Then, as in Section 10.1, we look for the optimal direct mechanism.

Denote by $\widetilde{\mathcal{W}}$ the set of forcing wage schemes $\{\tilde{w}_e : e \in E\}$. A direct mechanism is a pair

$$(e^{\dagger}, w^{\dagger}) : \Theta \to E \times \widetilde{\mathcal{W}}.$$

The principal chooses an incentive compatible direct mechanism to maximize profits. That is, the principal chooses $(e^{\dagger}, w^{\dagger})$ to maximize

$$E_\theta \int_{\underline{\pi}}^{\bar{\pi}} (\pi - \tilde{w}_{e^{\dagger}(\theta)}(\pi)) f(\pi \mid e^{\dagger}(\theta)) d\pi$$

subject to

$$\int_{\underline{\pi}}^{\bar{\pi}} v(\tilde{w}_{e^{\dagger}(\theta)}(\pi)) f(\pi \mid e^{\dagger}(\theta)) d\pi - c(e^{\dagger}(\theta), \theta)$$

$$\geq \int_{\underline{\pi}}^{\bar{\pi}} v(\tilde{w}_{e^{\dagger}(\hat{\theta})}(\pi)) f(\pi \mid e^{\dagger}(\hat{\theta})) d\pi - c(e^{\dagger}(\hat{\theta}), \theta) \qquad \forall \hat{\theta}, \theta$$

and

$$\int_{\underline{\pi}}^{\bar{\pi}} v(\tilde{w}_{e^{\dagger}(\theta)}(\pi)) f(\pi \mid e^{\dagger}(\theta)) d\pi - c(e^{\dagger}, \theta) \geq \bar{u}.$$

Since effort is observable, there is no value to imposing risk on the agent:

Lemma 13.4.1. *In the optimal mechanism, $\tilde{w}_{e^{\dagger}(\theta)}$ is a constant (i.e., it is independent of π.*

In light of this lemma, it is enough to look for direct mechanisms of the form

$$(e^{\dagger}, w^*) : \Theta \to E \times \mathbb{R}$$

maximizing

$$E_{\theta} \int_{\underline{\pi}}^{\bar{\pi}} \pi f(\pi \mid e^{\dagger}(\theta)) d\pi - w^*(\theta)$$

subject to

$$v(w^*\theta)) - c(e^{\dagger}(\theta), \theta) \geq v(w^*(\hat{\theta})) - c(e^{\dagger}(\hat{\theta}), \theta) \qquad \forall \hat{\theta}, \theta$$

and

$$v(w^*(\theta)) - c(e^{\dagger}(\theta), \theta) \geq \bar{u}.$$

Thus, this is conceptually identical to the model in Section 10.1, and the analysis of the two type case follows familiar lines (Problem 13.6.3).

13.5   A Hybrid Model

Suppose both effort and effort cost are unobservable. A direct mechanism is now the pair

$$(e^{\dagger}, w^{\dagger}) : \Theta \to E \times \mathbb{R}^{\Pi},$$

that is, for each report, the direct mechanism specifies an effort and a wage schedule (wage as a function of output). The principal chooses $(e^{\dagger}, w^{\dagger})$ to maximize

$$E_{\theta} \int_{\underline{\pi}}^{\bar{\pi}} (\pi - w^{\dagger}(\pi; \theta)) f(\pi \mid e^{\dagger}(\theta)) d\pi$$

subject to *incentive compatibility*

$$\int_{\underline{\pi}}^{\bar{\pi}} v(w^\dagger(\pi;\theta))f(\pi \mid e^\dagger(\theta))d\pi - c(e^\dagger(\theta),\theta)$$

$$\geq \int_{\underline{\pi}}^{\bar{\pi}} v(w^\dagger(\pi;\hat{\theta}))f(\pi \mid e)d\pi - c(e,\theta) \qquad \forall \hat{\theta}, \theta, e.$$

and *individual rationality*

$$\int_{\underline{\pi}}^{\bar{\pi}} v(w^\dagger(\pi;\theta))f(\pi \mid e^\dagger(\theta))d\pi - c(e^\dagger(\theta),\theta) \geq \bar{u}.$$

Incentive compatibility implies *truthful reporting*,

$$\int_{\underline{\pi}}^{\bar{\pi}} v(w^\dagger(\pi;\theta))f(\pi \mid e^\dagger(\theta))d\pi - c(e^\dagger(\theta),\theta)$$

$$\geq \int_{\underline{\pi}}^{\bar{\pi}} v(w^\dagger(\pi;\hat{\theta}))f(\pi \mid e^\dagger(\hat{\theta}))d\pi - c(e^\dagger(\theta),\theta) \qquad \forall \hat{\theta}, \theta,$$

and *obedience* (i.e., it is optimal for type θ to actually choose $e^\dagger(\theta)$)

$$\int_{\underline{\pi}}^{\bar{\pi}} v(w^\dagger(\pi;\theta))f(\pi \mid e^\dagger(\theta))d\pi - c(e^\dagger(\theta),\theta)$$

$$\geq \int_{\underline{\pi}}^{\bar{\pi}} v(w^\dagger(\pi;\theta))f(\pi \mid e)d\pi - c(e,\theta) \qquad \forall e.$$

Incentive compatibility also ensures that it is not profitable to both misreport and then not be obedient given the incorrect report.

Similar to the taxation principle, the principal's optimal mechanism can be implemented via a menu of wage contracts $\mathcal{M} := \{w^\dagger(\cdot;\hat{\theta}) : \hat{\theta} \in \Theta\}$. The principal offers the agent a menu $\mathcal{M}$, and the agent selects one of the wage contracts from $\mathcal{M}$. The menu is chosen so that the agent of type θ finds it optimal to choose $w^\dagger(\cdot;\theta)$, and then optimally chooses effort $e^\dagger(\theta)$.

Example 13.5.1 (Continuation of Example 13.3.2). We now suppose costs are unknown, so $\Theta = \{\theta', \theta''\}$, and costs are given by

$$c(e,\theta) = \begin{cases} 0, & e = 0, \\ c_L', & e = e_L, \theta = \theta', \\ c_H', & e = e_H, \theta = \theta', \\ c_L'', & e = e_L, \theta = \theta'', \\ c_H'', & e = e_H, \theta = \theta'', \end{cases}$$

with $c'_L < c''_L$ and $c'_H < c''_H$. Assume $c'_H > 2c'_L$ and $c''_H > 2c''_L$. Let α be the prior probability that the agent is low cost.

For α close to 1, the optimal mechanism is to offer the low-cost e_H inducing contract, and have the high cost agent reject the contract.

For α close to 0, the optimal mechanism is to offer the high-cost e_H inducing contract (a pooling contract), so that the low cost agent receives a benefit of $c''_H - c'_H > 0$.

For intermediate α, the optimal contract is a *screening* menu of contracts: the effort choice of the lost cost type θ' is undistorted (i.e., it equals the effort the principal would induce if θ' is common knowledge) but earns information rents, while the high cost type θ'' is distorted (to e_L), but has IR binding.

Parameter values:

$$\alpha = \tfrac{1}{2}, c'_L = 1, c'_H = 4, c''_L = 2, \text{ and } c''_H = 6.$$

For these parameter values, if

$$y > 17$$

each type would be induced to choose high effort if that type were common knowledge (i.e., (13.3.4) holds for both θ' and θ'').

The optimal screening contract: The high-cost contract $(\underline{v}'', \bar{v}'')$ is $(-c''_L, 3c''_L) = (-2, 6)$. Under this contract, the low-cost agent has expected utility from e_L of

$$c''_L - c'_L = 1 > 0.$$

(His utility from exerting high effort under this contract is $2c''_L - c'_H = 0.$[1])
Thus, the low-cost contract $(\underline{v}', \bar{v}')$ is

$$(3c'_L - 2c'_H + (c''_L - c'_L), \; 2c'_H - c'_L + (c''_L - c'_L)) = (-4, \; 8).$$

The high-cost agent's payoff under the low-cost agent's contract: If agent chooses e_H, the expected payoff is[2]

$$\tfrac{1}{4}(-4) + \tfrac{3}{4}(8) - c''_H = 5 - 6 < 0.$$

---

[1]More generally, high effort is suboptimal if $2c''_L - c'_H < c''_L - c'_L \iff c''_L + c'_L < c'_H$.
[2]This inequality uses single-crossing: If the high-cost agent chooses e_H, the expected payoff is

$$\tfrac{3}{4}(2(c'_H - c'_L) + c''_L) + \tfrac{1}{4}(-2(c'_H - c'_L) + c''_L) - c''_H$$
$$= c'_H - c'_L + c''_L - c''_H = (c''_L - c'_L) - (c''_H - c'_H) < 0.$$

If agent chooses e_L, the expected payoff is

$$\tfrac{1}{2}(-4) + \tfrac{1}{2}(8) - c_L'' = 2 - 2 = 0.$$

(This equality should not be a surprise.)

The expected wage payment of the low-cost contract (conditional on the agent having low cost) is

$$\tfrac{1}{4}\tfrac{1}{2}(-4) + \tfrac{3}{4}(8) = \tfrac{11}{2},$$

so the total cost of the screening menu is

$$\tfrac{1}{2}\tfrac{5}{4}c_L'' + \tfrac{1}{2}\tfrac{11}{2} = 4.$$

In order to verify optimality of the screening contract, we need to show that the screening contract is not dominated by either the high-cost pooling contract, or the low-cost only contract.

In the high-cost pooling contract, the principal offers the single contract $(-6,\ 10)$, at an expected cost of $\tfrac{27}{4}$, and so receives an expected profit of

$$\frac{3}{4}y - \frac{27}{4}.$$

In the low-cost only pooling contract, the principal offers the single contract $(-5,\ 7)$, which the the high-cost agent rejects. The expected cost of hiring the low-cost agent is $\tfrac{37}{8}$, and so the expected profit is

$$\frac{1}{2}\left\{\frac{3}{4}y - \frac{37}{8}\right\}.$$

Since $y > 17$, the high-cost pooling is more profitable than the low-cost only contract.

The screening contract loses revenue of $\tfrac{1}{8}y$, but saves costs of

$$\frac{27}{4} - 4 = \frac{11}{4},$$

and so the screening contract is optimal if

$$y < 22.$$

★

13.6   Problems

13.6.1. Define $\hat{u}(\pi) := u(\hat{w}(\pi))$, so that $\hat{u}$ is the promised agent utility from the wage function $\hat{w}$. Rewrite the problem of choosing $\hat{w}$ to minimize (13.3.1) subject to (13.3.2) and (13.3.3) to one of choosing $\hat{u}$ to minimize the expected cost of providing utility $\hat{u}$ subject to the appropriate modifications of (13.3.2) and (13.3.3). Verify that the result is a convex minimization problem and so the approach in the text is valid.

13.6.2. Prove Lemma 13.4.1.

13.6.3. Consider the model of Section 13.4.

(a) Prove that $V(e) := \int_{\underline{\pi}}^{\bar{\pi}} \pi f(\pi \mid e) d\pi$ is strictly increasing in e.

(b) Suppose $E = \mathbb{R}_+$ and V is convex and continuously differentiable, and $c(\cdot, \theta)$ is strictly increasing, concave, and continuously differentiable, and satisfies the *single-crossing condition*:

$$\frac{\partial^2 c}{\partial e \partial \theta} > 0.$$

Suppose the principal assigns probability $\alpha_H \in (0,1)$ to $\theta = \theta_H$ and complementary probability $\alpha_L := 1 - \alpha_H$ to $\theta_L < \theta_H$. Characterize the principal's optimal mechanism.

(c) Suppose $E = \{e_L, e_H\}$, with $e_L < e_H$. Suppose $\Theta = [\underline{\theta}, \bar{\theta}] \subset \mathbb{R}_+$, and the principal's beliefs are described by the probability distribution F with density f and F/f increasing in θ. Suppose $c(e_L, \theta) = 0$ for all θ and $c(e_H, \theta) = \theta$. Characterize the principal's optimal mechanism.

Chapter 14

Appendices

14.1 Proof of Theorem 2.4.1

It will simplify notation to assume (as we do in Chapter 5) that each action can only be taken at one information set (i.e., $A(h) \cap A(h') = \varnothing$ for all $h \neq h'$). With this assumption, behavior strategies can be defined as follows:

Definition 14.1.1. A behavior strategy *for player i is a function* $b_i : A_i \to [0,1]$ *such that, for all $h \in H_i$,*

$$\sum_{a \in A(h)} b_i(a) = 1.$$

We then have that the behavior strategy corresponding to a pure strategy s_i is given by

$$b_i(s_i(h)) = 1, \ \forall h \in H_i,$$

and

$$b_i(a) = 0, \ \forall a \notin \cup_{h \in H_i} \{s_i(h)\}.$$

Nature's move ρ is now denoted by b_0. Given $b = (b_0, b_1, \ldots, b_n)$, play begins with player $\iota(t_0)$ randomly selecting an action in $A(h)$ according to the distribution $\left(b_{\iota(h)}(a) \right)_{a \in A(h)}$. Define recursively $p^n(t) = p(p^{n-1}(t))$ for $t \neq t_0$, and let $\ell(t)$ solve $p^{\ell(t)}(t) = t_0$. The unique path to $z \in Z$ starts at $t_0 = p^{\ell(z)}(z)$ and is given by $\left(p^{\ell(z)}(z), p^{\ell(z)-1}(z), \ldots, p(z), z \right)$. The action leading to $p^{j-1}(z)$ is $\alpha \left(p^{j-1}(z) \right)$, which is taken by $\iota(p^j(z))$ with probability

$$b_{\iota(p^j(z))} \left(\alpha \left(p^{j-1}(z) \right) \right).$$

329

Thus, the probability that z occurs is

$$\mathbf{P}^b(z) = \prod_{j=1}^{\ell(z)} b_{\iota(p^j(z))}\left(\alpha\left(p^{j-1}(z)\right)\right).$$

If γ_i^h is the number of actions available to i at h, a mixed strategy for i is a point in the ξ_i ($= \prod_{h \in H_i} \gamma_i^h - 1$)-dimensional simplex. A behavior strategy, on the other hand involves specifying $\eta_i = \sum_{h \in H_i}(\gamma_i^h - 1)$ real numbers, one point on a $\left(\gamma_i^h - 1\right)$-dimensional simplex for each $h \in H_i$. In general, ξ_i is significantly larger than η_i.

Definition 14.1.2. *The mixed strategy σ_i corresponding to a behavior strategy b_i is given by*

$$\sigma_i(s_i) = \prod_{h \in H_i} b_i(s_i(h)).$$

Definition 14.1.3. *Two strategies for a player are* realization equivalent *if, for all specifications of strategies for the opponents, the two strategies induce the same distribution over terminal nodes.*

It is easy to check that a mixed strategy and the behavior strategy corresponding to it are realization equivalent.

Definition 14.1.4. *A node $t \in T$ is* reachable *under a mixed strategy $\sigma_i \in \Delta(S_i)$ for i if there exists σ'_{-i} such that the probability of reaching t under (σ_i, σ'_{-i}) is positive. An information set $h \in H_i$ is reachable under σ_i if some node $t \in h$ is reachable under σ_i.*

Remark 14.1.1. Perfect recall implies that if $h \in H_i$ is reachable for i under σ_i, then all $t \in h$ are reachable under σ_i. ◆

Define $R(\sigma_i) = \{h \in H_i : h$ is reachable for i under $\sigma_i\}$.

Definition 14.1.5. *The behavior strategy, b_i, generated by σ_i is*

$$b_i(a) = \begin{cases} \displaystyle\sum_{\substack{s_i \in S_i \text{ s.t.} \\ h \in R(s_i) \text{ and } s_i(h)=a}} \sigma_i(s_i) \Bigg/ \displaystyle\sum_{\substack{s_i \in S_i \text{ s.t.} \\ h \in R(s_i)}} \sigma_i(s_i) , & \text{if } h \in R(\sigma_i), \\[3em] \displaystyle\sum_{\substack{s_i \in S_i \text{ s.t.} \\ s_i(h)=a}} \sigma_i(s_i), & \text{if } h \notin R(\sigma_i), \end{cases}$$

for all $h \in H_i$ and for all $a \in A(h)$.

Note that $h \in R(\sigma_i)$ implies $\sum_{s_i \in S_i \text{ s.t. } h \in R(s_i)} \sigma_i(s_i) \neq 0$. If $h \in R(\sigma_i)$, $b_i(a)$ is the probability that a will be chosen, conditional on h being reached. The specification of $b_i(a)$ for $h \notin R(\sigma_i)$ is arbitrary.

Theorem 14.1.1 (Kuhn, 1953). *Fix a finite extensive game with perfect recall. For all mixed strategies, $\sigma_i \in \Delta(S_i)$, the behavior strategy b_i generated by σ_i is realization equivalent to σ_i.*

Proof. Fix $z \in Z$. Let $t_1, \ldots, t_\ell$ be i's decision nodes (in order) in the path to z, and a_k is the action at t_k required to stay on the path. Independence across players means that it is enough to show

$$\beta_{\sigma_i} := \Pr\{a_k \text{ is chosen at } t_k \forall k | \sigma_i\}$$
$$= \Pr\{a_k \text{ is chosen at } t_k \forall k | b_i\} =: \beta_{b_i}.$$

Let h^k be the information set containing t_k.

Suppose there exists k such that $h^k \notin R(\sigma_i)$. Then there exists $k' < k$ such that $\Pr(a_{k'} | \sigma_i) = 0$ and so $\beta_{\sigma_i} = 0$. Letting k' be the smallest first k such that $\Pr(a_{k'} | \sigma_i) = 0$, we have $b_i(a_{k'}) = 0$, and so $\beta_{b_i} = 0$.

Suppose $h^k \in R(\sigma_i)$ for all k. Then

$$\beta_{\sigma_i} = \sum_{\substack{s_i \in S_i \text{ s.t.} \\ s_i(h^k) = a_k, \ k \leq \ell}} \sigma_i(s_i).$$

Perfect recall implies that every information set owned by i is made reachable by a unique sequence of i's actions. Since $s_i(h^{k'}) = a_{k'} \ \forall k' < k$ implies $h^k \in R(s_i)$,

$$h^{k+1} \in R(s_i) \iff h^k \in R(s_i), \ s_i(h^k) = a_k.$$

Thus,

$$\left\{ s_i \in S_i : h^k \in R(s_i), \ s_i(h^k) = a_k \right\} = \left\{ s_i \in S_i : h^{k+1} \in R(s_i) \right\}$$

and

$$\beta_{b_i} = \prod_{k=1}^{\ell} b_i(a_k)$$

$$= \prod_{k=1}^{\ell} \sum_{\substack{s_i \in S_i \text{ s.t.} \\ h^k \in R(s_i) \text{ and } s_i(h^k) = a_k}} \sigma_i(s_i) \Bigg/ \sum_{\substack{s_i \in S_i \text{ s.t.} \\ h^{k+1} \in R(s_i)}} \sigma_i(s_i)$$

$$= \sum_{\substack{s_i \in S_i: \\ h^\ell \in R(s_i), s_i(h^\ell) = a_\ell}} \sigma_i(s_i) \Bigg/ \sum_{\substack{s_i \in S_i: \\ h^1 \in R(s_i)}} \sigma_i(s_i) \, .$$

But the denominator equals 1, so

$$\beta_{b_i} = \sum_{\substack{s_i \in S_i: \\ h^\ell \in R(s_i), s_i(h^\ell) = a_\ell}} \sigma_i(s_i)$$

$$= \sum_{\substack{s_i \in S_i: \\ s_i(h^k) = a_k, 1 \le k \le \ell}} \sigma_i(s_i) = \beta_{\sigma_i}.$$

∎

14.2 Trembling Hand Perfection

14.2.1 Existence and Characterization

This subsection proves the equivalence between Definition 2.5.1 and Selten's (1975) original definition.

Let $\eta : \cup S_i \to (0,1)$ be a function satisfying $\sum_{s_i \in S_i} \eta(s_i) < 1$ for all i. The *associated perturbed game*, denoted (G, η), is the normal form game $\{(R_1^\eta, v_1), \dots, (R_n^\eta, v_n)\}$ where

$$R_i^\eta = \{\sigma_i \in \Delta(S_i) : \sigma_i(s_i) \ge \eta(s_i), \quad \forall s_i \in S_i\}$$

and v_i is expected payoffs. Note that σ is a Nash equilibrium of (G, η) if and only if for all i, s_i, $s_i' \in S_i$,

$$v_i(s_i, \sigma_{-i}) < v_i(s_i', \sigma_{-i}) \implies \sigma_i(s_i) = \eta(s_i).$$

Definition 14.2.1 (Selten (1975)). *An equilibrium σ of a normal form game G is* (normal form) trembling hand perfect *if there exists a sequence $\{\eta_k\}_k$ such that $\eta_k(s_i) \to 0 \; \forall s_i$ as $k \to \infty$ and an associated sequence of mixed strategy profiles $\{\sigma^k\}_k$ with σ^k a Nash equilibrium of (G, η_k) such that $\sigma^k \to \sigma$ as $k \to \infty$.*

Theorem 14.2.1. *Every finite normal form game has a trembling hand perfect equilibrium.*

Proof. Clearly (G, η) has a Nash equilibrium for all η. Suppose $\{\eta_m\}$ is a sequence such that $\eta_m(s_i) \to 0 \; \forall s_i$ as $m \to \infty$. Let σ^m be an equilibrium of (G, η_m). Since $\{\sigma^m\}$ is a sequence in the compact set $\prod_i \Delta(S_i)$, it has a convergent subsequence. Its limit is a trembling hand perfect equilibrium of G. ∎

Remark 14.2.1. If σ is not trembling hand perfect, then there exists $\varepsilon > 0$ such that for all sequences $\{\eta_k\}_k$ satisfying $\eta_k \to 0$, there is a subsequence on which all Nash equilibria of the associated perturbed games are bounded away from σ by at least ε (i.e., for all K there exists $k \geq K$ such that for all σ^k equilibria of (G, η_k), we have $\left| \sigma^k - \sigma \right| \geq \varepsilon$).   ◆

Definition 14.2.2 (Myerson, 1978). *The mixed strategy profile σ is an ε-perfect equilibrium of G if it is completely mixed ($\sigma_i(s_i) > 0 \ \forall s_i \in S_i$) and satisfies*

$$s_i \notin BR_i(\sigma_{-i}) \implies \sigma_i(s_i) \leq \varepsilon.$$

Theorem 14.2.2. *Suppose σ is a strategy profile of the normal form game G. The following are equivalent:*

1. *σ is a trembling hand perfect equilibrium of G;*

2. *there exists a sequence $\{\varepsilon_k : \varepsilon_k \to 0\}$ and an associated sequence of ε_k-perfect equilibria converging to σ; and*

3. *there exists a sequence $\{\sigma^k\}$ of completely mixed strategy profiles converging to σ such that σ_i is a best reply to σ_{-i}^k, for all k.*

Proof. (1) $\Rightarrow$ (2). Take $\varepsilon_k = \max_{s_i \in S_i, i} \eta_k(s_i)$.

(2) $\Rightarrow$ (3). Let $\{\sigma^k\}$ be the sequence of ε_k-perfect equilibria. Suppose s_i receives positive probability under σ_i. We need to show that s_i is a best reply to σ_{-i}^k. Since $\varepsilon_k \to 0$ and $\sigma_i^k(s_i) \to \sigma_i(s_i) > 0$, there exists $k^*(s_i)$ such that $k > k^*(s_i)$ implies $\sigma_i^k(s_i) > \sigma_i(s_i)/2 > \varepsilon_k$. But σ^k is ε_k-perfect, so $s_i \in BR_i\left(\sigma_{-i}^k\right)$. The desired sequence is $\{\sigma^k : k > \bar{k}\}$, where $\bar{k} := \max\{k^*(s_i) : \sigma_i(s_i) > 0, \ i\}$.

(3) $\Rightarrow$ (1). Define η_k as

$$\eta_k(s_i) = \begin{cases} \sigma_i^k(s_i), & \text{if } \sigma_i(s_i) = 0, \\ 1/k, & \text{if } \sigma_i(s_i) > 0. \end{cases}$$

Since $\sigma^k \to \sigma$, there exists k' such that $k > k'$ implies $\sum_{s_i \in S_i} \eta_k(s_i) < 1$ for all i.

Let $m = \min_{i,s_i} \{\sigma_i(s_i) : \sigma_i(s_i) > 0\}$. There exists k'' such that, for all $k > k''$, $\left| \sigma_i^k(s_i) - \sigma_i(s_i) \right| < m/2$. Suppose $k > \max\{k', k'', 2/m\}$. Then $\sigma_i^k(s_i) \geq \eta_k(s_i)$. [If $\sigma_i(s_i) = 0$, then it follows immediately. If $\sigma_i(s_i) > 0$, then $\sigma_i^k(s_i) > \sigma_i(s_i) - m/2 > m/2 > 1/k$.]

Since σ_i is a best reply to σ_{-i}^k, if s_i is not a best reply to σ_{-i}^k, then $\sigma_i(s_i) = 0$. But this implies that σ^k is an equilibrium of (G, η_k) (since s_i is played with minimum probability).   ∎

14.2.2   Extensive Form Trembling Hand Perfection

Definition 14.2.3. *An equilibrium* $b := (b_1, \ldots, b_n)$ *of a finite extensive from game* Γ *is* extensive form trembling hand perfect *if there exists a sequence* $\{b^k\}_k$ *of completely mixed behavior strategy profiles converging to* b *such that for all players* i *and information sets* $h \in H_i$, *conditional on reaching* h, *for all* k, $b_i(h)$ *maximizes player* i's *expected payoff, given* b^k_{-i} *and* $b^k_i(h')$ *for* $h' \neq h$.

Theorem 14.2.3. *If* b *is an* extensive form trembling hand perfect *equilibrium of the finite extensive from game* Γ, *then it is subgame perfect.*

Proof. This is a corollary of Theorem 14.2.4.   ∎

Remark 14.2.2. When each player only has one information set (such as in Selten's horse, Example 5.1.1), trembling hand perfect in the normal and extensive form coincide. In general, they differ. Moreover, trembling hand perfect in the normal form need not imply subgame perfection (see problems 2.6.25 and 14.4.1).   ♦

Remark 14.2.3. Note that trembling hand perfect in the extensive form requires players to be sensitive not only to the possibility of trembles by the other players, but also their own trembles (at other information sets). The notion of *quasi-perfect* drops the latter requirement:

Definition 14.2.4 (van Damme, 1984). *A behavior strategy profile* b *is* quasi-perfect *if it is the limit of a sequence of completely mixed behavior profiles* b^n, *and if for each player* i *and information set* h *owned by that player, conditional on reaching* h, b_i *is a best response to* b^n_{-i} *for all* n.   ♦

Theorem 14.2.4. *Suppose* b *is a extensive form trembling hand perfect equilibrium of a finite extensive from game* Γ. *Then,* b *is sequentially rational given some consistent system of beliefs* μ, *and so is sequential.*

Proof. Suppose b is trembling hand perfect in the extensive form. Let $\{b^k\}_k$ be the trembles, i.e., the sequence of completely mixed behavior strategy profiles converging to b. Let μ^k be the system of beliefs implied by Bayes' rule by b^k (since b^k is completely mixed, μ^k is well-defined). Since the collection of systems of beliefs, $\prod_{h \in \cup_i H_i} \Delta(h)$, is compact, the sequence $\{\mu^k\}$ has a convergent subsequence with limit μ, and so μ is consistent. Moreover, a few minutes of reflection reveals that for each i, (b_i, b^k_{-i}) is sequentially rational at every information set owned by i, i.e., for all $h \in H_i$, given μ^k. Since best replies are hemicontinuous, for each i, b is sequentially rational at every information set owned by i, i.e., for all $h \in H_i$, given μ. That is, b is sequentially rational given μ.   ∎

14.3   Completion of Proof of Theorem 12.3.1

In this section, we derive the contradiction to

$$k_b + k_s = -\mathcal{S}$$

when the equalities $\underline{v}_b = \underline{v}_s$ and $\bar{v}_b = \bar{v}_s$ need not hold.
Since individual rationality requires $U_b^*(\underline{v}_b) \geq 0$ and $U_s^*(\bar{v}_s) \geq 0$,

$$U_b^*(\underline{v}_b) = \int_{\{v_s < \underline{v}_b\}} [\underline{v}_b - v_s]\, G(dv_s) + k_b,$$

and

$$U_s^*(\bar{v}_s) = \int_{\{v_b > \bar{v}_s\}} [v_b - \bar{v}_s]\, F(dv_b) + k_s,$$

we have

$$- k_b - k_s$$

$$\leq \int_{\underline{v}_s}^{\underline{v}_b} [\underline{v}_b - v_s]\, G(dv_s) + \int_{\bar{v}_s}^{\bar{v}_b} [v_b - \bar{v}_s]\, F(dv_b)$$

$$= \int_{\underline{v}_b}^{\bar{v}_b} \int_{\underline{v}_s}^{\underline{v}_b} [v_b - v_s]\, G(dv_s)F(dv_b) - \int_{\underline{v}_b}^{\bar{v}_b} \int_{\underline{v}_s}^{\underline{v}_b} [v_b - \underline{v}_b]\, G(dv_s)F(dv_b)$$

$$+ \int_{\underline{v}_s}^{\bar{v}_s} \int_{\bar{v}_s}^{\bar{v}_b} [v_b - v_s]\, F(dv_b)G(dv_s) - \int_{\underline{v}_s}^{\bar{v}_s} \int_{\bar{v}_s}^{\bar{v}_b} [\bar{v}_s - v_s]\, F(dv_b)G(dv_s)$$

$$< \mathcal{S} + \int_{\underline{v}_s}^{\underline{v}_b} \int_{\bar{v}_s}^{\bar{v}_b} [v_b - v_s]\, F(dv_b)G(dv_s)$$

$$- \int_{\underline{v}_b}^{\bar{v}_b} \int_{\underline{v}_s}^{\underline{v}_b} [v_b - \underline{v}_b]\, G(dv_s)F(dv_b) - \int_{\underline{v}_s}^{\bar{v}_s} \int_{\bar{v}_s}^{\bar{v}_b} [\bar{v}_s - v_s]\, F(dv_b)G(dv_s)$$

$$= \mathcal{S} + \int_{\underline{v}_s}^{\underline{v}_b} \int_{\bar{v}_s}^{\bar{v}_b} [v_b - v_s - (v_b - \underline{v}_b) - (\bar{v}_s - v_s)]\, F(dv_b)G(dv_s)$$

$$- \int_{\underline{v}_b}^{\bar{v}_s} \int_{\underline{v}_s}^{\underline{v}_b} [v_b - \underline{v}_b]\, G(dv_s)F(dv_b) - \int_{\underline{v}_b}^{\bar{v}_s} \int_{\bar{v}_s}^{\bar{v}_b} [\bar{v}_s - v_s]\, F(dv_b)G(dv_s)$$

$$= \mathcal{S} + \int_{\underline{v}_s}^{\underline{v}_b} \int_{\bar{v}_s}^{\bar{v}_b} [\underline{v}_b - \bar{v}_s]\, F(dv_b)G(dv_s)$$

$$- \int_{\underline{v}_b}^{\bar{v}_s} \int_{\underline{v}_s}^{\underline{v}_b} [v_b - \underline{v}_b]\, G(dv_s)F(dv_b) - \int_{\underline{v}_b}^{\bar{v}_s} \int_{\bar{v}_s}^{\bar{v}_b} [\bar{v}_s - v_s]\, F(dv_b)G(dv_s)$$

$$< \mathcal{S},$$

since $\underline{v}_b < \bar{v}_s$. That is,

$$k_b + k_s > -\mathcal{S},$$

our desired contradiction.

14.4   Problems

14.4.1. By explicitly presenting the completely mixed trembles, show that the profile $L\ell$ is *normal form* trembling hand perfect in the following game (this is the normal form from Example 2.3.4):

	ℓ	r
L	$2,0$	$2,0$
T	$-1,1$	$4,0$
B	$0,0$	$5,1$

Show that there is no *extensive form* trembling hand perfect equilibrium with that outcome in the first extensive form presented in Example 2.3.4.

References

ABREU, D., D. PEARCE, AND E. STACCHETTI (1990): "Toward a Theory of Discounted Repeated Games with Imperfect Monitoring," *Econometrica*, 58(5), 1041–1063. 194

ALIPRANTIS, C. D., AND K. C. BORDER (1999): *Infinite Dimensional Analysis: A Hitchhiker's Guide.* Springer-Verlag, Berlin, second edn. 102

AUMANN, R. J. (1987): "Correlated Equilibrium as an Expression of Bayesian Rationality," *Econometrica*, 55(1), 1–18. 83

AUMANN, R. J., Y. KATZNELSON, R. RADNER, R. W. ROSENTHAL, AND B. WEISS (1983): "Approximate Purification of Mixed Strategies," *Mathematics of Operations Research*, 8, 327–341. 72

BARON, D. P., AND R. B. MYERSON (1982): "Regulating a Monopolist with Unknown Costs," *Econometrica*, 50, 911–30. 313

BEN-EL-MECHAIEKH, H., AND R. W. DIMAND (2011): "A Simpler Proof of the Von Neumann Minimax Theorem," *American Mathematical Monthly*, 118(7), 636–641. 114

BENOIT, J.-P. (2000): "The Gibbard–Satterthwaite Theorem: A Simple Proof," *Economics Letters*, 69(3), 319–322. 284

BENOIT, J.-P., AND V. KRISHNA (1993): "Renegotiation in Finitely Repeated Games," *Econometrica*, 61(2), 303–323. 182

BERGEMANN, D., AND S. MORRIS (2005): "Robust Mechanism Design," *Econometrica*, 73(6), 1771–1813. 89

BERNHEIM, B. D. (1984): "Rationalizable Strategic Behavior," *Econometrica*, 52(4), 1007–1028. 39, 40

BHASKAR, V., G. J. MAILATH, AND S. MORRIS (2013): "A Foundation for Markov Equilibria in Sequential Games with Finite Social Memory," *Review of Economic Studies*, 80(3), 925–948. 221

BILLINGSLEY, P. (1995): *Probability and Measure*. John Wiley and Sons, New York, 3rd edn. 269

BORDER, K. C. (1985): *Fixed Point Theorems with Applications to Economics and Game Theory*. Cambridge University Press, Cambridge. 102

BÖRGERS, T. (2015): *An Introduction to Mechanism Design*. Oxford Univeristy Press. 263, 273

BRANDENBURGER, A., A. FRIEDENBERG, AND H. J. KEISLER (2008): "Admissibility in Games," *Econometrica*, 76(2), 307–352. 40

BUSCH, L.-A., AND Q. WEN (1995): "Perfect Equilibria in a Negotiation Model," *Econometrica*, 63(3), 545–565. 260

CARLSSON, H., AND E. VAN DAMME (1993): "Global Games and Equilibrium Selection," *Econometrica*, 61(5), 989–1018. 90

CARMICHAEL, H. L., AND W. B. MACLEOD (1997): "Gift Giving and the Evolution of Cooperation," *International Economic Review*, 38(3), 485–509. 191

CHO, I.-K., AND D. KREPS (1987): "Signaling Games and Stable Equilibria," *Quarterly Journal of Economics*, 102(2), 179–221. 147, 148, 155

CHO, I.-K., AND J. SOBEL (1990): "Strategic Stability and Uniqueness in Signaling Games," *Journal of Economic Theory*, 50(2), 381–413. 155

COASE, R. H. (1972): "Durability and Monopoly," *Journal of Law and Economics*, 15, 143–149. 229

CRAMTON, P., R. GIBBONS, AND P. KLEMPERER (1987): "Dissolving a Partnership Efficiently," *Econometrica*, 55(3), 615–632. 311

DEB, R., AND M. M. PAI (2017): "Discrimination via Symmetric Auctions," *American Economic Journal: Microeconomics*, 9(1), 275–314. 308

ELMES, S., AND P. J. RENY (1994): "On the Strategic Equivalence of Extensive Form Games," *Journal of Economic Theory*, 62(1), 1–23. 33

FARRELL, J., AND E. MASKIN (1989): "Renegotiation in Repeated Games," *Games and economic behavior*, 1(4), 327–360. 181

FERNANDEZ, R., AND J. GLAZER (1991): "Striking for a Bargain between Two Completely Informed Agents," *American Economic Review*, 81(1), 240–52. 260

FUDENBERG, D., AND D. K. LEVINE (1998): *The Theory of Learning in Games*. MIT Press, Cambridge, MA. 104, 108, 113

FUDENBERG, D., AND E. MASKIN (1986): "The Folk Theorem in Repeated Games with Discounting or Incomplete Information," *Econometrica*, 54(3), 533–554. 196

FUDENBERG, D., AND J. TIROLE (1991): "Perfect Bayesian Equilibrium and Sequential Equilibrium," *Journal of Economic Theory*, 53(2), 236–260. 132

GEANAKOPLOS, J. (2005): "Three Brief Proofs of Arrow's Impossibility Theorem," *Economic Theory*, 26(1), 211–215. 280

GERMANO, F., J. WEINSTEIN, AND P. ZUAZO-GARIN (2017): "Uncertain Rationality, Depth of Reasoning and Robustness in Games with Incomplete Information," unpublished, Universitat Pompeu Fabra, Washington University in St Loius, and University of the Basque Country. 92

GERSHKOV, A., J. K. GOEREE, A. KUSHNIR, B. MOLDOVANU, AND X. SHI (2013): "On the Equivalence of Bayesian and Dominant Strategy Implementation," *Econometrica*, 81(1), 197–220. 295

GIBBONS, R. (1992): *Game Theory for Applied Economists*. Princeton University Press, Princeton, NJ. 200

GILBOA, I., L. SAMUELSON, AND D. SCHMEIDLER (2014): "No-Betting-Pareto Dominance," *Econometrica*, 82(4), 1405–1442. 83

GOVINDAN, S., P. J. RENY, AND A. J. ROBSON (2003): "A Short Proof of Harsanyi's Purification Theorem," *Games and Economic Behavior*, 45(2), 369–374. 66

GREEN, J. R., AND J.-J. LAFFONT (1979): *Incentives in Public Decision Making*. North-Holland. 289

GUL, F. (1998): "A Comment on Aumann's Bayesian View," *Econometrica*, 66(4), 923–927. 83

GUL, F., H. SONNENSCHEIN, AND R. WILSON (1986): "Foundations of dynamic monopoly and the Coase conjecture," *Journal of Economic Theory*, 39(1), 155–190. 230

HARRIS, C. J., P. J. RENY, AND A. J. ROBSON (1995): "The Existence of Subgame-Perfect Equilibrium in Continuous Games with Almost
Perfect Information: A Case for Public Randomization," *Econometrica*,
63(3), 507–544. 115

HARSANYI, J. C. (1967): "Games with Incomplete Information Played
by Bayesian Players. Part I," *Management Science*, 14(3), 159–182. 81

———— (1968a): "Games with Incomplete Information Played by
Bayesian Players. Part II," *Management Science*, 14(5), 320–334. 81

———— (1968b): "Games with Incomplete Information Played by
Bayesian Players. Part III," *Management Science*, 14(7), 486–502. 81

———— (1973): "Games with Randomly Disturbed Payoffs: A New
Rationale for Mixed-Strategy Equilibrium Points," *International Journal
of Game Theory*, 2(1), 1–23. 64, 65, 66

HEIFETZ, A., AND W. KETS (2018): "Robust Multiplicity with a Grain
of Naiveté," *Theoretical Economics*, 13(1), 415–465. 92

HILLAS, J., AND E. KOHLBERG (2002): "Foundations of Strategic Equilibrium," in *Handbook of Game Theory with Economic Applications, Volume 3*, ed. by R. J. Aumann, and S. Hart, chap. 42. North Holland. 48

HU, J. (2014): "Reputation in the Presence of Noisy Exogenous Learning," *Journal of Economic Theory*, 153, 64–73. 242

JACKSON, M. O., L. K. SIMON, J. M. SWINKELS, AND W. R. ZAME
(2002): "Communication and Equilibrium in Discontinuous Games of
Incomplete Information," *Econometrica*, 70(5), 1711–1740. 102

JEHLE, G. A., AND P. J. RENY (2011): *Advanced Microeconomic Theory*. Pearson Education, third edn. 310

KANDORI, M., G. J. MAILATH, AND R. ROB (1993): "Learning, Mu-
tation, and Long Run Equilibria in Games," *Econometrica*, 61, 29–56.
52

KOHLBERG, E. (1990): "Refinement of Nash Equilibrium: The Main
Ideas," in *Game Theory and Applications*, ed. by T. Ichiishi, A. Neyman,
and Y. Tauman, pp. 3–45. Academic Press, San Diego. 48

KOHLBERG, E., AND J.-F. MERTENS (1986): "On the Strategic Stability
of Equilibria," *Econometrica*, 54(5), 1003–1037. 33, 47, 48, 148, 155

KREPS, D., AND R. WILSON (1982): "Sequential Equilibria," *Econometrica*, 50(4), 863–894. 33, 133

KREPS, D. M. (1988): *Notes on the Theory of Choice.* Westview Press, Boulder. 13

——— (1990): *A Course in Microeconomic Theory.* Princeton University Press, Princeton, NJ. 13

KRISHNA, V. (2002): *Auction Theory.* Academic press. 298, 307

KRISHNA, V., AND M. PERRY (1998): "Efficient Mechanism Design," Available at SSRN: http://ssrn.com/abstract=64934 or http://dx.doi.org/10.2139/ssrn.64934. 295

KUHN, H. W. (1953): "Extensive Games and the Problem of Information," in *Contributions to the Theory of Games,* ed. by H. W. Kuhn, and A. W. Tucker, vol. II of *Annals of Mathematical Studies 28.* Princeton University Press. 46, 331

LAFFONT, J.-J., AND D. MARTIMORT (2002): *The Theory of Incentives: The Principal-Agent Model.* Princeton University Press, Princeton. 320

LEONARD, R. J. (2010): *Von Neumann, Morgenstern, and the Creation of Game Theory: From Chess to Social Science, 1900–1960.* Cambridge University Press. viii

LIVSHITS, I. (2002): "On Non-Existence of Pure Strategy Markov Perfect Equilibrium," *Economics Letters,* 76(3), 393–396. 239

LUTTMER, E. G., AND T. MARIOTTI (2003): "The Existence of Subgame-Perfect equilibrium in Continuous Games with Almost Perfect Information: A Comment," *Econometrica,* 71(6), 1909–1911. 115

MACLEOD, W. B., AND J. M. MALCOMSON (1989): "Implicit Contracts, Incentive Compatibility, and Involuntary Unemployment," *Econometrica,* 57(2), 447–480. 191

MADRIGAL, V., T. C. C. TAN, AND S. R. D. C. WERLANG (1987): "Support Restrictions and Sequential Equilibria," *Journal of Economic Theory,* 43(2), 329–334. 136, 140

MAILATH, G. J. (1987): "Incentive Compatibility in Signaling Games with a Continuum of Types," *Econometrica,* 55(6), 1349–1365. 155

———— (1988): "On the Behavior of Separating Equilibria of Signaling Games with a Finite Set of Types as the Set of Types Becomes Dense in an Interval," *Journal of Economic Theory*, 44, 413–24. 158

———— (1998): "Do People Play Nash Equilibrium? Lessons From Evolutionary Game Theory," *Journal of Economic Literature*, 36(3), 1347–1374. 104, 109

MAILATH, G. J., AND Λ. POSTLEWAITE (1990): "Asymmetric Information Bargaining Problems with Many Agents," *Review of Economic Studies*, 57, 351–367. 312

MAILATH, G. J., AND L. SAMUELSON (2006): *Repeated Games and Reputations: Long-Run Relationships*. Oxford University Press, New York, NY. 168, 193, 195, 196, 200, 203, 207, 235

———— (2014): "Reputations in Repeated Games," in *Handbook of Game Theory, volume 4*, ed. by H. P. Young, and S. Zamir. North Holland. 235

MAILATH, G. J., L. SAMUELSON, AND J. M. SWINKELS (1993): "Extensive Form Reasoning in Normal Form Games," *Econometrica*, 61(2), 273–302. 33

———— (1997): "How Proper Is Sequential Equilibrium?," *Games and Economic Behavior*, 18, 193–218, Erratum, 19 (1997), 249. 33

MAILATH, G. J., AND E.-L. VON THADDEN (2013): "Incentive Compatibility and Differentiability: New Results and Classic Applications," *Journal of Economic Theory*, 148(5), 1841–1861. 155

MARX, L. M., AND J. M. SWINKELS (1997): "Order Independence for Iterated Weak Dominance," *Games and Economic Behavior*, 18(2), 219–245. 9, 28

MAS-COLELL, A., M. D. WHINSTON, AND J. GREEN (1995): *Microeconomic Theory*. Oxford University Press, New York, NY. 13, 100, 268, 289

MASKIN, E., AND J. TIROLE (2001): "Markov Perfect Equilibrium I. Observable Actions," *Journal of Economic Theory*, 100(2), 191–219. 221

MILGROM, P. R., AND R. J. WEBER (1985): "Distributional Strategies for Games with Incomplete Information," *Mathematics and Operations Research*, 10(4), 619–632. 72

MONDERER, D., AND D. SAMET (1989): "Approximating Common Knowledge with Common Beliefs," *Games and Economic Behavior*, 1(2), 170–190. 88

MORRIS, S. (2008): "Purification," in *The New Palgrave Dictionary of Economics Second Edition*, ed. by S. Durlauf, and L. Blume, pp. 779–782. Macmillan Palgrave. 66

MORRIS, S., AND H. S. SHIN (2003): "Global Games: Theory and Applications," in *Advances in Economics and Econometrics (Proceedings of the Eighth World Congress of the Econometric Society)*, ed. by M. Dewatripont, L. Hansen, and S. Turnovsky. Cambridge University Press. 91

MYERSON, R. B. (1978): "Refinements of the Nash Equilibrium Concept," *International Journal of Game Theory*, 7, 73–80. 333

——— (1981): "Optimal Auction Design," *Mathematics of Operations Research*, 6, 58–73. 307

——— (1991): *Game Theory: Analysis of Conflict*. Harvard Univ. Press., Cambridge, MA. 83

MYERSON, R. B., AND P. J. RENY (2015): "Sequential Equilibria of Multi-stage Games with Infinite Sets of Types and Actions," University of Chicago. 136

MYERSON, R. B., AND M. SATTERTHWAITE (1983): "Efficient Mechanisms for Bilateral Trading," *Journal of Economic Theory*, 28, 265–281. 296, 298, 301

NASH, JR., J. F. (1950a): "The Bargaining Problem," *Econometrica*, 18, 155–162. 244

——— (1950b): "Equilibrium Points in n-person Games," *Proceedings of the National Academy of Sciences USA*, 36(1), 48–49. 100

——— (1950c): "Non-cooperative Games," Ph.D. thesis, Princeton. 104

——— (1951): "Non-cooperative Games," *Annals of Mathematics*, 54(1), 286–95. 100

NÖLDEKE, G., AND E. VAN DAMME (1990): "Switching Away from Probability One Beliefs," Discussion Paper A-304, SFB-303, Department of Economics, University of Bonn. 136

OK, E. A. (2007): *Real Analysis with Economic Applications*. Princeton University Press. 100

OSBORNE, M. J. (2004): *An Introduction to Game Theory*. Oxford University Press, New York. vii

OSBORNE, M. J., AND A. RUBINSTEIN (1990): *Bargaining and Markets*. Academic Press, Inc., San Diego, CA. 256

———— (1994): *A Course in Game Theory*. The MIT Press, Cambridge, MA. 48

OWEN, G. (1982): *Game Theory*. Academic Press, second edn. 114

PEARCE, D. (1984): "Rationalizable Strategic Behavior and the Problem of Perfection," *Econometrica*, 52(4), 1029–50. 39, 40

POUNDSTONE, W. (1993): *Prisoner's Dilemma*. Anchor Books. 1

RENY, P. J. (2001): "Arrow's Theorem and The Gibbard-Satterthwaite Theorem: A Unified Approach," *Economics Letters*, 70, 99–105. 284

RILEY, J. G. (1979): "Informational Equilibrium," *Econometrica*, 47, 331–359. 153

ROCHET, J.-C. (1980): "Selection on an Unique Equilibrium Value for Extensive Games with Perfect Information," Unpublished, Université Paris IX-Dauphine, Centre de recherche de mathématiques de la décision. 28

ROSENTHAL, R. (1981): "Games of Perfect Information, Predatory Pricing and the Chain-Store Paradox," *Journal of Economic Theory*, 25(1), 92–100. 26

ROUSSEAU, J.-J. (1984): *A Discourse on Inequality*. Penguin. 51

ROYDEN, H. L., AND P. FITZPATRICK (2010): *Real Analysis*. Prentice Hall, New York, 4 edn. 268, 269

RUBINSTEIN, A. (1982): "Perfect Equilibrium in a Bargaining Model," *Econometrica*, 50(1), 97–109. 245, 255, 312

———— (1989): "The Electronic Mail Game: Strategic Behavior under Almost Common Knowledge," *American Economic Review*, 79(3), 385–391. 88

SAMUELSON, L. (1992): "Dominated Strategies and Common Knowledge," *Games and Economic Behavior*, 4(2), 284–313. 40

——— (1997): *Evolutionary Games and Equilibrium Selection.* MIT Press, Cambridge, MA. 104, 111

SCHWALBE, U., AND P. WALKER (2001): "Zermelo and the Early History of Game Theory," *Games and Economic Behavior*, 34(1), 123–137. 28

SELTEN, R. (1975): "Reexamination of the Perfectness Concept for Equilibrium Points in Extensive Games," *International Journal of Game Theory*, 4, 22–55. 121, 332

SHAKED, A., AND J. SUTTON (1984): "Involuntary Unemployment as a Perfect Equilibrium in a Bargaining Model," *Econometrica*, 52(6), 1351–1364. 259

SHAPIRO, C., AND J. STIGLITZ (1984): "Equilibrium Unemployment as a Worker Discipline Device," *American Economic Review*, 74(3), 433–444. 191

SIMON, L. K., AND W. R. ZAME (1990): "Discontinuous Games and Endogenous Sharing Rules," *Econometrica*, 58(4), 861–872. 116

SPENCE, A. M. (1973): "Job Market Signaling," *Quarterly Journal of Economics*, 87(3), 355–374. 149

STOKEY, N., AND R. E. LUCAS, JR. (1989): *Recursive Methods in Economic Dynamics.* Harvard University Press, Cambridge, MA. 175

TADELIS, S. (2013): *Game Theory: An Introduction.* Princeton University Press. vii

VAN DAMME, E. (1984): "A Relation between Perfect Equilibria in Extensive Form Games and Proper Equilibria in Normal Form Games," *International Journal of Game Theory*, 13, 1–13. 33, 334

——— (1991): *Stability and Perfection of Nash Equilibria.* Springer-Verlag, Berlin, second, revised and enlarged edn. 49

——— (2002): "Strategic Equilibrium," in *Handbook of Game Theory with Economic Applications, Volume 3*, ed. by R. J. Aumann, and S. Hart, chap. 41. North Holland. 48

VICKREY, W. (1961): "Counterspeculation, Auctions, and Competitive Sealed Tenders," *Journal of Finance*, 16(1), 8–37. 6

VOHRA, R. V. (2005): *Advanced Mathematical Economics.* Routledge, London and New York. 38, 102

WATSON, J. (2013): *Strategy: An Introduction to Game Theory.* WW Norton. vii

――― (2016): "Perfect Bayesian Equilibrium: General Definitions and Illustrations," unpublished, University of California at San Diego. 132

WEIBULL, J. W. (1995): *Evolutionary Game Theory.* MIT Press, Cambridge. 104

WEINSTEIN, J., AND M. YILDIZ (2007): "A Structure Theorem for Rationalizability with Application to Robust Predictions of Refinements," *Econometrica*, 75(2), 365–400. 91, 92

WILLIAMS, S. R. (1999): "A characterization of efficient, Bayesian incentive compatible mechanisms," *Economic Theory*, 14(1), 155–180. 295, 298

Index

Printed in the United States
By Bookmasters